Complex Analysis and Potential Theory

Complex Analysis and Potential Theory

Proceedings of the Conference Satellite to ICM 2006

Gebze Institute of Technology, Turkey 8 – 14 September 2006

Editors

Tahir Aliyev Azeroğlu
Gebze Institute of Technology, Turkey

Promarz M. Tamrazov
National Academy of Sciences, Ukraine

Gebze Institute of Technology

World Scientific

NEW JERSEY • LONDON • SINGAPORE • BEIJING • SHANGHAI • HONG KONG • TAIPEI • CHENNAI

Published by

World Scientific Publishing Co. Pte. Ltd.
5 Toh Tuck Link, Singapore 596224
USA office: 27 Warren Street, Suite 401-402, Hackensack, NJ 07601
UK office: 57 Shelton Street, Covent Garden, London WC2H 9HE

British Library Cataloguing-in-Publication Data
A catalogue record for this book is available from the British Library.

COMPLEX ANALYSIS AND POTENTIAL THEORY
Proceedings of the Conference Satellite to ICM 2006

ISBN-13 978-981-270-598-3
ISBN-10 981-270-598-8

Printed in Singapore by World Scientific Printers (S) Pte Ltd

PREFACE

The satellite to ICM (2006) International Conference on Complex Analysis and Potential Theory was held at the Gebze Institute of Technology in Gebze, Turkey during the period 8–14 September, 2006. The Proceedings involves the most of presentations delivered at the Conference by the participants, among of which many of the top-notch mathematicians were present.

Topics discussed include Grunsky inequalities and Moser‘s conjecture, speed of approximation to degenerate quasiconformal mappings, combinatorial Theorems of complex analysis, geometry of the general Beltrami equations, polyharmonic Dirichlet problem, isoperimetric inequalities for sums of reciprocal eigenvalues of the Laplacian, contour-solid theorems for finely meromorphic functions, residues on a Klein surface, functional analytic approach to the analysis of nonlinear boundary value problems, generalized quasiconformal mappings, properties of separately quasi-nearly subharmonic functions, implicit function theorem for Sobolev Mappings, The Martin boundary and the restricted mean value property for harmonic functions, approximate properties of the Bieberbach polynomials on the complex domains, asymptotic expansions of solutions of the heat equation with generalized functions initial data, Hausdorff operators, harmonic transfinite diameter and Chebyshev constants, analogues of the Mittag-Leffler and the Weierstrass theorems for harmonic differential forms on noncompact Riemannian manifolds, harmonic commutative Banach algebras and spatial potentials fields, parameter space of error functions and others. Besides, Prof. C. C. Yang proposed to include in the Proceedings a set of the articles devoted to so-called "open problems," i.e., the problems of great importance, but unsolved yet. This suggestion was approved by the scientific peers and accepted. In this connection, the Proceeding is composed of two Parts. The first one, Part A, involves the talks, which were presented and discussed at the Conference. The second one is devoted to the open problems completely. However, one can find some open problems in the first part as well, where they are given in passing with the main contents of the talks.

The articles published in this Proceedings are oriented on the active researchers, who works in these areas directly and in the adjacent fields of mathematics and would like to update recent developments in the field. This book will be useful for the Ph.D. and M.S. students as well as researchers who just start only or continue their activity in this area of mathematics and its applications in engineering.

We would like to thank all participants for their invaluable contributions. We acknowledge great efforts of our colleagues, Dr. Faik Mikailov, Tuğba Akyel and others who made major contribution to the organization of the meeting. Special thanks to Prof. Alinur Büyükaksoy, Rector of Gebze Institute of Technology, who supported every stages of preparation and holding of the meeting, and to the Scientific and the Technical Council of Turkey (TÜBİTAK) for the support of this conference.

Tahir Aliyev Azeroğlu
Department of Mathematics,
Gebze Institute of Technology,
Gebze, 41410 Kocaeli, Turkey

Promarz M. Tamrazov
Institute of Mathematics,
National Academy of Sciences
of Ukraine, Kiev, 01601, Ukraine

PARTICIPANTS

Olli Martio, University of Helsinki, Finland
Vladimir Mazya, Ohio State University, USA
Samuel Krushkal, Bar-Ilan University, Ramat Gan, Israel
Hakan Hedenmalm, Royal Institute of Technology, Stockholm, Sweden
Promarz M. Tamrazov, Institute of Mathematics of NAS, Ukraine
Chung Chun Yang, HKUST, Hong Kong
Vladimir Miklyukov, Volgograd State University, Russia
Reiner Kuhnau, Martin-Luther Universitat Halle-Wittenberg, Germany
Bodo Dittmar, Martin-Luther-Universitat Halle-Wittenberg, Germany
Bogdan Bojarski, Polish Academy of Sciences, Warsaw, Poland
Heinrich Begehr, I. Math. Inst., FU Berlin, Germany
Tahir Aliyev Azeroglu, Gebze Institute of Technology, Turkey
Sergei Favorov, Kharkov, Ukraine
Tatyana Shaposhnikova, Linkoping, Sweden
Lev Aizenberg, Bar-Ilan University, Ramat-Gan, Israel
Akif Gadjiev, Institute of Mathematics and Mechanics, Baku, Azerbaijan
Massimo Lanza De Cristoforis, Universita Degli Studi di Padova, Italy
Arturo Fernandez Arias, UNED, Madrid, Spain
Yurii Zelinskii, Institute of Mathematics of NAS, Ukraine
Anatoly Golberg, Bar-Ilan University, Ramat-Gan, Israel
A. V.Pokrovskii, Institute of Mathematics of NAS, Ukraine
Juhani Riihentaus, University of Joensuu, Joensuu, Finland
Sergiy Plaksa, Institute of Mathematics of NAS, Ukraine
Matti Vuorinen, University of Turku, Finland
Oleg F. Gerus, Zhytomyr, Ukraine
Shunsuke Morosawa, Kochi University, Japan
Victor V. Starkov, Petrozavodsk State University, Russian
Allami Benyaiche, Universitè Ibn Tofail, Kenitra, Morocco
Igor V. Zhuravlev, Volgograd, Russia
Elijah Liflyand, Bar-Ilan University, Ramat-Gan, Israel
Mubariz T. Karayev, Suleyman Demirel University, Isparta, Turkey
Kunio Yoshino, Sophia University, Tokyo, Japan
Eugenia Malinnikova, Trondheim, Norway
Daniyal M.Israfilov, Balikesir University, Turkey
Aydin Aytuna, Sabanci University, Turkey

Aydin Aytuna, Sabanci University, Turkey
Anatoliy Pogoruy, Zhytomyr, Ukraine
Andrey L. Targonskii, Institute of Mathematics of NAS, Ukraine
Yasuyuki Oka, Sophia University, Tokyo, Japan
Bulent N. Ornek, Gebze Institute of Technology, Turkey
Shahram Rezapour, Azarbaidjan University of Tarbiat Moallem, Iran
Hiroshige Shiga, Tokyo, Japan
Vyacheslav Zakharyuta, Istanbul, Turkey
H. Turgay Kaptanoglu, Ankara, Turkey
Olena Karupu, Institute of Mathematics of NAS, Ukraine
Mehmet Acikgoz, University of Gaziantep, Turkey
Yu.V. Vasil'eva, Institute of Mathematics of NAS, Ukraine
Coskun Yakar, Gebze Institute of Technology, Kocaeli, Turkey
Allaberen Ashyralyev, Fatih University, Istanbul, Turkey
Ali Sirma, Gebze Institute of Technology, Gebze, Kocaeli, Turkey
Mehmet Kucukaslan, Mersin University, Turkey
Peter Tien-Yu Chern, Kaohsiung, Taiwan

CONTENTS

PART A
TALKS

STRENGTHENED MOSER'S CONJECTURE AND FINSLER GEOMETRY OF GRUNSKY COEFFICIENTS

SAMUEL KRUSHKAL

Department of Mathematics,
Bar-Ilan University,52900 Ramat-Gan,Israel
and Department of Mathematical Sciences,
University of Memphis, Memphis, TN 38152, USA

The Grunsky and Teichmüller norms $\varkappa(f)$ and $k(f)$ of a univalent function f in a finitely connected domain $D \ni \infty$ with quasiconformal extension to $\widehat{\mathbb{C}}$ are related by $\varkappa(f) \le k(f)$. In 1985, Jürgen Moser conjectured that any univalent function in the disk $\Delta^* = \{z : |z| > 1\}$ can be approximated locally uniformly by functions with $\varkappa(f) < k(f)$. This conjecture has been recently proved by R. Kühnau and the author. We prove that approximation is possible in a much stronger sense and also give applications. One of them is the inversion of Ahlfors and Grunsky inequalities, which solves the old Kühnau problem on exact lower bound in the inverse inequality estimating $k(f)$ by $\varkappa(f)$.

Keywords: Quasiconformal, univalent function, Grunsky coefficient inequalities, universal Teichmüller space, subharmonic function, Strebel's point, Kobayashi metric, generalized Gaussian curvature, holomorphic curvature, Fredholm eigenvalues.

1. Grunsky inequalities and Fredholm eigenvalues

1.1.

The technique of Grunsky coefficient inequalities provides a powerful tool in solving various problems. We shall consider the Grunsky coefficients from several points of view.

These coefficients give rise to a noninvariant Finsler structure on the tangent bundle of the universal Teichmüller space $\mathbb{T}$ dominated by the canonical Finsler structure of this space. Estimating deviation of these structures from below provides interesting analytic and geometric consequences. The metric generated on $\mathbb{T}$ by this structure is naturally related to quasiconformal extensions of conformal maps of quasidisks and reflections across their boundaries and allows to establish the bounds for dilatations of such extensions.

This geometric approach is somewhat new. The exposition is mainly focussed on the presentation of main ideas and results; for complete details, see the author's recent papers.

The classical Grunsky theorem states that a holomorphic function $f(z) = z + const + O(z^{-1})$ in a neighborhood U_0 of $z = \infty$ is extended to a univalent holomorphic function on the disk $\Delta^* = \{z \in \widehat{\mathbb{C}} = \mathbb{C} \cup \{\infty\} : |z| > 1\}$ if and only if its Grunsky coefficients α_{mn} satisfy the inequalities $\left| \sum_{m,n=1}^{\infty} \sqrt{mn}\, \alpha_{mn} x_m x_n \right| \le 1$, where α_{mn} are generated by

$$\log \frac{f(z) - f(\zeta)}{z - \zeta} = - \sum_{m,n=1}^{\infty} \alpha_{mn} z^{-m} \zeta^{-n}, \quad (z, \zeta) \in (\Delta^*)^2, \tag{1}$$

$\mathbf{x} = (x_n)$ runs over the unit sphere $S(l^2)$ of the Hilbert space l^2 with $||x||^2 = \sum_1^\infty |x_n|^2$, and the principal branch of logarithmic function is chosen (cf.[Gr]). The quantity

$$\varkappa(f) := \sup \left\{ \left| \sum_{m,n=1}^{\infty} \sqrt{mn}\; mn x_m x_n \right| : \mathbf{x} = (x_n) \in S(l^2) \right\} \tag{2}$$

is called the **Grunsky norm** of f.

We denote by Σ the class of univalent holomorphic functions $f(z) = z + b_0 + b_1 z^{-1} + \ldots$ mapping Δ^* into $\widehat{\mathbb{C}} \setminus \{0\}$, and by $\Sigma(k)$ its subclass of f with k-quasiconformal extensions to the unit disk $= \{|z| < 1\}$ so that $f(0) = 0$. Let $\Sigma^0 = \bigcup_k \Sigma(k)$.

These functions are intrinsically connected with the **universal Teichmüller space** $\mathbb{T}$ modelled as a bounded domain in the Banach space $\mathbb{B}$ of holomorphic functions in Δ^* with norm $||\varphi||_{\mathbb{B}} = \sup_{\Delta^*} (|z|^2 - 1)^2 |(z)|$. All $\in \mathbb{B}$ can be regarded as the **Schwarzian derivatives**

$$S_f = (f''/f')' - (f''/f')^2/2$$

of locally univalent holomorphic functions in Δ^*. The points of $\mathbb{T}$ represent $f \in \Sigma^0$ whose **minimal dilatation**

$$k(f) := \inf\{k(w^\mu) = ||\mu||_{:}\ w^\mu|\partial\Delta^* = f\}$$

determines the Teichmüller metric on $\mathbb{T}$. Here $||\mu||_\infty = esssup_{\mathbb{C}}|\mu(z)|$.

Grunsky's theorem has been essentially strengthened for the functions with quasiconformal extensions, for which the Grunsky and Teichmüller norms of $f \in \Sigma$ are related as follows

$$\varkappa(f) \le k(f) \tag{3}$$

(see[Ku1]); on the other hand, by theorem of Pommerenke and Zhuravlev, any $f \in \Sigma$ with $\varkappa(f) \le k < 1$, has k_1-quasiconformal extensions to $\widehat{\mathbb{C}}$ with $k_1 = k_1(k) \ge k$ (see[Po] ,[Zh] ;[KK1] , pp. 82-84). An explicit not sharp bound $k_1(k)$ is given in[Ku4] . We will discuss this problem in the last section.

A point is that for a generic function $f \in \Sigma^0$, we have in (3) the strict inequality $\varkappa(f) < k(f)$ (see e.g.[KK2]). On the other hand, the functions with $\varkappa(f) = k(f)$ are crucial in many applications of the Grunsky inequality technique.

In 1985, Jürgen Moser conjectured that *the set of functions with $\varkappa(f) = k(f)$ is sparse in Σ^0 so that any function $f \in \Sigma$ is approximated by functions satisfying (1.8) uniformly on compact sets in Δ^**. This conjecture was recently proved in[KK2] .

A related conjecture, posed in[KK2] and which remains open, states that $f \in \Sigma^0$ satisfying (1.7) cannot be the limit functions of locally uniformly convergent sequences $\{f_n\} \subset \Sigma^0$ with $\varkappa(f_n) = k(f_n)$. It is proved only for the sequences of maps f_n which are asymptotically conformal on the unit circle $S^1 = \partial\Delta^*$.

In applications of Schwarzian derivatives, especially to Teichmüller spaces, one has to use a much stronger topology than locally uniform convergence. A question is how sparse is the set of derivatives $\varphi = S_f$ in $\mathbb{T}$ representing the maps with equal Grunsky and Teichmüller norms. The answer is given by the following key theorem.

Theorem 1. *The set of points $\varphi = S_f$, which represent the maps $f \in \Sigma^0$ with $\varkappa(f) < k(f)$, is open and dense in the space $\mathbb{T}$.*

1.2.

The first application of this theorem is to Fredholm eigenvalues theory. We consider the eigenvalues ρ_n of oriented quasiconformal Jordan curves $L \subset \widehat{\mathbb{C}}$ (quasicircles) which in the case of smooth curves coincide with eigenvalues of double-layer potential over L.

These values are intrinsically connected with Grunsky coefficients of the corresponding conformal maps, which is qualitatively expressed by Kühnau-Schiffer's theorem on reciprocity of $\varkappa(f)$ to the least positive Fredholm eigenvalue ρ_L. This value is defined for any quasicircle $L \subset \widehat{\mathbb{C}}$ by

$$\frac{1}{\rho_L} = \sup \frac{|\mathcal{D}_G(u) - \mathcal{D}_{G^*}(u)|}{\mathcal{D}_G(u) + \mathcal{D}_{G^*}(u)},$$

where G and G^* are, respectively, the interior and exterior of L; $\mathcal{D}$ denotes the Dirichlet integral, and the supremum is taken over all functions u continuous on $\widehat{\mathbb{C}}$ and harmonic on $G \cup G^*$ (cf.[Ku2] ,[Sc]).

A basic ingredient for estimating ρ_L is the well-known Ahlfors inequality

$$\frac{1}{\rho_L} \leq q_L \tag{4}$$

where q_L is the reflection coefficient of L, i.e., the minimal dilatation of quasiconformal reflections across L (cf.[1] ,[Kr4] ,[Ku2]).

It suffices to take the images $L = f^\mu(S^1)$ of the unit circle under quasiconformal self-maps of $\widehat{\mathbb{C}}$ with Beltrami coefficients $\mu = \partial_{\bar{z}} f / \partial_z f$ supported in the unit disk . Then q_L equals the minimal dilatation $k(f^\mu) = \|\mu\|$ of such maps, and the inequality (4) is reduced inequality (3). Applying Theorem 1, one obtains

Theorem 2 *The set of quasiconformal curves L, for which Ahlfors'inequality (4) is fulfilled in the strict form $1/\rho_L > q_L$, is open and dense in the strongest topology determined by the norm of the space $\mathbb{B}$.*

2. Sketch of the proof of Theorem 1

The openness follows from continuity of both quantities $\varkappa(f)$ and $k(f)$ as functions of the Schwarzian derivatives S_f on $\mathbb{T}$ (cf.[Sh]) , and the main counterpart of the proof concerns the density. The proof involves the density of Strebel points in $\mathbb{T}$ and relies on curvature properties of certain Finsler metrics on this space which are briefly presented below.

2.1.

Recall that if $f_0 := f^{\mu_0}$ is an extremal representative of its class $[f_0]$ with dilatation $k(f_0) = ||\mu_0||_\infty = \inf\{k(f^\mu) : f^\mu|S^1 = f_0|S^1\} = k$, and if there exists in this class a quasiconformal map f_1 whose Beltrami coefficient μ_{f_1} satisfies the strong inequality ${}_{A_r}|\mu_{f_1}(z)| < k$ in some annulus $A_r := \{z : r < |z| < 1\}$, then f_1 is called the **frame map** for the class $[f_0]$ and the corresponding point of the space $\mathbb{T}$ is called the **Strebel point**.

We use the following important properties of Strebel's points adopted to our case.

Proposition 1. *(i) If a class $[f]$ has a frame map, then the extremal map f_0 in this class is unique and either conformal or a Teichmüller map with Beltrami coefficient $\mu_0 = k|\psi_0|/\psi_0$ on , defined by an integrable holomorphic quadratic differential ψ on Δ and a constant $k \in (0,1)$*[12] .

(ii) The set of Strebel points is open and dense in $\mathbb{T}$.[GL]

We shall use the following construction exploited in[GL] . Suppose f_0 with Beltrami coefficient μ_0 is extremal in its class. Fix a number ϵ between 0 and 1, and take an increasing sequence $\{r_n\}_1$ with $0 < r_n < 1$ approaching 1. Put

$$\mu_n(z) = \begin{cases} \mu_0(z) & \text{if } |z| < r_n, \\ (1-\epsilon)\mu_0(z) & \text{otherwise,} \end{cases} \tag{5}$$

and let f_n be a quasiconformal map with Beltrami coefficient μ_n. Then, for sufficiently large n, f_n is a frame map for its class, and the dilatation k_n of the extremal map in the class of f_n approaches $k_0 = k(f_0)$.

2.2.

As well-known, the universal Teichmüller space $\mathbb{T}$ is the space of quasisymmetric homeomorphisms of the unit circle $S^1 = \partial\Delta$ factorized by Möbius maps. The canonical complex Banach structure on $\mathbb{T}$ is defined by factorization of the ball of coefficients

$$Belt(\Delta)_1 = \{\mu \in L_\infty() : \ \mu|\Delta^* = 0, \ ||\mu|| < 1\}, \tag{6}$$

letting $\mu, \nu \in Belt(\Delta)_1$ be equivalent if the corresponding maps $w^\mu, w^\nu \in \Sigma^0$ coincide on S^1 (hence, on $\overline{\Delta^*}$) and passing to Schwarzian derivatives S_{f^μ}. The defining projection $\phi_T : \mu \to S_{w^\mu}$ is a holomorphic map from $L_\infty(\Delta)$ to B. The equivalence class of a map w^μ will be denoted by $[w^\mu]$.

An intrinsic complete metric on the space is the **Teichmüller metric** defined by

$$\tau_{\mathbb{T}}(\phi_{\mathbb{T}}(\mu), \phi_{\mathbb{T}}(\nu)) = \frac{1}{2} \inf\{\log K\left(w^{\mu_*} \circ \left(w^{\nu_*}\right)^{-1}\right) : \ \mu_* \in \phi_{\mathbb{T}}(\mu), \nu_* \in \phi_{\mathbb{T}}(\nu)\}. \quad (7)$$

It is generated by the **Finsler structure** on the tangent bundle $\mathcal{T}(\mathbb{T}) = \mathbb{T} \times \mathbb{B}$ of $\mathbb{T}$ defined by

$$F_T(\phi_{\mathbb{T}}(\mu), \phi'_{\mathbb{T}}(\mu)\nu) = \inf\{\left\|\nu_*(1-|\mu|^2)^{-1}\right\|_\infty : \phi'_{\mathbb{T}}(\mu)\nu_* = \phi'_T(\mu)\nu; \ \mu \in Belt(\Delta)_1; \ \nu, \nu_* \in L_\infty(\mathbb{C})\}. \quad (8)$$

The space $\mathbb{T}$ as a complex Banach manifold has also the contractible invariant metrics. The **Kobayashi metric** $d_{\mathbb{T}}$ on $\mathbb{T}$ is the largest pseudometric d on $\mathbb{T}$ contracted by holomorphic maps $h: \ \Delta \to \mathbb{T}$ so that for any two points $\psi_1, \ \psi_2 \in \mathbb{T}$, we have

$$d_{\mathbb{T}}(\psi_1, \psi_2) \leq \inf\{d_\Delta(0, t) : \ h(0) = \psi_1, \ h(t) = \psi_2\},$$

where d_Δ is the **hyperbolic Poincaré metric** on Δ of Gaussian curvature -4, with the differential form

$$ds = \lambda(z)|dz| := |dz|/(1 - |z|^2). \quad (9)$$

The **Caratheodory metric** $c_{\mathbb{T}}$ is the least pseudometric on T, which does not increase the holomorphic maps $\Delta \to \mathbb{T}$. We shall use also the infinitesimal Kobayashi metric $\mathcal{K}_{\mathbb{T}}(\psi, v)$ defined on $\mathcal{T}(T)$.

The following result is a strengthened version of the fundamental Gardiner-Royden theorem for universal Teichmüller space.

Proposition 2.[Kr4] *The differential metric $\mathcal{K}_{\mathbb{T}}(\varphi, v)$ on the tangent bundle $\mathcal{T}(\mathbb{T})$ of the universal Teichmüller space $\mathbb{T}$ is logarithmically plurisubharmonic in $\varphi \in \mathbb{T}$, equals the canonical Finsler structure $F_{\mathbb{T}}(\varphi, v)$ on $\mathcal{T}(\mathbb{T})$ generating the Teichmüller metric of $\mathbb{T}$ and has constant holomorphic sectional curvature $\kappa_{\mathcal{K}}(\varphi, v) = -4$ on $\mathcal{T}(\mathbb{T})$.*

The generalized Gaussian curvature κ_λ of a upper semicontinuos Finsler metric $ds = \lambda(t)|dt|$ in a domain $\Omega \subset \mathbb{C}$ is defined by

$$\kappa_\lambda(t) = -\frac{\boldsymbol{\Delta} \log \lambda(t)}{\lambda(t)^2}, \quad (10)$$

where $\boldsymbol{\Delta}$ is the **generalized Laplacian**

$$\boldsymbol{\Delta}\lambda(t) = 4 \liminf_{r \to 0} \frac{1}{r^2}\left\{\frac{1}{2\pi}\int_0^{2\pi} \lambda(t + re^{i\theta})d\theta - \lambda(t)\right\}$$

(provided that $-\infty \leq \lambda(t) < \infty$), and the **sectional holomorphic curvature** of a Finsler metric on $\mathbb{T}$ the supremum of curvatures (10) over appropriate collections of holomorphic maps from the disk into $\mathbb{T}$.

Similar to C^2 functions, for which $\boldsymbol{\Delta}$ coincides with the usual Laplacian, one obtains that λ is subharmonic on Ω if and only if $\boldsymbol{\Delta}\lambda(t) \geq 0$; hence, at the points t_0 of local maximuma of λ with $\lambda(t_0) > -\infty$, we have $\boldsymbol{\Delta}\lambda(t_0) \leq 0$. For details see e.g.[Di], [EKK], [GL], [Ko], [15], [Ro1].

2.3.

Letting $A_1(\delta) = \{\psi \in L_1(\Delta) : \psi \}$, $A_1^2 = \{\psi \in A_1(\Delta) : \psi = \omega^2, \omega \}$, and

$$\langle \mu, \psi \rangle_\Delta = \iint_\Delta \mu(z)\psi(z)dxdy \quad (\mu \in L_\infty(\Delta),\ \psi \in L_1(\Delta),\ z = x + iy)$$

we have the following key result.

Proposition 3.[14] ,[Kr3] *The equality (1.6) holds if and only if the function f is the restriction to * of a quasiconformal self-map w^{μ_0} of $\widehat{\mathbb{C}}$ with Beltrami coefficient μ_0 satisfying the condition*

$$\sup |\langle \mu_0, \varphi \rangle_\Delta| = \|\mu_0\|_\infty, \tag{11}$$

where the supremum is taken over holomorphic functions $\varphi \in A_1^2(\Delta)$ with $\|\varphi\|_{A_1(\Delta)} = 1$.

If, in addition, the class $[f]$ contains a frame map (is a Strebel point), then μ_0 is of the form

$$\mu_0(z) = \|\mu_0\|_\infty |\psi_0(z)|/\psi_0(z) \quad \textit{with } \psi_0 \in A_1^2 \textit{ in } \Delta. \tag{12}$$

Geometrically the condition (11) means that the Caratheodory metric on the holomorphic extremal disk $\{\psi_{\mathbb{T}}(t\mu_0/\|\mu_0\|) : t \in \Delta\}$ in $\mathbb{T}$ coincides with the Teichmüller metric of this space.

For analytic curves $f(S^1)$ the equality (12) was obtained by a different method in[Ku3] .

2.4.

The proof of Theorem 1 involves generic holomorphic disks in $\mathbb{T}$ and a new Finsler structure on $\mathbb{T}$ determined by generalized Grunsky coefficients. The method of Grunsky inequalities is extended to bordered Riemann surfaces X with a finite number of the boundary components, in particular, to multiply connected domains on the complex plane though to a somewhat less extent (cf.[Gr] ,[Mi] ,[SS]).

In the general case, the generating function must be replaced by the bilinear differential

$$-\log \frac{f(z) - f(\zeta)}{z - \zeta} - R_X(z, \zeta) = \sum_{m,n=1}^{\infty} \beta_{mn}\ \varphi_m(z)\varphi_n() : \ X \times X \to \mathbb{C}, \tag{13}$$

where the surface kernel $R_X(z, \zeta)$ relates to conformal map $j_\theta(z, \zeta)$ of X onto the sphere $\mathbb{C}$ slit along arcs of logarithmic spirals inclined at the angle $\theta \in [0, \pi)$ to a ray issuing from the origin so that $j_\theta(\zeta, \zeta) = 0$ and $j_\theta(z) = z - z_\theta + const + O(1/(z - z_\theta))$ as $z \to z_\theta = j_\theta^{-1}(\infty)$. Here $\{\varphi_n\}_1^\infty$ is a canonical system of holomorphic functions on X such that (in a local parameter)

$$\varphi_n(z) = a_{n,n} z^{-n} + a_{n+1,n} z^{-n-1} + \dots \quad \text{with } a_{n,n} > 0, \quad n = 1, 2, \dots,$$

and φ'_n form a complete orthonormal system in $H^2(X)$.

We shall deal only with simply connected domains $X \ni \infty$ with quasiconformal boundaries. For any such domain, the kernel R_X vanishes identically on $X \times X$, and the expansion (13) assumes the form

$$\log[(f(z) - f(\zeta))/(z - \zeta)] = - \sum_{m,n=1}^{\infty} \frac{\alpha_{mn}}{f(z)^m f(\zeta)^n}, \tag{14}$$

where f denotes a conformal map of X onto the disk * so that $f(\infty) = \infty,\ f'(\infty) > 0$, and $\alpha_{mn} = \beta_{mn}/\sqrt{mn}$ are the normalized generalized Grunsky coefficients. These coefficients also depend holomorphically on Schwarzian derivatives S_f.

A theorem of Milin extending the Grunsky univalence criterion to multiply connected domains X states that a holomorphic function $f(z) = z + const + O(z^{-1})$ in a neighborhood of the infinite point $z = \infty$ is continued to a univalent function in the whole domain X if and only if the coefficients β_{mn} in (2.10) satisfy the inequality $\left| \sum_{m,n=1}^{\infty} \beta_{mn}\ x_m x_n \right| \le \|x\|^2$ for any point $x = (x_n) \in S(l^2)$ (see[Mi]). Accordingly, we have the generalized constant $\varkappa_X(f) = \sup \left| \sum_{m,n=1}^{\infty} \beta_{mn}\ x_m x_n \right|$ over $\mathbf{x} = (x_n) \in S(l^2)$, which coincides with (2) for $X = \Delta^*$.

2.5.

Now the proof of Theorem 1 continues as follows. In view of continuity of both functions $\widehat{\varkappa}(S_f) := \varkappa(f)$ and $\widehat{k}(S_f) := k(f)$ on $\mathbb{T}$ mentioned above, we need to establish only that each point $\varphi^* = S_{f^*}$ representing a function $f^* \in \Sigma^0$ is the limit point of a sequence $\{\varphi_n = S_{f_n}\} \subset \mathbb{T}$ with $\varkappa(f_n) = k(f_n)$.

We may assume that φ^* is a Strebel point and $\|\varphi^*\|_\infty < 2$. Its class $[f^*]$ contains a Teichmüller extremal map with Beltrami coefficient $\mu^*(z) = k^*|\psi^*(z)|/\psi^*(z)$ in Δ and $\mu^*(z) = 0$ in Δ^* determined by a holomorphic quadratic differential ψ^* which has in the unit disk Δ only zeros of even order. We fix $r \in (0, 1)$ and define a family of Beltrami coefficients $\mu_t = \mu(\cdot, t)$ depending on a complex parameter t (complexify (5)), letting

$$\mu(z,t) = \begin{cases} \mu^*(z), & |z| < r, \\ \left(1 - \frac{1}{1+t}\right)\mu^*(z), & r < |z| < 1, \end{cases}$$

and $\mu(z,t) = 0$ for $|z| > 1$. The admissible values of t are those for which $|\mu(z,t)| < 1$. This inequality holds, provided t ranges over the disk

$$\widetilde{\Delta}^*_a = \{t' \in\colon\ |t' + a| > R(a)\} \quad \text{with } a = a(k^*) = 1/[1-(k^*)^2] > 1,\ \ R(a) = a(a-1). \tag{15}$$

For $t = \infty$ and $t = 0$, we have, respectively, $\mu_\infty = \mu^*$ and $\mu_0(z) = \mu^*(z)$ if $|z| < r$, $\mu_0(z) = 0$ otherwise. We must establish that if the disk (15) contains a sequence $\{t_n\}$ going to infinity and such that

$$\lim_{n\to\infty} \|\mu(\cdot, t_n) - \mu(\cdot, \infty)\|_\infty = 0; \quad \varkappa(f^{\mu(\cdot,t_n)}) = k(f^{\mu(\cdot,t_n)}), \quad n = 1, 2, \dots ,$$

then

$$\varkappa(f^{\mu_t}) = k(f^{\mu_t}) \quad \text{for all } t \in \Delta_a^*. \tag{16}$$

Here μ_t is the extremal Beltrami coefficients in the classes $[f^{\mu(\cdot,t)}]$, $t \in \widetilde{}_a^*$, and the points $\varphi_t = \phi_{\mathbb{T}}(\mu_t)$ run over a holomorhic disk in $\mathbb{T}$, which we denote by Ω_a.

To establish (16), we construct on Ω_a a Finsler metric of generalized Gaussian curvatures at most -4 and compare it with the Kobayashi metric. The underlying fact is that coefficients $\widehat{\alpha}_{mn}(S_f)$ generate for each $x = (x_n) \in S(l^2)$ the holomorphic map

$$h_{\mathbf{x}}(\varphi) = \sum_{m,n=1}^{\infty} \sqrt{mn}\, \alpha_{mn}(\varphi) x_m x_n : \mathbb{T} \to \Delta. \tag{17}$$

Consider in the tangent bundle $\mathcal{T}(\mathbb{T}) = \mathbb{T} \times \mathbb{B}$ the holomorphic disks $\widetilde{\Delta}_a^*(v)$ covering the disk $\phi_{\mathbb{T}}(\widetilde{\Delta}_a^*)$ in $\mathbb{T}$. Their points are pairs (φ, v), where $v = \phi'_{\mathbb{T}}[\varphi]\mu \in \mathbb{B}$ is a tangent vector to $\mathbb{T}$ at the point φ, and μ runs over the ball

$$Belt(D_\varphi)_1 = \{\mu \in L_\infty(\mathbb{C}) : \ \mu | D_\varphi^* = 0, \ \|\mu\|_\infty < 1\}.$$

Here D and D_* denote the images of and Δ^* under $f = f_\varphi \in \Sigma^0$ with $S_f = \varphi$.

To get the maps $\Delta \to \mathbb{T}$ preserving the origins, we transform the functions (17) by the chain rule for Beltrami coefficients $w^\nu = w^{\sigma(\nu)} \circ (f^{\nu_0})^{-1}$, where

$$\sigma(\nu) \circ f^{\nu_0} = \nu - \nu_0 1 - \nu_0 \nu \frac{\partial_z f^{\nu_0}}{\partial_z f^{\nu_0}};$$

denote the composed maps by $g_x[\sigma_\varphi]$.

Using these maps, we pull-back the hyperbolic metric (9) onto the disks in $\mathcal{T}(\mathbb{T})$ covering Ω_a and get on these disks the conformal subharmonic metrics $ds = \lambda_{g_x[\sigma_\varphi(\nu)]}(t)|dt|$ on G_0, with

$$\lambda_{g_x[\sigma_\varphi(\nu)]} = g_x[\sigma_\varphi(\nu)]^*(\lambda) = \frac{|g'_x[\sigma_\varphi(\nu)]|}{1 - |g_x[\sigma_\varphi(\nu)]|^2}$$

with curvature -4 at nonsingular points. Consider the upper envelope of these metrics

$$\widehat{\lambda}_\varkappa(t) = \sup \widehat{\lambda}_{g_x[\sigma_\varphi]}(t),$$

with supremum over all $x \in S(l^2)$ and all $\sigma_\varphi \in Belt(\Delta)_1$. Its upper semicontinuous regularization

$$\lambda_\varkappa(t) = \limsup_{t' \to t} \widehat{\lambda}_\varkappa(t')$$

descends to an upper semicontinuous metric on Ω_a. Similar to,[Kr4] one derives the following basic properties of this metric.

Lemma 1. *(a) The metric $\lambda_{\varkappa}$ is a logarithmically subharmonic Finsler metric on Ω_a;*

(b) For small $\tau = (t - a(k))/R(k)$, we have

$$\lambda_{\varkappa}(\tau) = \varkappa(\tau\varphi^*) + o(\tau) \quad as\ \tau \to 0,$$

where $\varkappa(\tau\varphi^*)$ denotes the Grunsky constant of the map $f \in \Sigma^0$ with $S_{f|\Delta^*} = \tau\varphi^*$;

(c) The generalized Gaussian curvature of $\lambda_{\varkappa}$ satisfies $k_{\lambda_{\varkappa}} \leq -4$. Equivalently, $\log \lambda_{\varkappa} \geq 4\lambda_{\varkappa}^2$ *or* $u_{\varkappa} \geq 4e^{2u_{\varkappa}}$, *where* $u_{\varkappa} = \log \lambda_{\varkappa}$.

2.6.

The next step consists of comparison of $\lambda_{\varkappa}$ with restriction of the infinitesimal Kobayashi metric $\lambda_{\mathcal{K}}$ of T onto Ω_a. This restriction provides a logarithmically subharmonic metric of generalized Gaussian curvature -4.

To this end, we define on $\mathcal{T}(\mathbb{T})$ a new Finsler structure

$$F_{\varkappa}(\varphi_0, v) = \sup\{|dH_{@}(\varphi_0; \varphi_0)v| : x \in S(l^2)\}, \tag{18}$$

using the form

$$H_{@}(\varphi, \varphi_0) = \frac{h_x() - h_x(\varphi_0)}{1 - h_x(\varphi_0)h_x(\varphi)}.$$

It is dominated by the canonical Finsler structure (8). The structure (18) allows us to construct in a standard way on embedded holomorphic disks (Δ) the Finsler metrics $\lambda_g(t) = F_{\varkappa}(g(t), g'(t))$ and, accordingly, the corresponding distance

$$d_(\varphi_1, \varphi_2) = \inf \int_{\beta} \lambda_g(t) ds_t,$$

taking the infimum over C^1 smooth curves $\beta : [0,1] \to \mathbb{T}$ joining the points φ_1 and φ_2.

Lemma 2. *On any extremal disk $\Delta(\mu_0) = \{\phi_{\mathbb{T}}(t\mu_0) : t \in \Delta\}$, we have the equality*

$$\tanh^{-1}[\varkappa(S_{f^{r\mu_0}})] = \int_0^r \lambda_{\varkappa}(t) dt. \tag{19}$$

Taking into account that the disk Ω_a touches the Teichmüller disks $\Delta(\mu^*)$ and $\Delta(\mu_n)$ at points φ^* and $\varphi_n = \phi_{\mathbb{T}}(\mu(\cdot, t_n))$ and that λ does not depend on tangent unit vectors with one end at points of Ω_a, one derives that $\lambda_{\varkappa}$ relates to Kobayashi metric $\lambda_{\mathcal{K}}|\Omega_a$ as follows $\lambda_{\varkappa}(0) = \lambda_{\varkappa^*}(\varphi^*) = \lambda_{\mathcal{K}}(0)$, $\lambda_{\varkappa}(t_n) = \lambda_{\varkappa_n}(\varphi_n) = \lambda_{\mathcal{K}}(t_n)$ and $\lambda_{\varkappa}(t) \leq \lambda_{\mathcal{K}}(t)$ for $t \in \Omega_a \setminus \{0, t_n\}$. Hence, by the maximum principle of Minda[Min] , the metrics λ and $_{\mathcal{K}}$ must coincide on the entire disk Ω_a, which is equivalent to the desired equality (16).

In the same way, one obtains that $\varkappa() = k(\varphi)$ in a ball $B(\varphi^*,) \subset \mathbb{T}$ centered at φ^*. Moving this ball from the point φ^* along the segment $[-\varphi^*, ^*]$, one derives that this equality must hold for all points of a ball centered at the origin $\varphi = \mathbf{0}$. But the latter is impossible, since it contradicts the existence of points $\varphi = S_f$ in a neighborhood of $\mathbf{0}$ at which $\varkappa(f) < k(f)$. This contradiction proves the theorem.

2.7.

A similar approach allows to establish that the set of the Schwarzian derivatives $\varphi = S_f \in \mathbb{T}$ representing the functions $f \in \Sigma^0$ with $\varkappa(f) = k(f)$ is convex (cf.[Kr6]).

3. Inversion of Ahlfors and Grunsky inequalities

3.1.

The important problem on the sharp estimation of the dilatation $k(f)$ by Grunsky norm of f, or equivalently, by Fredholm eigenvalue of $f(S^1)$ was first stated by Kühnau in 1981 and remains open.

There was an explicit bound $k_1(k)$ for dilatations of quasiconformal extensions of $f \in \Sigma$ with $\varkappa(f) \leq k$ found in.[Ku4] It is given by

$$K_1 := \frac{1+k_1}{1-k_1} \leq max \left[\lambda^{3/2}\Big(\frac{1+\varkappa}{1-\varkappa}\Big),\ 2\lambda\Big(\frac{1+\varkappa}{1-\varkappa}\Big) - 1\right],$$

where $\overline{\lambda}(K) = \max w(1)$, and the maximum is taken over K-quasiconformal self-maps w of $\mathbb{C}$ with the fixed points $-1, 0, \infty$. The distortion function λ. For small $\varkappa$ there is a somewhat better estimate which is also not sharp.

3.2.

The following theorem solves the problem and has many other applications.
Theorem 3.[Kr5] *For $f \in \Sigma^0$ we have the estimate*

$$k(f) \leq \frac{3}{2\sqrt{2}}(f) = 1.07\ldots(f) \tag{20}$$

(and accordingly for Fredholm eigenvalues $\rho_{f(S^1)}$), which is asymptotically sharp as $\to 0$. The equality holds for the map

$$f_{3,t}(z) = \begin{cases} z(1+t/z^3)^{2/3} & if\ \ |z| > 1 \\ z[1+t(|z|/z)^3]^{2/3} & if\ \ |z| \leq 1 \end{cases} \tag{21}$$

with $t = const \in (0,1)$.

Note that the Beltrami coefficient of this map in the disk is $\mu_3(z) = t|z|/z$.
The proof of this theorem consists of several independent stages. It suffices to establish its assertion for $f \in \Sigma^0$ having Teichmüller extremal quasiconformal extensions onto , i.e., with Beltrami coefficient of the form $\mu_f(z) = k|\varphi(z)|/\varphi(z)$, where $k = const \in (0,1)$ and φ is an integrable holomorphic function in , which again means that f is represented in by a Strebel point $[f]$. Put $\mu^*(z) = \mu(z)/\|\mu\|_\infty$ to have $\|\mu^*\|_\infty = 1$.

3.3.

We first prove

Theorem 4. *For every function $f \in \Sigma^0$ with unique extremal extension f^{μ_0} to Δ, we have the sharp bound*

$$k(f^{\mu_0}) \le \frac{1}{\alpha(f^{\mu_0})} \min_{|t|=1} \varkappa(f^{t\mu_0}) \tag{22}$$

with

$$a(f^{\mu_0}) = \sup_{\varphi \in A_1^2, \|\varphi\|_{A_1}=1} |\langle \mu_0^*, \varphi \rangle_\Delta|$$

The proof of Theorem 4 is geometric and relies on properties of conformal metrics $ds = \lambda(z)|dz|$ on the disk Δ with $\lambda(z) \ge 0$ of **negative integral curvature** bounded from above. The curvature is understanding in the supporting sense of Ahlfors or, more generally, in the potential sense of Royden (see e.g.[Ro2]). For such metrics, we have the following result which underlies getting the desired lower estimate.

Lemma 3.[Ro2] *If a circularly symmetric conformal metric $\lambda(|z|)|dz|$ in Δ has curvature at most -4 in the potential sense, then $\lambda(r) \ge a(1 - a^2r^2)$, where $a = \lambda(0)$.*

On the extremal disk $\Delta(\mu_0^*) = \{\phi_{\mathbb{T}}(t\mu_0^*) : t \in \Delta\} \subset \mathbb{T}$ the infinitesimal Kobayashi-Teichmüller metric $\lambda_{\mathcal{K}}$ of *mathbbT* is isometrically equivalent to hyperbolic metric (9) on Δ of curvature -4. The Grunsky coefficients of $f \in \Sigma^0$ allows us to construct the holomorphic maps

$$\widetilde{h}_{\mathbf{x}}(t) := h_{\mathbf{x}}(\varphi_t) = \sum_{m,n=1}^{\infty} \sqrt{mn}\ \alpha_{mn}(\varphi_t)\ x_m x_n : \ \Delta \to \Delta,$$

where $\varphi_t = S_{f^{t\mu_0^*}}$ and $\mathbf{x} = (x_n) \in S(l^2)$. Then $\sup \{|\widetilde{h}_{\mathbf{x}}(t)| : \mathbf{x} \in S(l^2)\} = \varkappa(f^{t\mu_0^*})$. Pull-backing the hyperbolic metric to $\Delta(\mu_0^*)$ by applying these maps, we get conformal metrics

$$\lambda_{\widetilde{h}_{\mathbf{x}}}(t) := \widetilde{h}_{\mathbf{x}}^*(\lambda) = |\widetilde{h}'_{\mathbf{x}}(t)|/(1 - |\widetilde{h}_{\mathbf{x}}(t)|^2)$$

of Gaussian curvature -4 at noncritical points. Take their upper envelope $\widetilde{(}t) = \sup\{_{\widetilde{h}_{\mathbf{x}}}(t) : \mathbf{x} \in S(l^2)\}$ and pass to upper semicontinuous regularization

$$\lambda_{\varkappa}(t) = \limsup_{t' \to t} \lambda_{\varkappa}(t').$$

This yields a logarithmically subharmonic metric on Δ whose curvature in the supporting and in the potential sense both are less than or equal -4. Its circular mean

$$\mathcal{M}[\lambda_{\varkappa}](|t|) = (2\pi)^{-1} \int_0^{2\pi} \lambda_{\varkappa}(re^{i\theta}) d\theta$$

is a circularly symmetric metric with curvature also at most -4 in the potential sense.

To calculate the value of $\mathcal{M}[\lambda_{\varkappa}](0)$, one can use the standard variational method to the maps $f^\mu \in \Sigma^0$ and to their Grunsky coefficients, which yields

$$\mathcal{M}[\lambda_](0) = \lambda_{\varkappa}(0) = \alpha(f^{\mu_0}).$$

Thereafter, applying Lemma 3, we get

$$\mathcal{M}[\varkappa](r) \geq \alpha(f_0)/[1-\alpha(f_0)^2r^2]$$

and, integrating both sides of this inequality over a radial segment $[0,\varrho]$ with $\varrho = \|\mu_0\|$,

$$\int_0^{\varrho} \mathcal{M}[\varkappa](r)dr \geq \tanh^{-1}[\alpha(f^{\mu_0})\varrho] = \tanh^{-1}[\alpha(f^{\mu_0})k(f^{\varrho\mu_0^*})] = \tanh^{-1}[\alpha(f^{\mu_0})k(f^{\mu_0})].$$

On the other hand, by (19),

$$\int_0^{\varkappa} \lambda_{\varkappa}(t)dt = \tanh^{-1}\varkappa, \quad \varkappa = \varkappa(f^{\mu_0}).$$

Using these relations, one obtains the desired estimate (22).

To get (21), we have to estimate the quantity (6) from below. Applying Proposition 3 and Theorem 4, we can restrict ourselves by finding the minimal value of the functionals $l_\mu(\psi) = |\langle\mu^*,\varphi\rangle|$ on the set $\{\varphi \in A_1^2, \|\varphi\|_1 = 1\}$ for $\mu^* = |\psi|/\psi$ defined by integrable holomorphic functions in Δ of the form

$$\psi(z) = z^m(c_0 + c_1 z + \dots), \quad m = 1,3,5,\dots\ .$$

This minimum equals $\frac{2\sqrt{2}}{3}$ and is attained on the map (21).

3.4. *Geometric applications*

Theorem 3 has interesting geometric consequnces. The inequalities (3) and (20) result in

$$\varkappa(f) \leq k(f) \leq 3\varkappa(f)/(2\sqrt{2}), \tag{23}$$

and similarly for reciprocals of Fredholm eigenvalues of quasicircles $f(S^1)$.

Since $\varkappa(f) \leq c_{\mathbb{T}}(\mathbf{0}, S_f)$ and the universal Teichmüller space *mathbbT* is a homogeneous domain, one obtains from (26) the following inequalities estimating the behavior of invariant metrics on this space.

Theorem 5. *For any two points $\varphi_1, \varphi_2 \in \mathbb{T}$, their Caratheodory and Kobayashi distances are related by*

$$c_{\mathbb{T}}(\varphi_1,\varphi_2) \leq d_{\mathbb{T}}(\varphi_{1,2}) \leq \tanh\Big[\frac{3}{2\sqrt{2}}\tanh^{-1} c_{\mathbb{T}}(\varphi_1,\varphi_2)\Big].$$

3.5.

Note that the equality in (21) is attained by map (21) only asymptotically as $\varkappa \to 0$, while for small $\varkappa$, we have

$$k(f_{3,t}) = |t| = \varkappa^* - O(\varkappa^{*3}) < \varkappa^*, \quad \text{where } \varkappa^* = 3\varkappa(f_{3,t})/(2\sqrt{2}).$$

The approach exploited above can be extended to other important inequalities, for example, to exact estimating of the minimal dilatation $k_0(M)$ of quasiconformal extensions of M-quasisymmetric homeomorphisms of the circle.

References

Ah. L. Ahlfors, *Remarks on the Neumann-Poincaré integral equation*, Pacific J. Math. **2** (1952), 271-280.

Di. S. Dineen, *The Schwarz Lemma*, Clarendon Press, Oxford, 1989.

EKK. C.J. Earle, I. Kra and S.L. Krushkal, *Holomorphic motions and spaces*, Trans. Amer. Math. Soc. **944** (1994), 927-948.

GL. F.P. Gardiner and N. Lakic, *Quasiconformal Teichmüller Theory*, Amer. Math. Soc., 2000.

Gr. H. Grunsky, *Koeffizientenbedingungen für schlicht abbildende meromorphe Funktionen*, Math. Z. **45** (1939), 29-61.

Ko. S. Kobayashi, *Hyperbolic Complex Spaces*, Springer, New York, 1998.

Kr1. S.L. Krushkal, *Grunsky coefficient inequalities, Carathéodory metric and extremal quasiconformal mappings*, Comment. Math. Helv. **64** (1989), 650-660.

Kr2. S.L. Krushkal, *Plurisubharmonic features of the Teichmüller metric*, Publications de l'Institut Mathématique-Beograd, Nouvelle série **75(89)** (2004), 119-138.

Kr3. S.L. Krushkal, *Quasiconformal extensions and reflections*, Ch. 11 in: Handbook of Complex Analysis: Geometric Function Theory, Vol. II (R. Kühnau, ed.), Elsevier Science, Amsterdam, 2005, pp. 507-553.

Kr4. S.L. Krushkal, *Schwarzian derivative and complex Finsler metrics*, Contemporary Math. **382** (2005), 243-262.

Kr5. S.L. Krushkal, *Inversion of Ahlfors and Grunsky inequalities*, Georgian Math. J. **13** (2006), 495-500.

Kr6. S.L. Krushkal, *Rational approximation of holomorphic functions and geometry of Grunsky inequalities*, Contemporary Math., to appear.

KK1. S.L. Krushkal and R. Kühnau, *Quasikonforme Abbildungen - neue Methoden und Anwendungen*, Teubner-Texte zur Math., Bd. 54, Teubner, Leipzig, 1983.

KK2. S. L. Krushkal and R. Kühnau, *Grunsky inequalities and quasiconformal extension*, Israel J. Math. **152** (2006), 49-59.

Ku1. R. Kühnau, *Verzerrungssätze und Koeffizientenbedingungen vom Grunskyschen Typ für quasikonforme Abbildungen*, Math. Nachr. **48** (1971), 77-105.

Ku2. R. Kühnau, *Möglichst konforme Spiegelung an einer Jordankurve*, Jber. Deutsch. Math. Verein. **90** (1988), 90-109.

Ku3. R. Kühnau, *Wann sind die Grunskyschen Koeffizientenbedingungen hinreichend für Q-quasikonforme Fortsetzbarkeit?* Comment. Math. Helv. **61** (1986), 290-307.

Ku4. R. Kühnau, *Über die Grunskyschen Koeffizientenbedingungen*, Ann. Univ. Mariae Curie-Sklodowska, sect. A **54** (2000), 53-60.

Mi. I.M. Milin, *Univalent Functions and Orthonormal Systems*, Transl. of mathematical monographs, vol. 49, Transl. of Odnolistnye funktcii i normirovannie systemy, Amer. Math. Soc., Providence, RI, 1977.

Min. D. Minda, *The strong form of Ahlfors' lemma*, Rocky Mountain J. Math., **17** (1987), 457-461.

Po. Chr. Pommerenke, *Univalent Functions*, Vandenhoeck & Ruprecht, Göttingen, 1975.

Ro1. H.L. Royden, *Automorphisms and isometries of Teichmüller space*, Advances in the Theory of Riemann Surfaces (Ann. of Math. Stud., vol. 66), Princeton Univ. Press, Princeton, 1971, pp. 369-383.

Ro2. H.L. Royden, *The Ahlfors-Schwarz lemma: the case of equality*, J. Anal. Math. **46** (1986), 261-270.

Sc. M. Schiffer, *Fredholm eigenvalues and Grunsky matrices*, Ann. Polon. Math. **39** (1981), 149-164.

SS. M. Schiffer and D. Spencer, *Functionals of Finite Riemann Surfaces*, Princeton Univ. Press, Princeton, 1954.

Sh. Y.L. Shen, *Pull-back operators by quasisymmetric functions and invariant metrics on Teichmüller spaces*, Complex Variables **42** (2000), 289-307.

St. K. Strebel, *On the existence of extremal Teichmueller mappings*, J. Anal. Math **30** (1976), 464-480.

Zh. I.V. Zhuravlev, *Univalent functions and Teichmüller spaces*, Inst. of Mathematics, Novosibirsk, preprint, 1979, 23 pp. (Russian).

DECOMPOSITIONS OF MEROMORPHIC FUNCTIONS OVER SMALL FUNCTIONS FIELDS

CHUNG-CHUN YANG

Department of Mathematics
The Hong Kong Univ. of Sci. & Tech.
Clear Water Bay, Kowloon Hong Kong
mayang@ust.hk

PING LI

Department of Mathematics
Univ. of Sci. & Tech. of China
Hefei, Anhui, 230026
People's Republic of China
pli@ustc.edu.cn

Let $\mathcal{P}$ denote the polynomial ring of meromorphic functions f, with their coefficients being meromorphic functions of f. In this paper, we shall use Nevanlinns's distribution theory to discuss the decompositions of functions in $\mathcal{P}$. As an application of the results, we are able to show the non-existence of admissible solutions f and g of a certain class of functional equations of the form: $P(f) = Q(g)$, where both $P(f)$ and $Q(g)$ belong to $\mathcal{P}$.

1. Introduction and Results

Throughout the the text, we shall use $f(z)$ to denote a nonconstant meromorphic function, and use the standard notations in Nevanlinna's value distribution theory, such as $T(r, f)$, $N(r, f)$ and $m(r, f)$. The notation $S(r, f)$ is defined to be any quantity satisfying $S(r, f) = o(T(r, f))$ as $r \to \infty$ possibly outside a set of r of finite linear measure. For details of the theory and its notations, we refer the readers to [2]. A meromorphic function $a(z)$ ($\not\equiv \infty$) is called a small function of $f(z)$ provided that $T(r, a) = S(r, f)$. It is obvious that the set of all small functions of f is a field of functions.

Let $\mathcal{P}$ denote the polynomial ring of meromorphic functions f, with their coefficients being small meromorphic functions of f. When the coefficients of a polynomial $P(f) \in \mathcal{P}$ are constants, $P(f)$ can be decomposed as a product of linear functions in f. This result is no longer clear when the coefficients are replaced by small functions of f. However, we shall utilize recent results obtained by Yamanoi ([6] and [7]) regarding the generalized second main theorem and its applications to show that if $P(f)$ have few zeros, then $P(f)$ can be decomposed as a product of linear functions or polynomials of degree two in f over small functions field.

Theorem A ([5]). *Let $f(z)$ be a transcendental meromorphic function, and $P(f) \in \mathcal{P}$ with degree n, i.e.,*

$$P(f) = a_n f^n + a_{n-1} f^{n-1} + ... + a_1 f + a_0, \ a_n \not\equiv 0$$

If a_j $(j = 0, ..., n)$ are small functions of f and if $\overline{N}(r, 1/P(f)) = S(r, f)$, then there are only three possibilities below.

(*i*)

$$P(f) = a_n \left(f + \frac{a_{n-1}}{na_n} \right)^n;$$

(*ii*)

$$P(f) = a_n \left(\left(f + \frac{a_{n-1}}{na_n} \right)^2 + \alpha_0 \right)^\mu,$$

where $\alpha \not\equiv 0$ is a small function of f, and $\mu = n/2$ is a positive integer.

(*iii*)

$$P(f) = a_n \left(f + \frac{a_{n-1}}{na_n} - \lambda_1 \alpha_1 \right)^{\mu_1} \left(f + \frac{a_{n-1}}{na_n} - \lambda_2 \alpha_1 \right)^{\mu_2}$$

where $\alpha_1 \not\equiv 0$ is a small function of f, μ_1, μ_2 are positive integers, $\lambda_1, \lambda_2 \in \mathbb{C}$, $\lambda_1 \neq \lambda_2$, and $\mu_1 + \mu_2 = n$, $\lambda_1 \mu_1 + \lambda_2 \mu_2 = 0$.

Moreover, if f satisfy $\overline{N}(r, f) = S(r, f)$, then the first case above must hold.

Here we weaken the condition $\overline{N}(r, 1/P(f)) = S(r, f)$ in Theorem (A), and prove the following

Theorem 1. *Let $f(z)$ be a transcendental meromorphic function, and $P(f) \in P$ with degree n, i.e.,*

$$P(f) = a_n f^n + a_{n-1} f^{n-1} + ... + a_1 f + a_0, \ a_n \not\equiv 0 \tag{1}$$

If a_j $(j = 0, ..., n)$ are small functions of f, then either

$$\overline{N}\left(r, \frac{1}{f - c}\right) \leq \overline{N}\left(r, \frac{1}{P(f)}\right) + S(r, f) \tag{2}$$

holds for any $c \in \mathbb{C}$, or one of the three cases in Theorem A holds.

Theorem 2. *Let $f(z)$ be a transcendental meromorphic function, and $P(f)$ a polynomial in f defined by (1). Then either*

$$T(r,f) \leq \overline{N}\left(r, \frac{1}{P(f)}\right) + \epsilon T(r,f) + S(r,f) \tag{3}$$

holds for any positive number ϵ, or one of the three cases in Theorem A holds.

As applications of Theorem 2, we can obtain the following

Corollary 1. *Suppose that f and g are transcendental meromorphic functions, n and k two positive integers such that $k > n$. Let $P(f)$,$Q(g) \in \mathcal{P}$ with coefficients being small functions of f and g, respectively, and $P(f) = a_n f^n + a_{n-1} f^{n-1} + ... + a_1 f + a_0, \ \ a_n \not\equiv 0$.*

If

$$\overline{N}\left(r, \frac{1}{Q(g)}\right) \leq \frac{1}{k} T(r, Q(g)) + S(r,g), \tag{4}$$

and if f and g satisfy the following functional equation:

$$Q(g) = P(f) \tag{5}$$

then $P(f)$ assumes one of the form described in Theorem A.

Corollary 2. *Let $a_0, a_1, ..., a_n$ be given meromorphic functions. Suppose that f is a transcendental meromorphic functions with $a_0, a_1, ..., a_n$ being its small functions, and $P(f) \in P$ is as defined in Corollary 1. If there exist a positive integer $k > 2n$ and a nonconstant meromorphic function g satisfying the functional equation:*

$$g^k = P(f), \tag{6}$$

then

$$P(f) = a_n \left(f + \frac{a_{n-1}}{n a_n}\right)^n$$

Remark. The condition $k > 2n$ in Corollary 2 is necessary. For example, let h be the elliptic function which satisfy $(h')^2 = 4h^3 + 6h^2 + 4h + 1$, and let $g = 1 + 1/h$, $f = h'/h^2$, then $g^4 = f^2 - 1$.

Also, Corollary 2 shows that the functional equation (6) has no admissible transcendental meromorphic solution when $k > 2n \geq 4$, $a_{n-1} \equiv 0$ and $a_0 a_n \neq 0$. Here and in the sequel, we call a meromorphic function f an admissible solution of a differential or functional equation if the coefficients of the equation are small functions of f.

Corollary 3. *Let $f, g, P(f), Q(g)$ be as defined in Corollary 1. Suppose that k and n are two relatively prime positive integers and $k > 4$, $k \neq 6, 2 \leq n < k$. Then the* *functional equation* (5) *has no admissible transcendental meromorphic solutions* f *and* g *provided that* $a_{n-1} \not\equiv 0$ *and* $a_0 a_n \not\equiv 0$.

Remark. The assumption that $k > 4$ and $k \neq 6$ are necessary. In fact, let h_1 and h_2 are solutions of the equations $(h_1')^2 = h_1(h_1^2 + 3)$ and $(h_2')^2 = h_2^3 + 5$, respectively. Let $f_1 = h_1^2 + 2$, $g_1 = h_1'$ and $f2 = h_2^3 + 3$, $g_2 = h_2 h_2'$. Then we have $g_1^4 = (f_1 - 2)(f_1 + 1)^2$ and $g_2^6 = (f_2 - 3)^2(f_2 + 2)^3$

Suppose that f is a transcendental meromorphic function with $a_1, ..., a_n$ being distinct small functions of f. Then by the second fundamental theorem related to n small functions (see, [6]), the inequality

$$(n-2)T(r,f) \leq \sum_{j=1}^{n} \overline{N}\left(r, \frac{1}{f-a_j}\right) + \epsilon T(r,f) + S(r,f). \tag{7}$$

holds for any positive number ϵ. Let

$$P(f) = (f - a_1)(f - a_2)...(f - a_n), n \geq 3 \tag{8}$$

Then (7) can be rewritten as

$$(n-2)T(r,f) \leq \overline{N}\left(r, \frac{1}{P(f)}\right) + \epsilon T(r,f) + S(r,f). \tag{9}$$

Theorem 2 shows that (9) is still true if $P(f)$ is a polynomial in f of degree 3, and $P(f)$ has no factor of the form $(f - \alpha)^2$, where α is a small function of f.

It is natural to ask: whether or not the inequality (9) remains to be valid for arbitrary polynomials which can not be decomposed into form of (8)? We just realized that the question can be resolved, according to a very recent paper of Yamanoi (Corollary 6.1 in [7]). Also we would like to point out that both the methods used in the present paper and [7] are pertinent to discuss the existence or non-existence of admissible solutions f, g for functional equations of the form: $P(f) = Q(g)$, with $P(f), Q(g) \in P$ of higher degree, see [3]. Furthermore, it appears that both methods may not be useful to deal the similar question with P and Q of lower degree. For further investigations, we propose such a simple looking conjecture as follows:

Conjecture. *Suppose that* $a \in C$ *and* $b(z)$ *a nonconstant meromorphic function. Then the functional equation* $g^2 = f^3 + b(z)f + a$ *has no* admissible meromorphic solutions f and g.

2. Some Lemmas

Lemma 1 (Clunie's lemma, see [1]). *Suppose that* $f(z)$ *is meromorphic and transcendental in the complex plane* $\mathbb{C}$, *and*

$$f^n(z)P(f) = Q(f)$$

where $P(f)$ and $Q(f)$ are differential polynomials in f with the coefficients being small functions of f, and the degree of $Q(f)$ is at most n. Then

$$m(r, P(f)) = S(r, f)$$

Lemma 2. (Corollary 5 in [8]). *Let f be a transcendental meromorphic function, and n a positive integer. Let $Q_1(f)$ and $Q_2(f)$ be two differential polynomials in f not vanish identically. Suppose that F is a differential polynomials in f defined by*

$$F := f^n Q_1(f) + Q_2(f)$$

If $F \not\equiv 0$, then

$$(n - \gamma_{Q_2})T(r, f) \le \overline{N}\left(r, \frac{1}{F}\right) + \overline{N}\left(r, \frac{1}{f}\right) + (\Gamma_{Q_2} - \gamma_{Q_2} + 1)\overline{N}(r, f) + S(r, f)$$

where γ_{Q_2} and Γ_{Q_2} are the degree and weight of Q_2, respectively.

Lemma 3. ([4]). *Suppose that f is a nonconstant meromorphic function. If*

$$R(f) = \frac{a_0(z)f^p(z) + a_1(z)f^{p-1}(z) + ... + a_p(z)}{b_0(z)f^q(z) + b_1(z)f^{q-1}(z) + ... + b_q(z)}$$

is an irreducible rational polynomial in f with coefficients being small functions of f, and $a_0(z)b_0(z) \not\equiv 0$, then

$$T(r, R(f)) = max\{p, q\}T(r, f) + S(r, f).$$

Lemma 4. *Let f be a transcendental meromorphic function, and*

$$P = a_n f^{2n} + a_{n-1} f^{2n-2} + ... + a_1 f^2 + a_0,$$

where $a_0, ..., a_n$ are small functions of f, and $a_n \neq 0$. Then either

$$T(r, f) \le \overline{N}\left(r, \frac{1}{P}\right) + S(r, f), \tag{10}$$

or

$$P = a_n \left(f^2 + \frac{a_{n-1}}{na_n}\right)^n. \tag{11}$$

Proof. Let $g = f^2 + \frac{a_{n-1}}{na_n}$. Then

$$P = a_n g^n + b_{n-2} g^{n-2} + ... + b_1 g + b_0,$$

where $b_0, ..., b_{n-2}$ are polynomials in $a_0, ..., a_n$. If $b_{n-2}g^{n-2} + ... + b_1 g + b_0 \equiv 0$ then ([11]) holds. If $b_{n-2}g^{n-2} + ... + b_1 g + b_0 \not\equiv 0$, then by Lemma 2 , we have

$$2T(r, g) \le \overline{N}\left(r, \frac{1}{F}\right) + \overline{N}\left(r, \frac{1}{g}\right) + \overline{N}(r, g) + S(r, g),$$

which implies (10).

Lemma 5. *Let f be a transcendental meromorphic function, and*

$$P = a_n f^n + a_{n-1} f^{n-1} + ... + a_2 f^2 + 1,$$

where $a_2, ..., a_n$ $(n > 3)$ are small functions of f, and $a_n \not\equiv 0$. Then either

$$\overline{N}(r, f) \leq \overline{N}\left(r, \frac{1}{P}\right) + S(r, f), \ \overline{N}\left(r, \frac{1}{f}\right) \leq \overline{N}\left(r, \frac{1}{P}\right) + S(r, f), \tag{12}$$

or P satisfies one of the following two decompositions:

(i) There exist a small function α_0 of f and a positive integer $\mu = n/2$ such that

$$P = a_n (f^2 + \alpha_0)^\mu. \tag{13}$$

(ii) There exist a small function α_1 of f, positive integers μ_1, μ_2 and constants λ_1, λ_2 such that $n = \mu_1 + \mu_2$, $\lambda_1 \mu_2 + \lambda_2 \mu_1 = 0$, and

$$P = a_n (f + \lambda_1 \alpha_1)^{\mu_1} (f + \lambda_2 \alpha_1)^{\mu_2}. \tag{14}$$

Proof. If $a_2 \equiv 0$ then $P = f^3 \sum_{k=3}^{n} a_k f^{k-3} + 1$. By Lemma (2), we have

$$3T(r, f) \leq \overline{N}\left(r, \frac{1}{P}\right) + \overline{N}\left(r, \frac{1}{f}\right) + \overline{N}(r, f) + S(r, f),$$

which implies (12).

In the sequel, we suppose $a_2 \not\equiv 0$. Let $P_1 = a_n f^{n-2} + a_{n-1} f^{n-3} + ... + a_2$. Then P can be rewritten as $P = f^2 P_1 + 1$. By Lemma 2, we have

$$2T(r, f) \leq \overline{N}\left(r, \frac{1}{P}\right) + \overline{N}\left(r, \frac{1}{f}\right) + \overline{N}(r, f) + S(r, f) \tag{15}$$

which is equivalent to

$$m(r, f) + N_1(r, f) + m\left(r, \frac{1}{f}\right) + N_1\left(r, \frac{1}{f}\right) \leq \overline{N}\left(r, \frac{1}{P}\right) + S(r, f) \tag{16}$$

Let

$$a = f' P_1 + \frac{1}{2} f P_1' - \frac{1}{2} f P_1 \frac{P'}{P}, \tag{17}$$

and

$$G = 2 f' P_1 + f P_1'. \tag{18}$$

Then we have

$$a = \frac{G}{2P}, \quad af = \frac{1}{2} \cdot \frac{P'}{P}. \tag{19}$$

$$T(r, a) \leq \overline{N}\left(r, \frac{1}{P}\right) + S(r, f) \tag{20}$$

If $a \equiv 0$, then P is a constant. This is impossible. Therefore, $a \not\equiv 0$. It follows from the above equations that the poles of a must come from the zeros of P if they are not the poles of the coefficients of P. Thus $N(r, a) \leq \overline{N}(r, 1/P) + S(r, f)$. On the other hand, from the second equation in (19) and by Lemma 1, we have $m(r, a) \leq S(r, f)$. Hence

Suppose that z_0 is a simple pole of f, but not a zero or pole of the coefficients of P. At the neighborhood of z_0, we have Laurent expansion

$$f(z) = \frac{c}{z - z_0} + d + O(z - z_0) = \frac{c}{z - z_0}\left(1 + \frac{d}{c}(z - z_0) + O(z - z_0)^2\right), \tag{21}$$

where c, d are complex number and $c \neq 0$. Simple computation shows that

$$\frac{f'(z)}{f(z)} = \frac{-1}{z - z_0}\left(1 - \frac{d}{c}(z - z_0) + O(z - z_0)^2\right), \tag{22}$$

$$\frac{P'(z)}{P(z)} = \frac{-n}{z - z_0}\left(1 - B(z - z_0) + O(z - z_0)^2\right), \tag{23}$$

hold at the neighborhood of z_0, where

$$B = \frac{nda_n(z_0) + ca_n^{'}(z_0) + a_{n-1}(z_0)}{nca_n(z_0)}$$

Therefore,

$$\frac{P'(z)}{P(z)} - n\frac{f'(z)}{f(z)} = \frac{ca_n^{'}(z_0) + a_{n-1}(z_0)}{ca_n(z_0)} + O(z - z_0).$$

From (19), (21) and (23), we get $c = -\frac{c}{2a(z_0)}$ and thus

$$\frac{P'(z)}{P(z)} - n\frac{f'(z)}{f(z)} = \frac{a_n^{'}(z_0)}{a_n(z_0)} - \frac{2a(z_0)a_{n-1}(z_0)}{na_n(z_0)} + O(z - z_0),$$

which shows that z_0 is a zero of the following function

$$\varphi := \frac{P'}{P} - n\frac{f'}{f} - \frac{a_n^{'}}{a_n} + \frac{2aa_{n-1}}{na_n} \tag{25}$$

Let $R = a_{n-1}f^{n-1} + ... + a_2 f^2 + 1$.Then we have $P = a_n f^n + R$. Furthermore, we have

$$\frac{P'}{P} - n\frac{f'}{f} - \frac{a_n^{'}}{a_n} = \left\{ \frac{R'}{a_n f^n} - \frac{R}{a_n f^n}\left(n\frac{f'}{f} + \frac{a_n^{'}}{a_n}\right) / \left(1 + \frac{R}{a_n f^n}\right) \right\}$$

which shows that the multiple poles of f must be the zeros of $\frac{P'}{P} - n\frac{f'}{f} - \frac{a_n^{'}}{a_n}$ as long as they are not the zeros or poles of the coefficients of P.Note from (19) that such multiple poles of f are zeros of a, and hence they are zeros of φ. From (17), (18), (19) and (25), we have

$$a_2 f\varphi + na = fH \tag{26}$$

where

$$H := a_2\frac{P'}{P} - \frac{n}{2}P_1\frac{P'}{P} + nf^{'}\sum_{k=1}^{n-2} a_{k+2}f^{k-1} + \frac{n}{2}P_1' + \frac{2aa_2a_{n-1}}{na_n} - \frac{a_2a_n^{'}}{a_n} \tag{27}$$

From (26), we see that any pole of f is a zero of H as long as it is not a pole or zero of the coefficients of P. By Lemma 1, we have $m(r, H) = S(r, f)$. Therefore,

$$T(r, H) \leq \overline{N}\left(r, \frac{1}{P}\right) + S(r, f) \tag{28}$$

Let

$$H_1 := H - \frac{3na_3a}{2a_2} - \frac{2a_2a_{n-1}a}{na_n} - \frac{n}{2}a_2^{'} + \frac{a_2a_n^{'}}{a_n} \tag{29}$$

Clearly, $m(r, H_1) = S(r, f)$. From (17) and (27), we get $N(r, H_1) = N(r, 1/P) + S(r, f)$.
Therefore,

$$T(r, H_1) \leq \overline{N}\left(r, \frac{1}{P}\right) + S(r, f) \tag{30}$$

From (17), (18), (19) and (27), we deduce that

$$H_1 = 2a_2af - naP_1f + nf'\sum_{k=4}^{n} a_k f^{k-3} + na_3f' + \frac{n}{2}\sum_{k=2}^{n} a_k' f^{k-2}$$
$$+\frac{n}{2}\sum_{k=3}^{n}(k-2)a_k f^{k-3}f' - \frac{3na_3a}{2a_2} - \frac{n}{2}a_2'$$
$$= f\left[2a_2a - naP_1 + \frac{n}{2}\sum_{k=4}^{n} ka_k f^{k-4}f' + \frac{n}{2}\sum_{k=3}^{n} a_k' f^{k-3}\right] + \frac{3na_3}{2a_2}(a_2f' - a)$$

Since

$$a_2f' - a = a_2f' - f'P_1 - \frac{1}{2}fP_1' + \frac{1}{2}fP_1\frac{P'}{P}$$
$$= a_2f' - \sum_{k=2}^{n} a_k f^{k-2}f' - \frac{1}{2}fP_1' + \frac{1}{2}fP_1\frac{P'}{P}$$
$$= f\left[-\sum_{k=3}^{n} a_k f^{k-3}f' - \frac{1}{2}P_1' + afP_1\right],$$

it follows that

$$H_1 = f\left[2a_2a - naP_1 + \frac{n}{2}\sum_{k=4}^{n} ka_k f^{k-4}f' + \frac{n}{2}\sum_{k=3}^{n} a_k' f^{k-3}\right.$$
$$\left.-\frac{3na_3}{2a_2}\left(\sum_{k=3}^{n} a_k f^{k-3}f' + \frac{1}{2}P_1' - afP_1\right)\right].$$

Therefore, any zero of f is a zero of H_1 as long as it is not the zero or pole of the coefficients $a_2, ..., a_n$. Thus, if $H \not\equiv 0$ and $H_1 \not\equiv 0$, then

$$\overline{N}(r, f) \leq N(r, \frac{1}{H}) \leq \overline{N}(r, \frac{1}{P}) + S(r, f),$$

$$\overline{N}(r, \frac{1}{f}) \leq N(r, \frac{1}{H_1}) \leq \overline{N}(r, \frac{1}{P}) + S(r, f),$$

which prove (12).

In the following we distinguish two cases.

Case 1. $H \equiv 0$.

In this case, from (26) and the first equation in (19), we get

$$f' = \frac{2a}{n}f^2 + \frac{1}{n}\left(\frac{2aa_{n-1}}{na_n} - \frac{a_n'}{a_n}\right)f + \frac{a}{a_2}, \tag{31}$$

which can be rewritten as

$$f' = Q_2 a - \frac{a_n'}{na_n} f, \tag{32}$$

where

$$Q_2 := \frac{2}{n} f^2 + \frac{2a_{n-1}}{n^2 a_n} f + \frac{1}{a_2}. \tag{33}$$

Substituting $P_1 = \sum_{k=1}^{n} a_k f^{n-2}$ and (32) into (18), and from equation (19), we obtain

$$2aP = \sum_{k=2}^{n} a_k' f^{n-1} + \left(\sum_{k=2}^{n} k a_k f^{n-2}\right)\left(Q_2 a - \frac{a_n'}{na_n} f\right),$$

which is equivalent to

$$aQ = \sum_{k=2}^{n-1} \left(\frac{k a_k a_n'}{na_n} - a_k'\right) f^{k-1}, \tag{34}$$

where

$$Q := Q_2 \left(\sum_{k=2}^{n} k a_k f^{n-2}\right) - 2P. \tag{35}$$

From the definition of Q_2 and P, it follows from the above equation that

$$Q = \sum_{k=2}^{n-1} \left(\frac{k+1}{a_2} a_{k+1} + \frac{2k a_{n-1}}{n^2 a_n} - \frac{2(n-k+1)}{n} a_{k-1}\right) f^{k-1}, \tag{36}$$

where $a_1 = 0$.

If $Q \equiv 0$, then from (34) and (36) we get

$$\frac{k+1}{a_2} a_{k+1} + \frac{2k a_{n-1}}{n^2 a_n} - \frac{2(n-k+1)}{n} a_{k-1} \equiv 0,\ k = 2, ..., n-1. \tag{37}$$

$$\frac{k a_k a_n'}{na_n} - a_k' \equiv 0,\ k = 2, ..., n-1, \tag{38}$$

If $a_n \equiv 0$ then it follows from (37) that $a_{2k-1} \equiv 0$, $k = 1, 2, ...$ Note that if $a_n \not\equiv 0$, then n is even. By Lemma 4, we can obtain the conclusion easily.

In the sequel, we assume $a_n \not\equiv 0$.From (38), we see that there exists a nonzero constant A such that $a_n^{n-1} \equiv A a_{n-1}^n$. Since n and n-1 are relatively prime, thus there exists a meromorphic function h which is small with respect to f such that $a_n = h^n$. Therefore, by (38), we can find constants d_k $(k = 2, ..., n)$ such that

$$a_k = d_k h^k, \quad k = 2, ..., n.$$

Thus

$$P = \sum_{k=2}^{n} d_k (hf)^k + 1 = d_n (hf - r_1)...(hf - r_n),$$

where r_k, $k = 1, ..., n$ are roots of the equation $\sum_{k=2}^{n} d_k z^k + 1 = 0$. If this equation has at least three different roots r_1, r_2, r_3, then by Nevanlinna's second fundamental theorem, we have

$$T(r,f) \leq \sum_{k=1}^{3} \overline{N}\left(r, \frac{1}{f - r_k/h}\right) + S(r,f) \leq \overline{N}\left(r, \frac{1}{P}\right) + S(r,f),$$

which yields (12) by Nevanlinns's first fundamental theorem. If all r_k are equal to each other, then we have $P = a_n(f - r_k/h)^n$, which contradicts $a_1 = 0$ and $a_0 = 1$. If two and only two of r_k are different, then P assumes the form (14).

Now we consider the case: $Q \neq 0$. From (34), we can see that a is a rational function in f with coefficients being small functions of f. If this rational function can not be reduced to a small function of f, then by Lemma 3, we have

$$T(r,f) \leq T(r,a) + S(r,f) \leq \overline{N}(r, 1/P) + S(r,f),$$

which leads to (12). If this rational function can be reduced to a small function of f, then $T(r,a) \leq S(r,f)$. Thus from the second equation in (19) we get

$$\overline{N}(r, 1/P) \leq \overline{N}(r, 1/a) \leq S(r,f)$$

In terms of Theorem A, F has one of the three forms described in Theorem A. Since $a_1 = 0$ and $a_0 = 1$, the first form cannot occur. The second form and third form yield (13) and (14), respectively.

Case 2: $H_1 \equiv 0$

From (25), (26) and (29), we can deduce that

$$f' = \frac{2a}{n} f^2 - \frac{1}{2a_2}\left(\frac{3a_3 a}{a_2} + a_2'\right) f + \frac{a}{a_2} \tag{39}$$

By the arguments similar to that in Case 1, we can obtain

$$a\widetilde{Q} = \sum_{k=2}^{n} \left(a_k' - \frac{k a_2'}{2a_2} a_k\right) f^{k-1} \tag{40}$$

where

$$\widetilde{Q} := \sum_{k=3}^{n} \left(\frac{2(n-k+1)}{n} a_{k-1} + \frac{3ka_3}{2a_2^2} a_k - \frac{k+1}{a_2} a_{k+1} \right) f^{k-1} \tag{41}$$

with $a_{n+1} = 0$. By similar arguments used for Case 1, we still get the conclusion that Case 2 holds. The proof of Lemma 5 is thus completed.

Lemma 6. *Let $f(z)$ be a transcendental meromorphic function, and $P(f) \in \mathcal{P}$ defined by (1). Then either*

$$\overline{N}(r, f) \leq \overline{N}\left(r, \frac{1}{P(f)}\right) + S(r, f) \tag{42}$$

or one of the three cases in Theorem A holds.

Proof. Without loss of generality, we assume $a_{n-1} \equiv 0$, otherwise, we can consider the transformation $f = \widetilde{f} - a_{n-1}/na_n$. Thus

$$P(f) = a_n f^n + a_{n-2} f^{n-2} + ... + a_1 f + a_0 \tag{43}$$

If all $a_0, ..., a_{n-2}$ vanish identically, then $P(f) = a_n f^n$. This means that $P(f)$ satisfies the first case of Theorem A. Suppose that one of $a_0, ...,$ and a_{n-2} do not vanish identically. If $a_{n-2} \equiv 0$, then by Lemma 2, we have

$$2T(r, f) \leq \overline{N}\left(r, \frac{1}{P(f)}\right) + \overline{N}\left(r, \frac{1}{f}\right) + S(r, f)$$

which implies (42). In the sequel, we assume $a_{n-2} \not\equiv 0$.

If $a_0 \equiv 0$, then we can assume $a_0 \equiv ... a_{k-1} \equiv 0, a_k \neq 0, 1 \leq k \leq n-2$. Therefore, $P(f) = f^k \Phi$, where $\Phi = a_n f^{n-k} + a_{n-2} f^{n-k-2} + ... + a_k$. By Lemma 2, we get

$$2T(r, f) \leq \overline{N}\left(r, \frac{1}{\Phi}\right) + \overline{N}\left(r, \frac{1}{f}\right) + \overline{N}(r, f) + S(r, f)$$

Since $\overline{N}(r, 1/P(f)) = \overline{N}(r, 1/f) + \overline{N}(r, 1/\Phi) + S(r, f)$, the above equation also implies (42).

Suppose $a_0 \neq 0$. From (43), we have $P(f) = a_n f^n \Psi$, where

$$\Psi = b_0 h^n + b_1 h^{n-1} + ... + b_{n-2} h^2 + 1, \ h = \frac{1}{f}, \ b_i = a_i/a_n, \ i = 0, 1, ..., n-2.$$

Therefore, $\overline{N}(r, 1/P(f)) = \overline{N}(r, 1/\Psi) + S(r, h)$. By Lemma 5, we have

$$\overline{N}\left(r, \frac{1}{h}\right) \leq \overline{N}\left(r, \frac{1}{\Psi}\right) + S(r, h)$$

which is equivalent to (42), or the following two cases hold:

(i) There exists a small function $\widetilde{\alpha}_0 \not\equiv 0$ of h and positive integer μ such that $n = 2\mu$, and

$$\Psi = b_o(h^2 + \widetilde{\alpha}_0)^{\mu}$$

Hence

$$P(f) = a_n b_o(1 + \widetilde{\alpha}_0 f^2)^{\mu} = a_n(f^2 + \alpha_0)^{\mu}, \ \alpha_0 = \frac{1}{\widetilde{\alpha}_0}$$

which means that F satisfies the second case in Theorem A.

(*ii*) There exists a small function $\widetilde{\alpha}_1 \not\equiv 0$ of h and positive integers μ_1, μ_2, constants $\widetilde{\lambda}_1$and $\widetilde{\lambda}_2$ such that $n = \mu_1 + \mu_2, \widetilde{\lambda}_1\mu_2 + \widetilde{\lambda}_2\mu_1 = 0$, and

$$\Psi = b_o(h + \widetilde{\lambda}_1\widetilde{\alpha}_1)^{\mu_1}(h + \widetilde{\lambda}_2\widetilde{\alpha}_1)^{\mu_2}$$

Hence

$$\begin{aligned} P(f) &= a_n b_o(1 + \widetilde{\lambda}_1\widetilde{\alpha}_1 f)^{\mu_1}(1 + \widetilde{\lambda}_2\widetilde{\alpha}_1 f)^{\mu_2} \\ &= a_n(f - \lambda_1\alpha_1)^{\mu_1}(f - \lambda_2\alpha_1)^{\mu_2} \end{aligned}$$

where $\alpha_1 = -1/(\widetilde{\lambda}_1\widetilde{\lambda}_2\widetilde{\alpha}_1)$, $\lambda_1 = \widetilde{\lambda}_2$, $\lambda_2 = \widetilde{\lambda}_1$. This means that $P(f)$ satisfies the third case in Theorem A. This also completes the proof of Lemma 6.

3. Proof of Theorem 1

Let $g = f - c$. Then $P(f)$ can be expressed as

$$P(f) = b_o g^n + b_1 g^{n-1} + ... + b_{n-1}g + b_n \tag{44}$$

where $b_k (k = 0, 1, ..., n)$ is a polynomial in $a_0, ..., a_n$, and

$$b_0 = a_n,\ b_n = a_n c^n + a_{n-1}c^{n-1} + ... + a_1 c + a_0 \tag{45}$$

If $b_n \equiv 0$, then from (44) we get

$$\overline{N}\left(r, \frac{1}{g}\right) \leq \overline{N}\left(r, \frac{1}{P(f)}\right) + S(r, f) \tag{46}$$

i.e., (2) holds. Suppose $b_n \not\equiv 0$. We have $P(f) = g^n\Psi$, where

$$\Psi := b_n h^n + b_{n-1}h^{n-1} + ... + b_1 h + b_0 \tag{47}$$

and $h = 1/g$. It is obvious that

$$\overline{N}\left(r, \frac{1}{P(f)}\right) = \overline{N}\left(r, \frac{1}{\Psi}\right) + S(r, f) \tag{48}$$

By Lemma 6, we have

$$\overline{N}(r, h) \leq \overline{N}\left(r, \frac{1}{\Psi}\right) + S(r, h)$$

which implies (2), or one of the following three cases holds:

(A)

$$\Psi = b_n\left(h + \frac{b_{n-1}}{nb_n}\right)^n$$

(B)

$$\Psi = b_n\left(\left(h + \frac{b_{n-1}}{nb_n}\right)^2 + \alpha_0\right)^\mu$$

where $\alpha_0 \not\equiv 0$ is a small function of h, and $\mu = n/2$ is a positive integer.

(C)

$$\Psi = b_n \left(h + \frac{b_{n-1}}{nb_n} - \lambda_1 \alpha_1\right)^{\mu_1} \left(h + \frac{b_{n-1}}{nb_n} - \lambda_2 \alpha_1\right)^{\mu_2}$$

where $\alpha_1 \not\equiv 0$ is a small function of h, μ_1, μ_2 are positive integers, $\lambda_1, \lambda_2 \in \mathbb{C}$, $\lambda_1 \not\equiv \lambda_2$,and $\mu_1 + \mu_2 = n$, $\lambda_1\mu_1 + \lambda_2\mu_2 = 0$. If needed, by a suitable transformation, we can see that $P(f)$ must satisfy one of the three cases in Theorem A. This also completes the proof of Theorem 1.

4. Proof of Theorem 2

If $P(f)$ does not assume any form in (i), (ii), or (iii) of Theorem A, then by Lemma 1, we see that (2) holds for any $c \in \mathbb{C}$. For any positive number ϵ, we can select a positive integer k such that $k > 2/\epsilon$. Let $c_1, c_2, ..., c_k$ be distinct values in $\mathbb{C}$. By Nevanlinna's second fundamental theorem, we have

$$(k-2)\,T(r,f) \leq \sum_{j=1}^{k} \overline{N}\left(r, \frac{1}{f-c_j}\right) + S\,(r,f) \leq k\overline{N}\left(r, \frac{1}{P(f)}\right) + S(r,f)$$

which yields (3). This completes the proof of Theorem 2.

5. Proof of Corollary 1

Suppose that $f, g, P(f)$ and $Q(g)$ are as defined with (5) being satisfied.If $P(f)$ does not satisfy any form described in Theorem A, then by Theorem 2 we have

$$T(r,f) \leq \bar{N}\left(r, \frac{1}{P(f)}\right) + \varepsilon T(r,f) + S(r,f).$$

On the other hand, by (5) and (4), we have $T(r,Q(g)) = nT(r,f) + S(r,f)$, and

$$\bar{N}\left(r, \frac{1}{P(f)}\right) = \bar{N}\left(r, \frac{1}{Q(f)}\right) \leq T(r, Q(g)) + S(r,g) \leq \frac{n}{k}T(r,f) + S(r,f).$$

Therefore,

$$\left(1 - \frac{n}{k}\right) T(r,f) \leq \varepsilon T(r,f) + S(r,f),$$

which is impossible for $\varepsilon < 1 - n/k$. Hence $P(f)$ must satisfy one of the form described in Theorem A.

6. Proof of Corollary 2

Since $g^k = P(f)$, we have $kT(r,g) = nT(r,f) + S(r,f)$. If $T(r,f) \leq \frac{3}{2}\overline{N}\,(r, 1/P(f)) + S(r,f)$, then we have

$$T(r,f) \leq \frac{3}{2}\bar{N}\left(r, \frac{1}{g}\right) + S(r,f) \leq \frac{3}{2}T(r,g) + S(r,f).$$

Therefore, the above equation leads to $T(r,f) \leq (3n/2k)T(r,f) + S(r,f)$,i.e.,$T(r,f) = S(r,f)$, a contradiction.

If $P(f)$ satisfies (ii) in Theorem A, then we have $g^k = a_n\left(f^2+\alpha_0\right)^{\mu}$, where $\alpha_0 \not\equiv 0$ is small function of f, and $n = 2\mu$.Therefore, the multiplicities of poles of f are at least k/n, and the multiplicities of zeros of $f^2+\alpha_0$ are least k/μ. By Nevanlinna's second fundamental theorem, we have

$$\begin{aligned} T(r,f^2) &\leq \bar{N}(r,f) + \bar{N}\left(r,\frac{1}{f^2+\alpha_0}\right) + \bar{N}\left(r,\frac{1}{f}\right) + S(r,f) \\ &\leq \frac{n}{k}T(r,f) + \frac{\mu}{k}T(r,f^2) + T(r,f) + S(r,f), \end{aligned}$$

which implies $T(r,f) \leq (2n/k)T(r,f) + S(r,f)$.This contradicts $k > 2n$. Hence F can not satisfy (ii) in Theorem A.

By a similar argument, we can show that $P(f)$ can not satisfy (iii) in Theorem A. Hence by Theorem 2, $P(f)$ must satisfy (i) in Theorem A. This also completes the proof of Corollary 2.

7. Proof of Corollary 3

Suppose that (f,g) is a pair of admissible solutions of (5). Since $a_{n-1} \equiv 0$ and $a_0a_n \not= 0$, we see immediately that F can not satisfy (i) in Theorem A.

If $P(f)$ satisfies (iii) in Theorem A, then we have

$$g^k = a_n\left(f-\lambda_1\alpha_1\right)^{\mu_1}\left(f-\lambda_2\alpha_1\right)^{\mu_2}, \tag{50}$$

where $\alpha_1 \not\equiv 0$ is a small function of f, μ_1,μ_2 are positive integers, $\lambda_1,\lambda_2 \in \mathbb{C}, \lambda_1 \neq \lambda_2$,and $\mu_1+\mu_2 = n$, $\lambda_1\mu_1 + \lambda_2\mu_2 = 0$.

Let dj be the greatest common divisor of k and $\mu_j (j = 1,2)$. From (50), we see that the multiplicity of any zero of $f - \lambda_j\alpha_1$ is at least k/d_j. Since k and n are maturely prime, the multiplicity of any pole of f is at least k. By Nevanlinna's second fundamental theorem, we have

$$\begin{aligned} T(r,f) &\leq \bar{N}(r,f) + \bar{N}\left(r,\frac{1}{f-\lambda_1\alpha_1}\right) + \bar{N}\left(r,\frac{1}{f-\lambda_2\alpha_1}\right) + S(r,f) \\ &\leq \frac{1}{k}N(r,f) + \frac{d_1}{k}N\left(r,\frac{1}{f-\lambda_1\alpha_1}\right) + \frac{d_2}{k}N\left(r,\frac{1}{f-\lambda_2\alpha_1}\right) + S(r,f) \\ &\leq \frac{1+d_1+d_2}{k}T(r,f) + S(r,f), \end{aligned}$$

which implies $k \leq d_1+d_2+1$. Since $\mu_1+\mu_2 = n < k$ and $d_j \leq \mu_j$, we have $n < k \leq n+1$. Therefore, $k = n+1$. If $d_1 < \mu_1$ or $d_2 < \mu_2$, then $d_1+d_2 < \mu_1+\mu_2 = n$, which yields $k \leq n$, a contradiction to the assumption. Hence $d_j = \mu_j$, $j = 1,2$. Without loss of generality, we assume $d_1 \leq d_2$. Therefore, $k \leq 2d_2+1$. Note that $n > 3$ and k/d_2 is an integer. We deduce that $k = 2d_2$. And thus $d_2 = d_1+1$, which shows that d_1 and d_2 are relatively prime. Note that d_1 can divide k. Thus, there are two possibilities: $d_1 = 1$ or $d_1 = 2$. Therefore, $k = 4$ or $k = 6$, which contradicts the assumption. Hence F can not satisfy (iii) in Theorem A. A similar arguments shows that $P(f)$ can not satisfy (ii) in Theorem A as well. By Theorem 2, we see that (3) holds for any positive number ε. Therefore,

$$T(r,f) \leq \bar{N}\left(r, \frac{1}{g}\right) + \varepsilon T(r,f) + S(r,f) \leq T(r,g) + \varepsilon T(r,f) + S(r,f).$$

From (5), we get $T(r,g) = \frac{n}{k}T(r,f) + S(r,f)$. It follows from the above inequality that $\left(1 - \frac{n}{k}\right) T(r,f) \leq \varepsilon T(r,f) + S(r,f)$, which is impossible for $\varepsilon < 1 - n/k$. This also completes the proof of Corollary 3.

References

1. Clunie, J., On integral and meromorphic functions, J. London Math. Soc., 37(1962), 17–27.
2. Hayman, W., Meromorphic Functions, Oxford, Clarendon Press, 1964.
3. Li, P. and Yang, C.-C., Admissible Solutions of Functional Equations of Diophantine Type, preprint.
4. Mohon'ko, A., The Nevanlinna characteristics of certain meromorphic functions, Teor. Funktsiĭ Funktsional. Anal. i Prilozhen., 14(1971), 83–87.
5. Mues, E. and Steinmetz, N., The theorem of Tumura-Clunie for meromorphic function, J. London Math. Soc., 23(1981), no. 2, 113–122.
6. Yamanoi, K., The second main theorem for small functions and related problems, Acta Math. 192(2004), no. 2, 225–294.
7. Yamanoi, K., The second main theorem for small functions and related problems, Acta Math. 192(2004), no. 2, 225–294.
8. Yang, C.-C. and Yi, H.-X., Quantitative estimations on the zeros of differential polynmials, Science in China, Series A , 1(1993), no. 23, 9–20.

SPEED OF APPROXIMATION TO DEGENERATE QUASICONFORMAL MAPPINGS AND STABILITY PROBLEMS

V.M. MIKLYUKOV

Volgograd St. University, Russia

There are many results concerning existence and uniqueness for degenerate Beltrami systems (see Pesin [1969], Lehto [1970], Miklyukov and Suvorov [1972], Kruglikov [1973], Belinskii [1974], David [1988], Tukia [1991], Brakalova, Jenkins [1998], Gutlyanskii, Martio, Sugawa and Vuorinen [2001], Martio and Miklyukov [2002] at al.) The proofs are based on an existence and uniqueness theorem for nondegenerate quasiconformal mappings and on approximations of the degenerate Beltrami system

$$w_{\bar{z}} = \mu(z)\, w_z\,, \qquad \operatorname{ess\,sup} |\mu(z)| = 1\,,$$

by nondegenerate systems

$$w_{n,\bar{z}} = \mu_n(z)\, w_{n,z}\,, \qquad \operatorname{ess\,sup} |\mu_n(z)| < 1\,, \qquad n = 1, 2, \ldots .$$

A principal part of the proofs consists of convergence of $\{w_n\}$ to a homeomorphic mapping $w = w_\mu(z)$. The different methods providing convergence imply different existence theorems.

We study the speed of convergence

$$w_n \to w_\mu \quad \text{as} \quad n \to \infty\,.$$

This problem is connected with the problem of stability for conformal maps in the class of quasiconformal maps.

1. Isothermal coordinates

Let $D \subset \mathbb{R}^2$ be a domain, and let Ω be a two-dimensional surface in $\mathbb{R}^m$, $m \geq 3$, defined by a monotone $W^{1,2}_{\mathrm{loc}}$ vector function

$$\xi = f(x_1, x_2) = (f_1(x), \ldots, f_m(x)) : D \to \mathbb{R}^m \tag{1}$$

which realizes a homeomorphic map of D onto a set $f(D)$ whose metric (and topology) is induced from $\mathbb{R}^m$.

The vector function f is differentiable a.e. on D. Below we assume that

$$\operatorname{rank}(df) = 2 \quad \text{a.e. on } D\ . \tag{2}$$

At each point $x \in D$, where f is differentiable, the condition (2) is equivalent to

$$g = g_{11}g_{22} - g_{12}^2 > 0\,,$$

where

$$g_{ij} = \left\langle \frac{\partial f}{\partial x_i}, \frac{\partial f}{\partial x_j} \right\rangle \qquad (i,j = 1,2)$$

are the coefficients of the first quadratic form Ω.

If

$$g_{11}(x) = g_{22}(x)\,, \quad g_{12}(x) = 0 \quad \text{a.e. on} \quad D\,, \tag{3}$$

then (x_1, x_2) are the isothermal coordinates on Ω. In this case, we have

$$ds_\Omega^2 = \lambda^2(x)\,(dx_1^2 + dx_2^2)\,, \quad \lambda^2(x) = g_{11}(x) = g_{22}(x)\,.$$

The assumption (3) means that $f : D \to \Omega$ preserves a.e. angles between the curves and that f is conformal a.e. in the usual sense.

Let D be a simply connected domain in $\mathbb{R}^2$ and Ω be a surface given over D by a vector function (1) satisfying (2). Let $x = (x_1, x_2) \in D$ be a point, at which f is differentiable and (2) holds. The condition (2) implies that at this point $df \neq 0$, the metric ds_Ω does not degenerate, and every infinitesimal circle with respect to ds_Ω centered at x is an infinitesimal ellipse with respect to the Euclidean metric. Denote by $p(x) \geq 1$ and $0 \leq \theta(x) < \pi$ the characteristics of such ellipse, i.e., the ratio p of the lengths of its great and small axes and the angle θ between the great axis and the direction $\overrightarrow{Ox_1}$.

A calculation gives

$$p(x) = \frac{g_{11}(x) + g_{22}(x)}{2\sqrt{g_{11}(x)g_{22}(x) - g_{12}^2(x)}} + \left(\frac{(g_{11}(x) + g_{22}(x))^2}{4\,(g_{11}(x)g_{22}(x) - g_{12}^2(x))} - 1 \right)^{1/2}.$$

The characteristic $\theta(x)$ is well defined at each point where $p(x) > 1$.

Definition 1.1. *A homeomorphic $W^{1,2}_{\text{loc}}$ map $f : D \subset \mathbb{R}^2 \to \mathbb{R}^2$ is called quasiconformal with M.A. Lavrentiev characteristics $(p(x), \theta(x))$, if it transformes a.e. on D the infinitesimal ellipses with characteristics $(p(x), \theta(x))$ onto infinitesimal circles* [a].

A mapping $f : D \to \mathbb{R}^2$ is called q-quasiconformal, if $\operatorname{ess\,sup}_{x \in D} p(x) \leq q$.

Fix a number sequence $\{Q_n\}_{n=1}^{\infty}$ such that

$$Q_n \geq 1 \quad \text{for all} \quad n = 1, 2, \ldots \quad \text{and} \quad Q_n \to \infty \quad \text{as} \quad n \to \infty. \tag{4}$$

[a] This means that for a.e. $x \in D$ the differential $df(x) : \mathbb{R}^2 \to \mathbb{R}^2$ has characteristics $(p(x), \theta(x))$.

For $n = 1, 2, \ldots$, we set

$$p_n(x) = \begin{cases} p(x) \text{ if } p(x) \le Q_n \\ \\ 1 \text{ if } p(x) > Q_n. \end{cases}$$

By well-known results of the theory of quasiconformal mappings, there exists a quasiconformal map $\xi = w_n(x)$ of D onto an appropriate domain $\Delta_n \subset \mathbb{R}^2$ with the characteristics equal to (p_n, θ) a.e. on D. This domain Δ_n either is $\mathbb{R}^2$ if $D = \mathbb{R}^2$ or it is a simply connected proper subdomain of $\mathbb{R}^2$.

We denote by $B(a, r)$ the disc of radius $r > 0$ with center at $a \in \mathbb{R}^2$ and by $S(a, r)$ its boundary.

Fix two points a_0, $a_1 \in D$. By auxiliary conformal transformations of the plane of variables $\xi = (\xi_1, \xi_2)$, we attain that the domains Δ_n are the discs $B_n = B(0, R_n)$, where $1 < R_n < \infty$, and the maps w_n satisfy

$$w_n(a_0) = (0,0) \quad \text{and} \quad w_n(a_1) = (1,0) \quad (n = 1, 2, \ldots). \tag{5}$$

We call a sequence of maps $F_n = f \circ w_n^{-1} : B_n \to \mathbb{R}^m$ with $w_n : D \to B_n$ satisfying (5) the *canonical sequence of representations* of the surface Ω corresponding to $\{Q_n\}$ with (4).

Fix a sequence of canonical representations $F_n : B_n \to \mathbb{R}^m$ of a surface Ω.

Definition 1.2. *We say that a sequence $\{F_n\}$ converges locally uniformly to a canonical homeomorphism $F : B(0, R) \to \mathbb{R}^m$ if:*

(*i*) *the domains B_n converge as $n \to \infty$ to a domain $B = B(0, R)$ as their kernel with respect to the point $\xi = 0$ (in other words, there exists $R = \lim\limits_{n\to\infty} R_n$, $1 \le R \le \infty$);*

(*ii*) *the sequence $\{F_n\}$ converges as $n \to \infty$ to the map $F : B(0, R) \to \mathbb{R}^m$ uniformly in each subdomain $U \subset\subset B(0, R)$;*

(*iii*) *the sequence of inverse maps $F_n^{-1} : \Omega \to \mathbb{R}^2$ converges as $n \to \infty$ to the map $F^{-1} : \Omega \to B(0, R)$ uniformly on each subdomain $\Omega_1 \subset\subset \Omega$.*

Definition 1.3. *Let $D \subset \mathbb{R}^2$ be a domain. We say that a function $P : D \to \mathbb{R}$ is $W^{1,2}$-majorized in D if there exists a function $K \in W^{1,2}(D)$ such that*

$$P(x) \le K(x) \quad a.e. \text{ on } \quad D. \tag{6}$$

A function P is called locally $W^{1,2}$-majorized in D, if it is $W^{1,2}(D')$-majorized on each subdomain $D' \subset\subset D$.

The bounded functions provide simplest examples of $W^{1,2}$ majorized functions. Let $P : D \to \mathbb{R}$ be a bounded function defined on a domain D, $\mathcal{H}^2(D) < \infty$. Here we can choose $K = \sup_{x\in D} P(x)$. It is clear, that $K \in W^{1,2}(D)$ and (6) holds everywhere on D.

It is not known, how reach is the class of $W^{1,2}$ majorized functions in the general case. We will discuss this question later.

The following theorem is proved in my paper [2004].

Theorem 1.1 *Let Ω be a two-dimensional surface defined by a vector function (1) over a simply connected bounded domain $D \subset \mathbb{R}^2$ and satisfying (2). Let $\{Q_n\}$ be a sequence satisfying (4). Suppose that for a.e. $x \in D$,*

$$\mathcal{P}(x) = \frac{g_{11}(x) + g_{22}(x)}{2\sqrt{g_{11}(x)\, g_{22}(x) - g_{12}^2(x)}} \le K(x)\,, \quad K(x) \in W^{1,2}_{\mathrm{loc}}(D)\,. \qquad (7)$$

Then there exists an isothermal coordinate system $\xi = (\xi_1, \xi_2)$ on Ω defined by a canonical homeomorphism $y = F(\xi)$ of a disc $B(0,R) \subset \mathbb{R}^2$ onto Ω that is a limit of the canonical sequence of representations $F_n(\xi) : B(0,R_n) \to \Omega$ of Ω corresponding to $\{Q_n\}$.

Moreover, there exists a limit $\lim_{n\to\infty} R_n = R$, $1 < R < \infty$, and the sequence $\{F_n\}$ converges to $F(\xi)$ locally uniformly in the disc $B = B(0,R)$.

The canonical homeomorphism $F : B \to \Omega$ is unique.

We have obtained also some explicit two-sided estimates for the radius R in terms of the distance $|a_1 - a_0|$, integral

$$\int_D |\nabla K(x)|^2\, dx_1 dx_2$$

and the infimum of length arcs separating x_1 from a_0 with the ends on ∂D.

The differential properties of the canonical homeomorphism

$$F : B(0,R) \to \mathbb{R}^m$$

and of its inverse F^{-1} are described by certain properties of f and f^{-1}. We will study them separately.

2. Canonical homeomorphisms

Let Ω be a surface of the form (1). Since the metric on $f(D)$ is induced from $\mathbb{R}^m$, 2-dimensional Hausdorff measure $\mathcal{H}^2(E)$, $E \subset f(D)$, is defined on Ω. If $\Omega_1 \subset \mathbb{R}^n$ and $\Omega_2 \subset \mathbb{R}^m$ are two-dimensional surfaces and $h : \Omega_1 \to \Omega_2$ is a homeomorphism, then we say that a map h has the Lusin N-property, if for every subset $E \subset \Omega_1$, $\mathcal{H}^2(E) = 0$, we have $\mathcal{H}^2(h(E)) = 0$.

Let Ω be a surface of the form (1) satisfying (2). Suppose, in addition, that the vector function $f \in W^{1,2}_{\text{loc}}(D)$. Since this map f is homeomorphic, then we have for every subset $E \subset D$ the equality

$$\mathcal{H}^2\left(f(E)\right) = \int_E \sqrt{g_{11}g_{22} - g_{12}^2}\, dx_1 dx_2 \,. \tag{8}$$

It follows from (2) and (8) that

$$\mathcal{H}^2\left(f(E)\right) = 0 \quad \text{if and only if} \quad \mathcal{H}^2\left(E\right) = 0 \tag{9}$$

and that the maps f, f^{-1} have the Lusin N-property.

Since f is differentiable, by (9) a.e. on Ω, there exists a tangent plane $T_y(\Omega)$. Thus, we may define a.e. on Ω the gradient (connection) ∇_Ω.

Indeed, let $\varphi : \Omega \to \mathbb{R}$ be a function and $y \in \Omega$ be a fixed point, in which there exists a tangent plane $T_y(\Omega)$. By the standard scheme of definition of connection on surfaces, we extend φ onto a some neighborhood of the point y in $\mathbb{R}^m$. By the symbol $\tilde{\varphi}$ we denote a new function. Put

$$\nabla_\Omega \varphi(y) = \left(\overline{\nabla}\tilde{\varphi}(y)\right)^T ,$$

where $\overline{\nabla} = \nabla_{\mathbb{R}^m}$ is the standard differentiation in $\mathbb{R}^m$ and u^T means the orthogonal projection of a vector u onto the tangent plane $T_y(\Omega)$.

Fore an open bounded set $U \Subset \Omega$, we denote $U \Subset \Omega$ by $W^{1,p}(U)$, $p \geq 1$, the collection of functions $\varphi : U \to \mathbb{R}$ with the following property: *for every point $x \in U$ there exists a neighborhood $V \subset \mathbb{R}^m$ and a sequence of C^1-functions $\varphi_k : V \to \mathbb{R}$ such that the contraction $\tilde{\varphi}_k = \varphi_k|_{V\cap U}$ converges to φ uniformly and, in addition,*

$$\sup_{k\geq 1} \int_{V\cap U} |\nabla_\Omega \tilde{\varphi}_k|^p \, d\,\text{area}_\Omega \leq q < \infty$$

for some $q = \text{const} < \infty$. By $W^{1,p}_{\text{loc}}(\Omega)$ we denote the set of the functions $\varphi : \Omega \to \mathbb{R}$ belonging to $W^{1,p}(U)$ for every open subset $U \Subset \Omega$.

By Theorem 1.1, a canonical homeomorphism $F : B(0,R) \to \Omega$ is differentiable a.e. in the disc $B(0,R)$. If $f \in C^k(D)$, $k \geq 1$, then F defines the isothermal coordinates of the class C^k on Ω, and F is a conformal mapping from D onto Ω in the standard sense.

On the other hand, conformal mappings of nonregular surfaces were studied only in very special cases. H.A. Schwarz [1869] proved that every tetrahedron and cube may be mapped onto the unit sphere (see W. Gresky [Inaugural – Dissertation, Leipzig 1928] and C. Caratheodory [Conformal representation, 1932]). On conformal maps of general polyhedral surfaces (see R. Kühnau [1970], D. Ells, B. Fuglede [Harmonic Maps between Riemannian Polyhedra, 2001]) and of manifolds with bounded curvature (see Yu.G. Reshetnyak [1989] and also T. Toro [1994], S. Müller and V.Sverák [1995]). A generalized Christoffel – Schwarz formula for piecewise conformal homeomorphisms and the related questions were considered by I.M. Grudskii [1986], [1989].

In the case of $W^{1,2}_{\text{loc}}$-surfaces we can consider the conformality at the points where f is differentiable, i.e. the conformality of $f : D \to \Omega$ almost everywhere in the domain D. The notion of conformality is not defined in general, since one encounters substantial difficulties while trying to make sense of conformal maps of non-regular (singular) surfaces. This issue is similar, for example, to the well-known problem of removability of singular sets in the flat case.

As in the planar case, we can study different classes of mappings conformal outside of the singular set and having different properties 'in the large' (see, for example, D.E. Menchoff [1936], Yu.Yu. Trohimchuk [1963], E.P. Dolzhenko [1963], S.Ya. Khavinson [1963], G. David and P. Mattila [2000], et al.)

The following theorem connects the properties of a map $f : D \to \mathbb{R}^m$ with the properties of its canonical homeomorphism F.

Theorem 2.1 *Let Ω be a surface given by a vector function (1) over a simply connected bounded domain $D \subset \mathbb{R}^2$ and satisfying (2). Suppose that a function $\mathcal{P}$ defined by (7) is $W^{1,2}$-majorized on D. Then we have:*

(i) the canonical homeomorphism $F : B(0,R) \to \Omega$ and the inverse homeomorphism $F^{-1} : \Omega \to B(0,R))$ have the Lusin N-property if and only if $f : D \to \Omega$ (respectively, $f^{-1} : \Omega \to D$) have this property;

(ii) if $f \in W^{1,2}_{\text{loc}}(D)$, then $F \in W^{1,1}_{\text{loc}}(B(0,R))$;

(iii) if $f^{-1} \in W^{1,\alpha}_{\text{loc}}(\Omega)$ for a number α, $2 \le \alpha \le \infty$, and for every subdomain $D_1 \Subset D$:

$$\int\limits_{D_1} \left(g_{11}g_{22} - g_{12}^2\right)^{\beta} dx_1 dx_2 < \infty\,, \quad \beta = \frac{\alpha}{\alpha - 2}\,, \tag{10}$$

then the inverse map $F^{-1} : \Omega \to B(0,R)$ belongs to $W^{1,1}_{\text{loc}}(\Omega)$.

3. Main Theorem

In addition to Theorem 1.1, we shall establish the estimate for speed approximations of the canonical homeomorphism $F : B(0,R) \to \mathbb{R}^m$ by maps F_n.

For this purpose, we use the well-known concept of stability of conformal mapping in the class of mappings with bounded Dirichlét integral and show that F_n and F are close and can be obtained one from another by a small deformation.

There are many results describing such closeness. Below we formulate the simplest statement. Assume that the surface $\Omega \subset \mathbb{R}^m$ has the conformal hyperbolic type and the vector function f in (2) is chosen so that $f : B(0,R) \to \mathbb{R}^m$ for some $0 < R < \infty$.

Assume that $p(x) : B(x,R) \to [1,\infty)$ is a measurable function. Fix a number sequence $\{Q_n\}$ with (4). and put

$$I_n \quad = \{x \in B(x,R) : p(x) \geq Q_n\}\,, \quad n = 1,2,\dots\,,$$

$$p_n(x) = 1 \quad \text{for} \quad x \in I_n\,,$$

$$p_n(x) = p(x) \quad \text{for} \quad x \in B(x,R) \setminus I_n\,.$$

Assume also that $F(x) = f \circ w(x)$, $F_n(x) = f \circ w_n(x)$ and $w(x) : B(0,R) \to B(0,R)$ have Lavrentiev's characteristics (p,θ) and (p_n,θ) respectively.

Assuming that the mappings $w(\xi)$ and $w_n(\xi)$ are automorphisms of disk $B(0,R)$, $R > 1$, normalized by conditions 5, we have the following theorem.

Theorem 3.1. *Suppose that the functions*

$$P_n(x) = \begin{cases} p^2(x) - 1 \text{ for } & x \in I_n\,, \\ \\ 0 \qquad\quad \text{for } & x \in D \setminus I_n \end{cases}$$

have majorants $K_n(x)$ *of the class* $W_0^{1,2}(B(0,R))$, $n = 1,2,\dots$.

Let

$$\delta_n = \left(\int\limits_{B(0,R)} \left| \nabla K_n^{1/2}(x) \right|^2 p(x)\, dx_1 dx_2 \right)^{1/2} . \tag{11}$$

There are $N \geq 1$ *and* $\varepsilon_0 > 0$ *such that for all* $n \geq N$ *and* $0 < \delta_n < \varepsilon_0$

$$|h(x) - x| \leq C'\, \Delta(R\sqrt{2}\delta_n)\,, \quad h = w_n \circ w^{-1}(x)\,, \quad x \in B(0,R)\,, \tag{12}$$

and for every measurable set $E \subset B(0,R)$ *the following estimate holds*

$$|\mathcal{H}^2(h(E)) - \mathcal{H}^2(E)| \leq C''\, \Delta_1(R\sqrt{2}\delta_n)\,. \tag{13}$$

Here

$$\Delta(\varepsilon) \equiv \frac{2\,(2\pi\, k)^{\frac{1}{2}}}{\ln^{\frac{1}{2}}\, \varepsilon} + \frac{M}{2\sqrt{\pi}}\sqrt{\varepsilon}\,, \quad \varepsilon < \beta\,,$$

$$\Delta_1(\varepsilon) \equiv C\,\Delta(\varepsilon) + \sqrt{2\pi} R\varepsilon\,,$$

$$\beta = \left[\frac{1}{2} \min \left\{ 1,\, \frac{1}{4} R,\, \frac{1}{16} R^2 \right\} \right]^2 ,$$

$$k = \int\limits_{B(0,R)} |h'(x)|^2 dx_1 dx_2 \,,$$

$$M = \text{const} \le 8\pi^2 R \,.$$

The proof of this theorem is based on stability theorems of conformal maps in the class of maps with bounded Dirichlét integral. The stability problem goes back to M.A. Lavrentiev [1935], where the first qualitative solution was given for $(1+\varepsilon)$-quasiconformal mappings of a disk onto a disk. The quantitative aspects of the Lavrentiev result were revised by P.P. Belinskii [1953] – [1960]. The further development of this problems is connected with papers of S.L. Krushkal [1967], G.D. Suvorov [1968], J. Lawrynowicz [1969], V.I. Kruglikov, V.M. Miklyukov [1972], I.A. Volynec [1977] ($n = 2$) and with papers Yu.G. Reshetnyak [1961], [1967], V.M. Miklyukov [1969], P.P. Belinski [1973], A.P. Kopylov [2002] and others ($n > 2$).

In the proof of Theorem 3.1, we used the corresponding results of V.I. Kruglikov, V.M. Miklyukov [1972] and I.A. Volynec [1977].

4. Remarks on $W^{1,2}$ majorized functions

Until now, no complete description of the $W^{1,2}$ majorized functions has been obtained. We notice here only some their limit properties, arising from the theory of functions with generalized derivatives.

A. On the set P_∞. It follows from the definition of $W^{1,2}$ functions that the *restrictions of the $W^{1,2}$ majorized functions P onto almost all horizontal and vertical intervals are locally bounded.*

Assume that a domain D is the disk and the function $P(x)$ defined by (7) is a $W^{1,2}$ majorized in D. In other words, there exists a function $K(x) \in W^{1,2}(D)$, for which

$$P(x) \le K(x) \quad \text{for a.e.} \quad x \in D.$$

Denote by $L_1^2(\mathbb{R}^2)$ the set of the functions $\varphi(x)$ which are as the Bessel potentials

$$\varphi(x) = \int\limits_{\mathbb{R}^2} G_1(|x-y|)\, u(y)\, dy_1 dy_2,$$

of the functions $u \in L^2(\mathbb{R}^2)$ with the Bessel kernel $G_1(t)$ of the order 1 (see, e.g., Adams and Hedberg[34]). Since D is a disk, there exists by Theorem **1.2.3** of[34] for every function $\varphi(x) \in W^{1,2}(D)$ a function $\varphi^*(x) \in L_1^2(\mathbb{R}^2)$ such that a.e. $\varphi^*(x) = \varphi(x)$. Thus, since $K \in W^{1,2}(D)$, there is a function $u \in L^2(\mathbb{R}^2)$ such that

$$K(x) = \int_{\mathbb{R}^2} G_1(|x-y|)\, u(y)\, dy_1 dy_2 \quad \text{a.e. on} \quad D.$$

By Theorem **6.2.1**,[34] the set

$$\{x \in D : \lim_{\zeta \to x} K(\zeta) = \infty\}$$

has zero conformal capacity, and therefore the following statement holds.

Theorem 4.1. *If P is $W^{1,2}$-majorized, the set*

$$P_\infty = \{x \in D : \lim_{r \to 0} \frac{1}{r^2} \int_{B(x,r)} P(y)\, d\mathcal{H}^n = \infty\}$$

is of zero capacity.

Problem I. Describe the class of functions $p(x) : D \to [1, \infty)$ that are $W^{1,2}$-majorized in D.

B. The class is not empty. We privide the conditions for characteristic $p(x)$ which ensure that the quantities $\delta_n \to 0$ and, consequently, the class of mappings described by Theorem 3.1 is not empty.

Namely, the following statement holds.

Theorem 4.2. *Let $p(x) : D \to [1, \infty]$ be a continuous function of the class $W^{1,2}(D)$. Then the quantities δ_n defined by*

$$\delta_n = \left(\int_D \left| \nabla K_n^{1/2}(x) \right|^2 p(x)\, dx_1 dx_2 \right)^{1/2}$$

satisfy the condition

$$\lim_{n \to \infty} \delta_n = 0\,.$$

Sketch of the proof. Let $(p(x), \theta(x))$ be the Lavrentiev's characteristics in D, and let $p : D \to [1, +\infty]$ be continuous. For a given sequence $Q_n \to \infty$, we find the sets $I_n \subset D$ and the functions P_n involved in Theorem 3.1.

Since p is continuous, the sets I_n are closed. Suppose that for some $Q_n > 1$, the set $I_n \Subset D$. Without loss of generality, we may assume that $D \setminus I_n$ is a domain.

For every function K_n we have

$$\delta_n^2 \le \int_D \left|\nabla K_n^{1/2}(x)\right|^2 p(x)\,dx_1dx_2$$

$$= \int_{D\setminus I_n} \left|\nabla K_n^{1/2}(x)\right|^2 p(x)\,dx_1dx_2 + \int_{I_n} |\nabla K_n(x)|^2 \,\frac{p(x)}{K_n(x)}\,dx_1dx_2$$

$$\le Q_n \int_{D\setminus I_n} \left|\nabla K_n^{1/2}(x)\right|^2 dx_1dx_2 + \int_{I_n} |\nabla K_n(x)|^2 \,\frac{1+K_n}{K_n}\,dx_1dx_2$$

$$\le Q_n \int_{D\setminus I_n} \left|\nabla K_n^{1/2}(x)\right|^2 dx_1dx_2 + \left(1+\frac{1}{Q_n}\right)\int_{I_n} |\nabla K_n(x)|^2 \,dx_1dx_2\,.$$

Let $\varphi : D \setminus I_n$ be a function which is extremal for the condenser capacity $(I_n, \partial D; D)$, that is, $\varphi = 0$ on ∂D, $\varphi = 1$ on I_n and

$$\int_{D\setminus I_n} |\nabla\varphi(x)|^2 \,dx_1dx_2 = \operatorname{cap}(I_n, \partial D; D)\,.$$

We choose K_n of the form

$$K_n(x) = \begin{cases} Q_n\,\varphi^2(x) & \text{for} \quad x \in D \setminus I_n \\ p(x) & \text{for} \quad x \in I_n\,. \end{cases}$$

This function has the desired properties and by the proved above,

$$\delta_n^2 \le Q_n^2 \operatorname{cap}(I_n, \partial D; D) + \left(1+\frac{1}{Q_n}\right)\int_{I_n} |\nabla p(x)|^2 \,dx_1dx_2\,.$$

Thus, if

$$\lim_{n\to\infty} Q_n^2 \operatorname{cap}(I_n, \partial D; D) = 0\,, \tag{14}$$

then $\delta_n \to 0$.

We show that the assumption $p(x) \in W^{1,2}(D)$ implies (14). Indeed, let

$$A_{n_0,n} = \int_{I_{n_0}\setminus I_n} |\nabla p(x)|^2 \,dx_1dx_2 < \infty\,.$$

Suppose that the set P_∞ is not empty. We choose $Q_{n_0} > 1$ so that the set I_{n_0} lies strongly interior to D. Fix $Q_n > Q_{n_0}$. Consider the condenser $(D \setminus I_{n_0}, I_n; D)$. The function

$$\varphi = \frac{1}{Q_n - Q_{n_0}}\,(p(x) - Q_{n_0})$$

is admissible in the variational problem for the capacity of the condenser. Therefore,

$$\operatorname{cap}(I_n, \partial D; D) \leq \operatorname{cap}(D \setminus I_{n_0}, I_n; D \leq \frac{A_{n_0,n}}{(Q_n - Q_{n_0})^2}.$$

This yields

$$Q_n^2 \operatorname{cap}(I_n, \partial D; D) \leq \frac{A_{n_0,n} Q_n^2}{(Q_n - Q_{n_0})^2}. \tag{15}$$

Setting $Q_{n_0} = \frac{1}{2} Q_n$, we get that the condition (14) holds. Thus, $\delta_n \to 0$ as $n \to \infty$, and the theorem is proved. □

Problem II. Describe the class of continuous functions $p(x) : D \to [1, \infty)$ for which

$$\liminf_{n \to \infty} \delta_n = 0.$$

References

1. I.N. PESIN: *Mappings which are quasiconformal in the mean.* - Dokl. Akad. Nauk SSSR **187**, 740–742, 1969 (in Russian).
2. O. LEHTO: *Homeomorphisms with a given dilatation.* - Proceedings of the Fifteenth Scandinavian Congress (Oslo, 1968), Springer-Verlag, 55–73, 1970.
3. O. LEHTO: *Remarks on generalized Beltrami equations and conformal mappings.* - Proceedings of the Romanian-Finnish Seminar on Teichmüller Spaces and Quasiconformal Mappings (Braşov, 1969), Publ House of the Acad. of the Socialist Republic of Romania, Bucharest, 203–214, 1971.
4. V.M. MIKLYUKOV and G.D. SUVOROV: *On the existence and uniqueness of quasiconformal mappings with unbounded characteristics* (in Russian). - Investigations in the theory of functions of a complex variable and its applications, Inst. Mat. Akad. Nauk Ukr., Kiev, 45–53, 1972.
5. V. I. KRUGLIKOV: *Existence and Uniqueness of Quasiconformal "in Mean" Mappings* (in Russian). - Metric questions of functions theory and mappings, n. 4, 123–147, 1973.
6. P. P. BELINSKII: *General properties of quasiconformal mappings* (in Russian). - Novosibirsk, Nauka, Sibirskoe otdelenie, 1974.
7. G. DAVID: *Solutions de l'équation de Beltrami avec* $\|\mu\|_\infty = 1$. - *Ann. Acad. Sci. Fenn. Ser. A I Math.* **13**, 25–70, 1988.
8. P. TUKIA: *Compactness properties of μ-homeomorphisms.* - *Ann. Acad. Sci. Fenn. Ser. A I Math.* **16**, 47–69, 1991.
9. M.A. BRAKALOVA and J.A. JENKINS: *On solutions of the Beltrami equation.* - *J. Anal. Math.* **76**, 67–92, 1998.
10. V. GUTLYANSKII, O. MARTIO, T. SUGAWA and M. VUORINEN: *On the Degenerate Beltrami Equation.* Trans. Amer. Math. Soc., v. 357, n. 3, 2005, 875-900.
11. O. MARTIO, V.M. MIKLYUKOV: *On existence and uniqueness of degenerate Beltrami equation.* - Complex Var. Theory Appl. 49, n. 7-9, 2004, 647–656.,
12. V.M. MIKLYUKOV: *Isothermic coordinates on singular surfaces.* - Mat. Sb. **195**, n. 1, 2004, 69–88.
13. C. CARATHÉODORY: *Conformal representation.* - Cambridge at the University Press, 1932.

14. J. EELLS and B. FUGLEDE: *Harmonic maps between Riemannian polyhedra.* - Cambridge Tracts in Mathematics **142**, Cambridge Univ. Press, UK, 2001.
15. YU.G. RESHETNYAK: *Two-dimensional manifolds of bounded curvature,* - Modern problems of math. Fundamental directions, v. 70, Itogi nauki i thechniki, VINITI, Moscow: 1989, 8-189.
16. T. TORO: *Surfaces with generalized second fundamental form in L^2 are Lipschitz manifolds.* - J. Differential Geometry, v. 39, 1994, 65-101.
17. S. MÜLLER, V. SVERÁK: *On surfaces of finite total curvature.* - J. Differential Geometry, v. 42, 1995, 229-258.
18. I.M. GRUDSKII: *Construction of inner coordinates in composite Riemannian surfaces.* "Differential, integral equations and complex analysis", Elista, 1986, 30-45.
19. I.M. GRUDSKII: *Christoffel – Schwarz formula for polyhedral surfaces.* - DAN SSSR, v. 307, n. 1, 1989, 15-17.
20. D.E. MENCHOFF: *Les conditions de monogénéité.* - Act. sci. et ind. v. 329, 1936, 1-52.
21. YU.YU. TROHIMCHUK: *Continuous mappings and conditions of monogeneity.* - GIFML, Moskow, 1963.
22. E.P. DOLZHENKO: *On "erasure" of singularities of analytic functions.* - Uspechi math. nauk, v. 18, n. 4, 1963, 135—142.
23. S.YA. KHAVINSON: *On erasure of singularities.* - Litov. math. sb., III, 1, 1963, 271—287.
24. G. DAVID and P. MATTILA: *Removable sets for Lipschitz harmonic functions in the plane.* - Revista Matem. Iberoamericana, v. 16, n. 1, 2000, 137—215.
25. M.A. LAVRENTIEV: *Sur une classe de representations continue.* - Math. sb., v. 42, n. 4, 1935, 407—424.
26. P.P. BELINSKII: *On distortion under quasiconformal mappings.* - DAN SSSR, v. 91, n. 5, 1953, 997—998.
27. P.P. BELINSKII: *On measure area under quasiconformal mappings* . - DAN SSSR, v. 121, n. 1, 1958, 16—17.
28. P.P. BELINSKII: *Solution of extremal problems of the quasiconformal mappings theory with the variational method.* - Sib. math. j., v. 1, n. 3, 1960, 303—330.
29. S.L. KRUSHKAL': *On mappings ε-quasiconformal in "means".* - Sib. math. j., v. 8, n. 4, 1967, 798—806.
30. G.D. SUVOROV: *Equistability of canformal mappings of closed domains.* - Ukr. math. j., v. 20, n. 1, 1968, 78—84.
31. J. LAWRYNOWICZ: *On a class of quasiconformal mappings with invariant boundary points, II, Applications and generalizations.* - Ann. polon. math., v. 21, n. 3, 1969.
32. V.I. KRUGLIKOV, V.M. MIKLJUKOV: *Stability theorems for BL-mappings.* - Dopov. AN USSR. ser. A, 1972, n. 5, 421-423 (in Úkráinian); In sb. " Metr. vopr. teor. funct. i otobr.", vyp. III, "Naukova Dumka", Kiev, 1971, 55-70 (in Russian).
33. I.A. VOLYNEC: *On distortion under BL-mappings.* - Sib. math. j., v. 18, n. 6, 1977, 1259-1270 (in Russian).
34. D.R. ADAMS and L.I. HEDBERG: *Function Spaces and Potential Theory.* - Springer-Verlag, Berlin – Heidelberg – New York etc., 1996.
35. V.M. MIKLYUKOV: *Conformal mapping of nonregular surface and its application.* - Volgograd State University Press, Volgograd, 2005 (in Russian).
36. H.A. SCHWARZ: *Conforme Abbildung der Oberfläche eines Tetraeders auf die Oberfläche einer Kugel.* - J reine angew. Math., v. 70, 1869, 121-136.
37. REINER KÜHNAU: *Trianguliere Riemannsche Mannigfaltigkeiten mit ganz-linear Bezugssubstitutionen und quasikonforme Abbildungen mit stückweise kpnstanter komplexer Dilatation.* - Mathematische Nachrichten, Band 46, Heft 1-6, 1970, 243 -261.

38. WALTER GRESKY: *Konforme Abbildung der Oberfläche eines rektangulären Hexaeders auf die Kugeloberfläche.* - Inaugural – Dissertation zur Erlangung der Doktorwürde der Hohen Philosophischen Fakultät der Universität Leipzig, Weida i. Thür. 1928, 1-74.

GRUNSKY INEQUALITIES, FREDHOLM EIGENVALUES, REFLECTION COEFFICIENTS

REINER KÜHNAU

FB Mathematik und Informatik
Martin-Luther-Universität Halle-Wittenberg
D-06099 Halle-Saale, Germany
reinhard.kuehnau@mathematik.uni-halle.de

We give the exact domain of variability of a fixed Grunsky functional in the class of all hydrodynamically normalized schlicht conformal mappings of the exterior of the unit disk for which the unit circle transforms onto a quasicircle with a given Fredholm eigenvalue.

Keywords: Univalent functions, Grunsky coefficients, quasiconformal map, Fredholm eigenvalues.

1. Introduction

Today a central role in Geometric Function Theory plays the Grunsky functional. We will restrict ourself here to the simply-connected case. Then we consider in the complex plane the class of all schlicht conformal mappings $w = f(z)$ of $|z| > 1$ with the hydrodynamical normalization

$$f(z) = z + \frac{b_1}{z} + \frac{b_2}{z^2} + \dots . \tag{1}$$

The corresponding Grunsky coefficients a_{kl} are defined by the development

$$\log \frac{f(z) - f(\zeta)}{z - \zeta} = - \sum_{k,l=1}^{\infty} \frac{a_{kl}}{z^k \zeta^l}, \tag{2}$$

and with a given fixed system of complex numbers $x_1, \dots, x_n$ $(n \geq 1, \sum |x_k|^2 > 0)$ the Grunsky functional G_f is defined as

$$G_f = \frac{\sum_{k,l=1}^{n} a_{kl} x_k x_l}{\sum_{k=1}^{n} \frac{|x_k|^2}{k}}. \tag{3}$$

The starting point of the theory was the famous

Theorem 0 [Gr], [P]. *For every n and all fixed $x_1, \dots, x_n$ the exact domain of variability of the functional G_f is the closed unit disk:*

$$|G_f| \leq 1 . \tag{4}$$

This theorem solves the extremal problem $|G_f| \to \max$ if we fix the constants x_k and vary the mappings $f(z)$.

It arises the following *complementary question*: What is the solution of the extremal problem $|G_f| \to \max$ if we conversely fix the mapping $f(z) \in \Sigma$ and vary the systems x_k? Surprisingly, the maximal value for $|G_f|$ is not always again 1. We have $\max_{x_k} |G_f| < 1$ for fixed f if and only if the image C of $|z| = 1$ is a quasicircle [Ku1], [P] (Theorem 9.12). More precisely, we have

$$\max_{x_k} |G_f| = \frac{1}{\lambda}, \tag{5}$$

where $\lambda = \lambda_C \geq 1$ denotes the Fredholm eigenvalue of C; cf. [Ku2,3], [S], [Kr3,4]. The Fredholm eigenvalue λ is defined (cf., e.g.,[Ku6]) as the greatest constant ≥ 1 for which the inequality

$$\frac{\lambda-1}{\lambda+1}\,\mathcal{D}[h_2] \;\leq\; \mathcal{D}[h_1] \;\leq\; \frac{\lambda+1}{\lambda-1}\,\mathcal{D}[h_2] \tag{6}$$

is satisfied with the Dirichlet integral $\mathcal{D}[...]$ for all pairs h_1, h_2 (both not constant), where h_1 is (real and) harmonic in the interior of C, h_2 (real and) harmonic in the exterior (including $\zeta = \infty$), with continuity and $h_1 = h_2$ at C. Because of the invariance of Dirichlet integrals under conformal mappings we have after a Möbius transformation of C the same Fredholm eigenvalue. This invariance property has also the so-called reflection coefficient Q_C of C. This means the smallest dilatation bound ≥ 1 in the class of all quasiconformal reflections at C. Basic is the Ahlfors inequality

$$\frac{1}{\lambda_C} \leq q_C := \frac{Q_C - 1}{Q_C + 1}. \tag{7}$$

For the question of equality here cf. [Kr2,3,4], [Ku5]. These cases with equality in (7) are in some sense sparse [KK2]. We have $\lambda_C = \infty$ or $Q_C = 1$ only in the trivial case of a circle C. Sometimes, we also write Q_f and q_f instead of Q_C and q_C.

Now we study here also the following subclasses of Σ (cf. [Ku3,5]).

Definition. *Let $\Sigma(\kappa)$ be the class of those mappings of Σ which have a schlicht and continuous extension to the whole plane which is $Q = \frac{1+\kappa}{1-\kappa}$-quasiconformal for $|z| < 1$. - Let $\Sigma\langle\kappa\rangle$ be the class of those mappings of Σ for which the image of $|z| = 1$ is a quasicircle C whose Fredholm eigenvalue λ_C satisfies $\frac{1}{\lambda_C} \leq \kappa$.*

Because of (7) it holds $\Sigma(\kappa) \subset \Sigma\langle\kappa\rangle$; cf. [Ku3]. Up to now a simple criterion (beside this definition) for functions of Σ to belong to the class $\Sigma(\kappa)$ does not exist. Contrary to this, we have with the Grunsky functional in form of (5) a simple criterion for functions of the class $\Sigma\langle\kappa\rangle$.

For every fixed system x_k the extremal problem $|G_f| \to \max$ in the class $\Sigma(\kappa)$ was solved in [Ku1](cf. also [Kr1],Chapter 4, § 5, [KK1], p.111) in form of the following Theorem.

Theorem 1. *For every **fixed** system $x_1, \ldots, x_k$ the exact domain of variability of the functional G_f in the class $\Sigma(\kappa)$ is the closed disk defined by*

$$|G_f| \leq \kappa \, . \tag{8}$$

To every boundary point corresponds exactly one extremal function. For example, the extremal function $w = f(z)$ with $G_f = \kappa$ satisfies

$$\sum_{k=1}^{n} \frac{1}{k} x_k F_k(w) = \begin{cases} \sum_{k=1}^{n} \frac{1}{k}(x_k z^k - \kappa \overline{x_k} z^{-k}) \text{ for } |z| \geq 1, \\ \\ \sum_{k=1}^{n} \frac{1}{k}(x_k z^k - \kappa \overline{x_k}\, \overline{z}^k) \quad \text{for } |z| \leq 1. \end{cases} \tag{9}$$

Note that we obtain $e^{2i\theta} G_f$ instead of G_f if we replace all x_k by $e^{i\theta} x_k$. Therefore we can restrict ourself in the discussion of equality in (8) to the point with $G_f = \kappa$.

We remark here further that, contrary to the class $\Sigma(\kappa)$, in the classical Grunsky case for the class Σ (cf. Theorem 0), corresponding to the limit case $\kappa \to 1$, to a boundary point of the domain of variability (4) can exist more than one extremal function. This fact was already mentioned by H. Grunsky itself [Gr] (last page) but is today not well-known or mostly forgotten or treated stepmotherly; cf. also remarks in [P] (p.61), [KK1] (end of p.109 and p.108/109). In the literature there is missing a complete discussion of those parameter systems x_k for which the corresponding extremal function in the class Σ is unique.

Contrary to the class $\Sigma(\kappa)$, there is hitherto only a small number of extremal problems which are completely solved in the class $\Sigma\langle\kappa\rangle$. The reason is that a variational formula does not exist for this class. What concerns the extremal problem $|G_f| \to \max$ in the class $\Sigma\langle\kappa\rangle$ we have of course by (5) for every fixed system x_k the estimate $|G_f| \leq \kappa$. Our aim is now to add here that it is impossible *for every fixed* system x_k to improve this estimate. Observe the subtle distinction to the assertion (5).

Theorem 2. *For every **fixed** system $x_1, \ldots, x_k$ the exact domain of variability of the functional G_f in the class $\Sigma\langle\kappa\rangle$ is again the closed disk defined by*

$$|G_f| \leq \kappa \, . \tag{10}$$

Again equality holds for the mapping (9) and its modifications by replacing all x_k by $e^{i\theta} x_k$. The image of $|z| = 1$ under these mappings is always a closed analytic Jordan curve C with $\frac{1}{\lambda_C} = q_C = \kappa$.

It remains here as an open question: Are there further extremal mappings ?

The proof of Theorem 2 uses an analysis of the mappings (9). Theorem 2 also immediately yields some new examples of quasicircles C for which we can give the exact value of the reflection coefficient and of the Fredholm eigenvalue; cf. older collections of such examples, e.g. in [Kr3,4], [Ku4,6,7], [W].

Finally, in Section 4 we add the analogous theorem for Golusin's functional. (What concerns the question of uniqueness of the extremal functions there are analogous remarks possible, as in the case of Grunsky's functional; cf. also [KK1], p. 108/109.)

2. Proof of Theorem 2

First we observe that indeed, by (5), for every system x_k the inequality (10) is true. Then we remark that, by Theorem 1, every boundary point of the disk (10) is attained by the mappings (9) and the mentioned modifications. These mappings are not only contained in $\Sigma(\kappa)$ but also in $\Sigma\langle\kappa\rangle$ because of (7). It remains to show that the mappings (9) transform the unit circle $|z| = 1$ onto a closed analytic Jordan curve. Namely, because the corresponding quadratic differential

$$\left(\sum_{k=1}^{n} \frac{1}{k} x_k F_k'(w)\right)^2 dw^2$$

has at the inside of C only zeroes of even order it would follow by [Ku5] (cf. also [Kru3,4]) that we have for these mappings $\kappa_f = q_f$, which will end the proof.

With the abbreviation

$$\varphi(z) = \sum_{1}^{n} \frac{1}{k} x_k z^k \tag{11}$$

the equation (9) becomes the form

$$\Phi(w) = \begin{cases} \varphi(z) - \kappa\overline{\varphi(1/\overline{z})} & \text{for } |z| \geq 1, \\ \varphi(z) - \kappa\overline{\varphi(z)} & \text{for } |z| \leq 1\,. \end{cases} \tag{12}$$

(By the way, $i\overline{\varphi(1/\overline{z})}$ corresponds to the complex eigenfunctions of C; cf. [Ku5], Satz 1.)

In (12) the left-hand side $\Phi(w)$ is a polynomial of degree n because we can assume $x_n \neq 0$. As the image of the whole w-plane appears a Riemann surface $\mathfrak{R}$ with n sheets. Therefore this $\mathfrak{R}$ also appears as image of the whole z-plane by the right-hand side of (12).

Now we show that for every given z_0 with $|z_0| = 1$ the mapping (9) transforms all sufficiently small arcs of $|z| = 1$ with center at z_0 onto an analytic arc. For this purpose we start with the development of $\varphi(z)$ in the neighborhood of z_0 :

$$\varphi(z) = a_0 + a_m(z - z_0)^m + \ldots,\quad a_m \neq 0\,,$$

with some $m \geq 1$. We denote here and in the following by $\dots$ higher powers in the corresponding power series. This yields

$$\Phi(w) = \begin{cases} (a_0 - \kappa\overline{a_0}) + [(a_m - \kappa\overline{a_m}(-z_0^2)^{-m}](z-z_0)^m + \dots & \text{for } |z| \geq 1, \\ [a_0 + a_m(z-z_0)^m + \dots] - \kappa\overline{[a_0 + a_m(z-z_0)^m + \dots]} & \text{for } |z| \leq 1\,. \end{cases} \tag{13}$$

Here we have

$$|a_m - \kappa\overline{a_m}(-z_0^2)^{-m}| \geq |a_m| - \kappa|\overline{a_m}| = (1-\kappa)|a_m| > 0,$$

therefore $a_m - \kappa\overline{a_m}(-z_0^2)^{-m} \neq 0$. This means that, roughly speaking, the half of the neighborhood of z_0 in $|z| \geq 1$ transforms onto one half of the m-sheeted neighborhood of $a_0 - \kappa\overline{a_0}$. The same holds for the other half in $|z| \leq 1$. We therefore obtain, altogether, a complete m-sheeted neighborhood of $a_0 - \kappa\overline{a_0}$, thus also by the mapping $\Phi(w)$. This means an equation of the form, e.g. for $|z| \geq 1$,

$$(a_0 - \kappa\overline{a_0}) + b_m(w-w_0)^m(1+\dots) = (a_0-\kappa\overline{a_0}) + (a_m - \kappa\overline{a_m}(z_0^2)^{-m})(z-z_0)^m(1+\dots) \tag{14}$$

with some $b_m \neq 0$. It follows with some $C \neq 0$

$$(w-w_0)^m(1+\dots) = C(z-z_0)^m(1+\dots),$$

$$(w-w_0)(1+\dots) = C^{1/m}(z-z_0)(1+\dots),$$

$$w - w_0 = C^{1/m}(z-z_0)(1+\dots).$$

This means, as desired, that this mapping is indeed in a sufficiently small neighborhood of z_0 analytic with a derivative $\neq 0$.

Therefore, we have indeed for the mappings (9) the desired equality $\kappa_f = q_f$. This means that all boundary points of the disk $|G_f| \leq \kappa$ are attained by mappings of the class $\Sigma\langle\kappa\rangle$. It is then trivial that also all interior points of this disk are attained. To see this we have only to consider the mappings (9) after replacing κ by a smaller value. (This idea is called in another context "Grötzsch's argument".)

3. Examples of mappings (9)

The representation (9) of the extremal functions contains unknown parameters, namely some coefficients b_k. These are hidden in the Faber polynomials. (This phenomenon of such unknown "accessory" parameters is a circumstance which often can be observed in Geometric Function Theory; cf., for example, also the Schwarz-Christoffel formulas, or the representation of extremal functions obtained by the variational method. For "general reasons", in our case by Theorem 1, it is obvious that there is always a solution for these accessory parameters.)

In our case, for a given fixed set x_k, the determination of these accessory parameters has to use the fact that we must obtain (if $x_n \neq 0$) the same n-sheeted Riemann surface $\mathfrak{R}$ (mentioned also in the proof of Theorem 2), therefore also the same ramification points, by both sides of (9) [a]. Of course, a concrete discussion is practicable only in some simple cases because in (9) polynomials are involved. We will give here such examples up to the concrete determination of the corresponding quasicircles C, with $\frac{1}{\lambda_C} = q_C = \kappa$. As usual, the corresponding extremal quasiconformal (= möglichst konforme) reflection at C is contained in (9); cf. [Ku6].

(i) The simplest case $n = 1$ leaves us with the well-known mapping

$$w = \begin{cases} z - \kappa e^{-2i\alpha}\frac{1}{z} & \text{for } |z| \geq 1, \\ \\ z - \kappa e^{-2i\alpha}\overline{z} & \text{for } |z| \leq 1 \,. \end{cases} \tag{15}$$

We obtain for every real α an ellipse C with semi-axis $1+\kappa$ and $1-\kappa$, and with $\frac{1}{\lambda_C} = q_C = \kappa$.

(ii) In the case $n = 2$ we can now assume $|x_2| = 1$. We will restrict ourself to the case $|x_1| \leq 1$. (The remaining case $|x_1| > 1$ needs a more complicate discussion.) This means that

$$\varphi'(z) = x_1 + x_2 z$$

has its zero $z_0 = -\frac{x_1}{x_2}$ in the unit disk $|z| \leq 1$. Because the left-hand side of (12),

$$\Phi(w) = x_1 w + \frac{1}{2}\, x_2(w^2 - 2b_1),$$

has its zero of the derivative at $w_0 = -\frac{x_1}{x_2} = z_o$, the equality

$$\Phi(w_0) = \varphi(z_0) - \kappa\,\overline{\varphi(z_o)}$$

yields

$$-\frac{x_1^2}{x_2} + \frac{1}{2}x_2\left(\frac{x_1^2}{x_2^2} - 2b_1\right) = -\frac{x_1^2}{x_2} + \frac{1}{2}x_2\frac{x_1^2}{x_2^2} - \kappa\overline{\left(-\frac{x_1^2}{x_2} + \frac{1}{2}x_2\frac{x_1^2}{x_2^2}\right)}\,,$$

$$2b_1 = -\kappa\overline{x_1^2}.$$

Therefore the equation (9) becomes

$$x_2(w - z_0)^2 = \begin{cases} x_2(z - z_0)^2 - \kappa\overline{x_2}\left(\frac{1}{z} - \overline{z_0}\right)^2 & \text{for } |z| \geq 1, \\ \\ x_2(z - z_0)^2 - \kappa\overline{x_2\,(z - z_0)^2} & \text{for } |z| \leq 1 \,. \end{cases} \tag{16}$$

With (16) and $z_0 = -\frac{x_1}{x_2}$ the parameter problem, mentioned at the very beginning of this section, is completely solved in the case $n = 2$, $|z_0| \leq 1$. Particularly, we obtain for $|z| = 1$ the corresponding image C with $\frac{1}{\lambda_C} = q_C = \kappa$ by the chain of elementary mappings

$$\mathfrak{z} = x_2(z - z_0)^2, \quad \mathfrak{w} = \mathfrak{z} - \kappa\bar{\mathfrak{z}}, \quad w = z_0 + \sqrt{\overline{x_2}\mathfrak{w}}. \tag{17}$$

---

[a] Especially, this means that the derivative of $\varphi(z)$ and of $\varphi(z) - \kappa\overline{\varphi(1/\overline{z})}$ must have the same number of zeroes for $|z| \geq 1$. This also follows by Rouché's theorem.

As a result, we obtain a family of quasicircles C, depending on the parameter x_1, x_2, and with explicitly known values $\frac{1}{\lambda_C} = q_C$.

In the special case $|z_0| = 1$ we have a cardioid in the $\mathfrak{z}$-plane. We repeat that the final quasicircle C is always analytic.

Finally, we leave the friendly reader with the following general inverse problem.

Problem. *Is it possible to produce any quasicircle C with $\frac{1}{\lambda_C} = q_C$ as the image of the unit circle by the special mappings (9), generalized by a suitable infinite case, perhaps also with an additional approximation procedure ?*

Here one has to bear in mind also the cases of quasicircles C without a corresponding quadratic differential for the extremal quasiconformal reflection; cf., for example, the case of a square C [Ku6]; further examples e.g. in [Kr3,4], [Ku4], [W].

4. Analogue: Golusin's inequalities

Completely similar to Theorem 2 there holds (always with the usual branch of the logarithm)

Theorem 3. *For every **fixed** system of complex numbers $\gamma_1, \ldots, \gamma_n$ $(n \geq 1, \sum |\gamma_n|^2 > 0)$ and every fixed and distinct points $z_1, \ldots, z_n$ with $|z_k| > 1$, in the class $\Sigma\langle\kappa\rangle$ the domain of variability of Golusin's functional*

$$G_f^* = \left(\sum_{k,l=1}^{n} \gamma_k \gamma_l \log \frac{f(z_k) - f(z_l)}{z_k - z_l} \right) \Bigg/ \left(\sum_{k,l=1}^{n} \overline{\gamma_k}\, \gamma_l \log \frac{1}{1 - \frac{1}{\overline{z_k} z_l}} \right) \tag{18}$$

is the closed disk defined by

$$|G_f^*| \leq \kappa \ . \tag{19}$$

Equality holds for the mapping which is uniquely and defined implicitly by

$$\sum_{k=1}^{n} \gamma_k \log \frac{f(z) - f(z_k)}{z - z_k} = \begin{cases} -\kappa \sum_{k=1}^{n} \overline{\gamma_k} \log(1 - \frac{1}{\overline{z_k} z}) & \text{for } |z| \geq 1, \\ \\ -\kappa \sum_{k=1}^{n} \overline{\gamma_k} \log(1 - \frac{\overline{z}}{\overline{z_k}}) & \text{for } |z| \leq 1\,, \end{cases} \tag{20}$$

and its modifications by replacing all γ_k by $e^{i\theta}\gamma_k$.

The image of $|z| = 1$ under these mappings is always a closed analytic Jordan curve C with $\frac{1}{\lambda_C} = q_C = \kappa$.

Again it remains as an open question: Are there further extremal mappings ?

Proof. We can write (20) again in form of (12), now with

$$\varphi(z) = \sum_{k=1}^{n} \gamma_k \log(z - z_k)$$

and a corresponding $\Phi(w)$. Therefore we can use for the proof of Theorem 3 the same ideas as in Section 3.

The great similarity of Theorem 2 and Theorem 3 is not a marvel because of the equivalence of the analogous theorems for the class $\Sigma(\kappa)$; cf. [Ku1], [KK1], [P].

References

Gr. H. Grunsky, *Koeffizientenbedingungen für schlicht abbildende meromorphe Funktionen*, Math. Z. **45** (1939), 29 - 61.

Kr1. S. L. Krushkal', *Quasiconformal mappings and Riemann surfaces,* V. H. Winston & Sons, Washington, D. C./ John Wiley & Sons, New York etc. 1979 (Russian original: Izdat. "Nauka", Sibirsk. Otd., Novosibirsk 1975).

Kr2. S. L. Krushkal, *On the Grunsky coefficient conditions*, Siberian Math. J. **28** (1987), 104 - 110.

Kr3. S. L. Krushkal, *Variational principles in the theory of quasiconformal maps*, Handbook of Complex Analysis: Geometric Function Theory, Vol.2 (ed. R. Kühnau), Elsevier, Amsterdam etc. 2005, 31 - 98.

Kr4. S. L. Krushkal, *Univalent holomorphic functions with quasiconformal extensions (variational approach)*, Handbook of Complex Analysis: Geometric Function Theory, Vol.2 (ed. R. Kühnau), Elsevier, Amsterdam etc. 2005, 165 - 241.

KK1. S. L. Kruschkal und R. Kühnau, *Quasikonforme Abbildungen - neue Methoden und Anwendungen,* B. G. Teubner Verlagsgesellschaft, Leipzig 1983 (Russian edition: Izdat. "Nauka", Sibirsk. Otd., Novosibirsk 1984).

KK2. S. Krushkal and R. Kühnau, *Grunsky inequalities and quasiconformal extension*, Israel J. of Math. **152**, 49 - 59 (2006).

Ku1. R. Kühnau, *Verzerrungssätze und Koeffizientenbedingungen vom GRUNSKYschen Typ für quasikonforme Abbildungen*, Math. Nachr. **48** (1971), 77 - 105.

Ku2. R. Kühnau, *Zu den Grunskyschen Coeffizientenbedingungen*, Annales Academiæ Scientiarum Fennicæ, Ser.A.I. Math. **6** (1981), 125 - 130.

Ku3. R. Kühnau, *Quasikonforme Fortsetzbarkeit, Fredholmsche Eigenwerte und Grunskysche Koeffizientenbedingungen*, Annales Academiæ Scientiarum Fennicæ, Ser.A.I. Math. **7** (1982), 383 - 391.

Ku4. R. Kühnau, *Zur Berechnung der Fredholmschen Eigenwerte ebener Kurven*, Zeitschr. angew. Math. Mech. (ZAMM) **66** (1986), 193 -200.

Ku5. R. Kühnau, *Wann sind die Grunskyschen Koeffizientenbedingungen hinreichend für Q-quasikonforme Fortsetzbarkeit ?*, Comment. Math. Helvetici **61** (1986), 290 - 307.

Ku6. R. Kühnau, *Möglichst konforme Spiegelung an einer Jordankurve*, Jahresber. d. Deutsch. Math.-Verein. **90** (1988), 90 - 109.

Ku7. R. Kühnau, *Quasiconformal reflection coefficient and Fredholm eigenvalue of an ellipse of hyperbolic geometry*, Publ. de l'Inst. Math, Nouv. Sér. (Beograd), **75 (89)** (2004), 77 - 86.

P. Ch. Pommerenke, *Univalent functions, with a chapter on quadratic differentials by Gerd Jensen,* Vandenhoeck & Ruprecht, Göttingen 1975.

S. M. Schiffer, *Fredholm eigenvalues and Grunsky matrices*, Ann. Polon. Math. **39** (1981), 149 - 164.

W. St. Werner, *Spiegelungskoeffizient und Fredholmscher Eigenwert für gewisse Polygone*, Annales Academiæ Scientiarum Fennicæ, Math. **22** (1997), 165 - 186.

SUMS OF RECIPROCAL EIGENVALUES

BODO DITTMAR

Institut für Mathematik Martin-Luther-Universität Halle-Wittenberg
bodo.dittmar@mathematik.uni-halle.de

Contents

1. Introduction

Eigenvalues and extremal problems for eigenvalues of partial differential equations with various boundary conditions are on the one hand important for people interested in applications such as acoustics, theoretical physics, quantum mechanics, solid mechanics and on the other hand this field is one of the most fascinating of mathematical analysis. The purpose of the lecture are formulas and isoperimetric inequalities for sums of reciprocal eigenvalues of the Laplacian and related eigenvalue problems. For the eigenvalues of the fixed membrane problem it was proven by Pólya and Schiffer 1953[25]

$$\frac{1}{\dot{r}^2}\sum_{i=1}^{n}\frac{1}{\lambda_i} \geq \Big(\sum_{i=1}^{n}\frac{1}{\lambda_i^{(o)}}\Big),$$

where λ_j are the eigenvalues of a simply connected and bounded domain in the plane with the maximum of the conformal radius $\dot{r}$ and $\lambda_j^{(o)}$ are the eigenvalues of the unit disk. This result has been generalized and extended by many authors. For a fuller treatment we refer the reader to.[1,7,12] The eigenvalue problems which we consider here are eigenvalue problems of Green's functions and a simply consequence of the well-known Expansion Theorem of the classical theory of integral equation are formulas for the sum over the squares of all reciprocals. After conformal transplantation into the unit disk this formulas are given in terms of the coefficients of the Taylor series for the conformal mapping. This formulas allow on the one hand to calculate the exact value of the $\sum_1^\infty \lambda_j^{-2}$ in some cases and on the other hand a variational characterization for the sum of the first n reciprocal eigenvalues has been obtained. Basing on this variational characterizations we are able to prove isoperimetric results for sums of reciprocals eigenvalues. We consider the fixed and free membrane eigenvalue problem and the Stekloff and mixed Stekloff eigenvalue problem. Among others the above inequality is proven for the free membrane eigenvalues.

2. Membrane problems

Let D be a simply connected domain in the plane, which is bounded. We consider the following eigenvalue problems

$$\begin{aligned} \Delta u + \lambda u &= 0 \text{ in } D, \\ u &= 0 \text{ on } \partial D, \end{aligned} \tag{1}$$

and

$$\begin{aligned} \Delta v + \mu v &= 0 \text{ in } D, \\ \frac{\partial v}{\partial n} &= 0 \text{ on } \partial D \end{aligned} \tag{2}$$

n stands for the normal to ∂D, λ and μ for the eigenvalue parameters. It is well-known that there exists an infinity of eigenvalues with finite multiplicity

$$0 < \lambda_1 \leq \lambda_2 \leq \lambda_3 ..., \tag{3}$$

$$0 = \mu_1 < \mu_2 \leq \mu_3 ... \, . \tag{4}$$

The eigenvalues are the stationary values of the Rayleigh quotient

$$\lambda_j = \min_{L_j} \max_{L_j} R[v] = \frac{\iint_D |\nabla v|^2 \, dxdy}{\iint_D v^2 \, dxdy} \tag{5}$$

where L_j ranges over all j-dimensional subspaces of the Sobolev space $H_0^1(D)$ in problem (1) and $H^1(D)$ in problem (2).

2.1. *Fixed membrane*

The basic isoperimetric inequality on vibrating membranes with fixed boundary is that of Rayleigh, Faber and Krahn[1,12,19,24]

Theorem 1.(Faber, Krahn) Among all membranes of given area, the circle hast the lowest first eigenvalue.

Proofs were given independently by Faber and Krahn in the early 1920s.. A counterpart of the Faber-Krahn inequality was derived using conformal mappings by Pólya and Szegö[24] .

Theorem 2.(Pólya and Szegö 1951) For all simply connected domains in the plane holds

$$\lambda_1 \dot{r}^2 \leq j^2, \tag{6}$$

where j denotes the first positive root of the Bessel function J_0, and $\dot{r}$ means the maximum of the inner conformal radius of D. Equality holds if and only if D is a disk. Of course, the products $\lambda_n^2 \dot{r}$ and $\mu_n^2 \dot{r}$ depend only on the shape of the domain

D, not on its size. For higher values of n the products $\lambda_n^2 \dot{r}$ and $\mu_n^2 \dot{r}$ do not attain the maximum for the disk and so do λ_n and μ_n for domains with the same $\dot{r}$. But for domains with a sufficiently high symmetry the circle yields the maximum of $\lambda_n^2 \dot{r}$ and $\mu_n^2 \dot{r}$, more precisely[22] .

Theorem 3.(Pólya 1955) If the domain D is convex and possesses a symmetry of the order not lower than $2n+1$, then

$$\lambda_1^2 \dot{r}, \lambda_2^2 \dot{r}, ..., \lambda_{l(n)}^2 \dot{r}, \mu_1^2 \dot{r}, \mu_2^2 \dot{r}, ..., \mu_{m(n)}^2 \dot{r}, \tag{7}$$

where $l(n)$ and $m(n)$ are strictly increasing functions of n, will be inferior to the corresponding quantity belonging to a circle.

On the other hand, from Weyl's well-known asymptotic result[28]

$$\lim_{n\to\infty} \frac{\lambda_n}{n} = \lim_{n\to\infty} \frac{\mu_n}{n} = \frac{4\pi}{A}, \tag{8}$$

where $A = |D|$ is the area of D, it follows that for sufficiently large n it holds $\lambda_n(D) < \lambda_n(U_1)$ and $\mu_n(D) < \mu_n(U_1)$ if D is the conformal image of the unit disk U_1 under a normalized conformal map such that $\dot{r}$ is the same because then $|D| > |U_1|$. Pólya and Schiffer derived an interesting result. They use conformal transplantation and convexity arguments in order to prove[25]

Theorem 4.(Pólya and Schiffer 1953) For any n

$$\frac{1}{\dot{r}^2} \sum_{i=1}^{n} \frac{1}{\lambda_i} \geq \Big(\sum_{i=1}^{n} \frac{1}{\lambda_i^{(o)}} \Big) \tag{9}$$

where $\lambda_i^{(o)}$ denotes the eigenvalues for the unit disk. Equality holds if and only if D is the unit disk.

There are generalizations given by C. Bandle and closely related papers by J. Hersch[1,7] .

An easy consequence of the Expansion Theorem is the following formula for the sum of all reciprocals[3]

Theorem 5.(Dittmar 2002) Let f be a conformal mapping from the unit disk U_1 onto the domain D with the area A, then it holds

$$\int_{U_1}\int_{U_1} G^2(z,\zeta)|f'(z)|^2|f'(\zeta)|^2 dA_z dA_\zeta = \sum_{j=1}^{\infty}\frac{1}{\lambda_j^2}, \tag{10}$$

where $G(z,\zeta)$ denotes Green's function of the unit disk.

From this theorem it follows a formula[11] in terms of the coefficients of the Taylor series for the conformal mapping which is similar to a formula for the torsional rigidity given in [25, p.330]

Theorem 6.(Hantke 2006) For the eigenvalues of the fixed membrane problem holds

$$\sum_{j=1}^{\infty}\frac{1}{\lambda_j^2} = \sum_{m=0}^{\infty}\sum_{l=0}^{\infty}(A_{m,l}+B_{m,l})a_{0,m}a_{0,l} + \tag{11}$$

$$\sum_{k=1}^{\infty}\sum_{m=1}^{\infty}\sum_{l=1}^{\infty}(C_{k,m,l}+D_{k,m,l}+\sum_{k=2}^{\infty}\sum_{m=1}^{\infty}\sum_{l=1}^{\infty}E_{k,m,l}, \tag{12}$$

where the coefficients A, B, C, D, E are known and

$$|f'(r,\varphi)|^2 = \sum_{n=0}^{\infty}a_{0,n}r^n\sum_{m=1}^{\infty}\sum_{n=1}^{\infty}(a_{m,n}\cos m\varphi + b_{m,n}\sin m\varphi)r^m. \tag{13}$$

It is worth pointing out that in some cases it is possible to calculate (10) exactly.[11]

Examples

1. Disk

Let D_1 be the unit disk then holds
$\sum_{j=1}^{\infty}\frac{1}{\lambda_j^2} = \frac{4}{2\cdot 4^3} + \sum_{n=1}^{\infty}\frac{8}{4(2n+4)(4n+4)(2n+2)} = \frac{\pi^2}{48} - \frac{5}{32}$.
For the unit disk see also [3, p.51].

2. Cardioid

$f(z) = z + z^2/2$ maps the unit disk onto a cardioid. It follows from the formula above $\sum_{j=1}^{\infty}\frac{1}{\lambda_j^2} = \frac{3}{64}\pi^2 - \frac{551}{1536}$. Similar results are given in[11] for the image of the unit disk by $f_n(z) = z + \frac{1}{n}z^n, n = 3, 4, \ldots$.
For the cardioid follows in the same matter also the value $11\pi/48$ for the torsional rigidity[11] .

Starting with (10) it follows[3]

Corollary 1. Let f be the conformal mapping from the unit disk U_1 onto the domain D with the area $A = \int_U |f'(z)|^2 dA < \infty$, then

$$\int_{U_1}\int_{U_1} G^2(z,\zeta)|f'(z)|^2|f'(\zeta)|^2 dA_z dA_\zeta = \sum_{1}^{\infty}\frac{\int_{U_1}(\nabla G_j)^2 dA}{\lambda_j^{(o)}}, \tag{14}$$

with

$$G_j(\zeta) = \int_{U_1} G(z,\zeta) u_j^{(o)}(z) |f'(z)|^2 dA_\zeta, \tag{15}$$

where $u_j^{(o)}$ is the j-th eigenfunction of U_1.

The following lemmas are essential tools in order to give isoperimetric results[3,5]

Lemma 1. Let $f(z) = a_1 z + a_2 z^2 + ...$ be an univalent conformal map of the unit disk onto U_1. We denote by $u_j^{(o)}$ and $\lambda_j^{(o)}$ the j-th eigenfunction and eigenvalues of the unit disk. If $u_j^{(o)}$ is radial, then it holds

$$\int_{U_1} (\nabla G_j)^2 dA \geq \frac{|a_1|^4}{\lambda_j^{(o)}}. \tag{16}$$

If $u_j^{(o)}$ is not radial, then let $u_j^{(o)}$ and $u_{j+1}^{(o)}$ be the eigenfunctions whose sum of the squares is radial and both of them have the same eigenvalue $\lambda_j^{(o)}$, then it holds

$$\int_{U_1} (\nabla G_j)^2 dA + \int_{U_1} (\nabla G_{j+1})^2 dA \geq \frac{2|a_1|^4}{\lambda_j^{(o)}}. \tag{17}$$

Equality occurs in both inequalities if and only if $f(z) = a_1 z$.

Lemma 2. For the eigenvalues of the fixed membrane problem holds

$$\sum_1^n \frac{1}{\lambda_j} \geq \sum_2^n \int_U \int_U G(z,\zeta) |f'(z)|^2 |f'(\zeta)|^2 v_j(z) v_j(\zeta) dA_z dA_\zeta, \tag{18}$$

where $\int_U |f'(z)|^2 v_i(z) v_j(z) dA_z = \delta_{i,j}, i,j = 1,2,...,n, v_j \in L^2(U)$.

Using this variational characterization we obtain following[5] the Pólya-Schiffer result (9) and with a result from Hardy, Littlewood and Pólya[5] follows[18]

Theorem 7.(Laugesen, Morpurgo 1998) Let $\Phi(a)$ be convex and increasing for $a \geq 0$. Then

$$\sum_{j=1}^n \Phi\Big(\frac{1}{\lambda_j(D)}\Big) \geq \sum_{j=1}^n \Phi\Big(\frac{1}{\lambda_j(U_1)}\Big), \tag{19}$$

where n is either a positive integer or $+\infty$, and D is the conformal image of the unit disk U_1 under the conformal mapping f with $|f'(0)| = 1$, with strict inequality unless D equals the unit disk.

2.2. *Free membrane*

In the free membrane case the situation is more complicated. Encouraged by a conjecture of Kornhauser and Stakgold it was proven by Szegö 1954 that of all simply connected domains of a given area, the circle yields the maximum value of μ_2. Szegö and Weinberger noticed that the same proof causes the disk to minimize

$$\frac{1}{\mu_2} + \frac{1}{\mu_3} \tag{20}$$

of all simply connected domains of a given area.

Generalizations are given by C. Bandle 1972, T. Gasser and J. Hersch 1968[1,7] and N. Nadirashvili 1997.[20]

In order to derive a formula similar to (10) we introduce a special symmetric function which will be the kernel in an integral equation for the eigenfunctions of Problem (2).[5]

Lemma 3. Let $N_f(z,\zeta)$ be the following symmetric function depending on an univalent conformal map f

$$N_f(z,\zeta) = A\ N(z,\zeta) + H_f(z) + H_f(\zeta), z, \zeta \in U_1, \tag{21}$$

where

$$H_f(z) = -\int_{U_1} |f'(\zeta)|^2 N(z,\zeta) dA_\zeta, z \in U_1 \tag{22}$$

and $A = \int_{U_1} |f'(z)|^2 dA_z < \infty$. Then

$$\Delta_z N_f(z,\zeta) = |f'(z)|^2, z \neq \zeta, z \in U_1, \tag{23}$$

$$\frac{\partial N_f(z,\zeta)}{\partial n_z} = 0 \text{ on } \partial U_1, \tag{24}$$

$$H_f(z) = \frac{1}{4}|f(z)|^2 + h(z), \tag{25}$$

where $h(z)$ is a harmonic function in U_1 with $\frac{\partial h}{\partial n} = A/(2\pi) - 1/4\frac{\partial |f|^2}{\partial n}$ on ∂U_1 for a sufficiently smooth $f(z)$ and n is the outward pointing normal.

In particular

$$H_{f\equiv z}(z) = \frac{1}{4}|z|^2. \tag{26}$$

Now we are able to give an integral equation for the eigenfunctions $v_j|f'|$ with the eigenvalues μ_j/A [5] .

Theorem 8. For the eigenfunctions of the free membrane Problem (2) it holds, if D is sufficiently smoothly bounded

$$v_j(\zeta) = \frac{\mu_j}{A}\int_{U_1} N_f(z,\zeta) v_j(z)|f'(z)|^2 dA_z, j = 2, 3, \dots , \tag{27}$$

where $A = \int_{U_1} |f'(z)|^2 dA < \infty$ and N_f referring to Lemma 2.2. The kernel $N_f(z,\zeta)|f'(z)||f'(\zeta)|$ has only the eigenfunctions $v_j(z)|f'(z)|$ with the eigenvalues μ_j/A in the space $V_f = L^2(U_1) \cap \{u : \int_{U_1} u|f'(z)|dA = 0\}$.

A consequence of Theorem 2.2 is the following

Theorem 9.(Dittmar 2002) If D is a simply connected sufficiently smoothly bounded domain with the area $A = \int_{U_1} |f'(z)|^2 dA_z < \infty$. Then for the eigenvalues of the free membrane it holds

$$\int_U \int_U \left(N_f(z,\zeta) - AC\right)^2 |f'(z)|^2 |f'(\zeta)|^2 dA_z dA_\zeta = A^2 \sum_{j=2}^{\infty} \frac{1}{\mu_j^2}, \tag{28}$$

where $C = -\frac{1}{A^2} \int_U \int_U N(z,\zeta) |f'(z)|^2 |f'(\zeta)|^2 dA_z dA_\zeta$.

The following lemmas are helpful in order to get isoperimetric results[3] .

Corollary 2. If D is a simply connected sufficiently smoothly bounded domain with the area $A = \int_{U_1} |f'(z)|^2 dA_z < \infty$. Then

$$\int_{U_1} \int_{U_1} \left(N_f(z,\zeta) - AC\right)^2 |f'(z)|^2 |f'(\zeta)|^2 dA_z dA_\zeta = \sum_{j=2}^{\infty} \frac{\int_{U_1} (\nabla F_j)^2 dA}{\mu_j^{(o)}}, \tag{29}$$

with

$$F_j(\zeta) = \int_{U_1} N_f(z,\zeta) |f'(z)|^2 v_j^{(o)}(z) dA_z. \tag{30}$$

Lemma 4. For a radial eigenfunction $v_m^{(o)}(z) = \alpha_0 J_0(k_{m,n} r)$, we have for every conformal map $f(z) = a_1 z + a_2 z^2 + ...$

$$A \int_{U_1} v_m^{(o)2} |f'(z)|^2 dA - \left(\int_{U_1} v_m^{(o)} |f'(z)|^2 dA\right)^2 \geq \pi |a_1|^4. \tag{31}$$

Let

$$\begin{aligned} v_m^{(o)}(z) &= \alpha_m J_m(k_{m,n} r) cos m\varphi, \\ v_{m+1}^{(o)}(z) &= \alpha_m J_m(k_{m,n} r) sin m\varphi, z = re^{i\varphi}, \end{aligned} \tag{32}$$

be eigenfunctions belonging to the same eigenvalue $\mu_m^{(o)}$. Then it holds for every $f(z) = a_1 z + a_2 z^2 + ...$

$$\begin{aligned} &A \int_{U_1} (v_m^{(o)2} + v_{m+1}^{(o)\,2}) |f'(z)|^2 dA - \\ &\left(\int_{U_1} v_m^{(o)} |f'(z)|^2 dA\right)^2 - \left(\int_{U_1} v_{m+1}^{(o)} |f'(z)|^2 dA\right)^2 \geq 2\pi |a_1|^4. \end{aligned} \tag{33}$$

Equality occurs in both inequalities if and only if $f(z) = a_1 z$.

The preceding lemmas lead to the following isoperimetric result.

Theorem 10.(Dittmar 2002) For all simply connected domains with the maximal conformal radius 1, the unit disk and only the unit disk yields the minimum of

$$A^2 \sum_{j=2}^{\infty} \frac{1}{\mu_j^2}, \tag{34}$$

where A is the area of the domain.
Remark 1. 1. We can also calculate the minimum of the preceding theorem. It follows

$$\int_U \int_U N^2(z,\zeta) dA_z dA_\zeta = -\frac{25}{32} + \frac{5}{48}\pi^2. \tag{35}$$

2. A still open conjecture of Pólya is $\mu_n < 4\pi n A^{-1} < \lambda_n$. It was proven by Kröger in 1992 that $\mu_n < 8\pi n A^{-1}$. Using Krögers result it is easy to check that a minimizer of $A^2 \sum \mu_j^{-2}$ must have an area less than $2.163..\pi$

For the finite case a variational characterization is also given.[5]

Theorem 11.

$$\sum_{j=2}^{n} \frac{1}{\mu_j} \geq \sum_{j=2}^{n} \int_U \int_U N(z,\zeta) |f'(z)|^2 h_j(z) |f'(\zeta)|^2 h_j(\zeta) dA_z dA_\zeta,$$

with $\int_U h_j |f'(z)|^2 dA_z = 0, \int_U h_i h_j |f'(z)|^2 dA_z = \delta_{ij}$. We take the functions $u_j = v_j^{(o)} + c_j, j = 2, ..., n$, with $c_j A = -\int_U v_j^{(o)} |f'(z)|^2 dA_z$, such that $\int_U u_j |f'(z)|^2 dA = 0$.

A consequence is the following isoperimetric result

Theorem 12.(Dittmar 2005) Let D be a domain with the area $A < \infty$, for any $n \geq 2$, we have

$$\frac{A}{\dot{r}^4} \sum_{j=2}^{n} \frac{1}{\mu_j} \geq \pi \sum_{j=2}^{n} \frac{1}{\mu_j^{(o)}},$$

where $\dot{r}$ denotes the maximal value for the inner conformal radius of D and $\mu_j^{(o)}$ denote the eigenvalues of the unit disk. Equality occurs if and only if $D = U$.

It is worth pointing out that a slight change in the proof in[5] gives more, namely the following[6]

Theorem 13.(Dittmar 2006) Let D be a domain with the area $A < \infty$, for any $n \geq 2$, we have

$$\frac{1}{\dot{r}^2} \sum_{j=2}^{n} \frac{1}{\mu_j} \geq \sum_{j=2}^{n} \frac{1}{\mu_j^{(o)}},$$

where $\dot{r}$ denotes the maximal value for the inner conformal radius of D and $\mu_j^{(o)}$ denote the eigenvalues of the unit disk. Equality occurs if and only if $D = U$.

3. Stekloff eigenvalue problems

We consider now the following classical eigenvalue problem

$$\Delta u = 0 \text{ in } D,$$
$$\frac{\partial u}{\partial n} = \nu u \text{ on } \partial D, \tag{36}$$

An infinity of eigenvalues also exists with finite multiplicity

$$0 = \nu_1 < \nu_2 \leq \nu_3 \leq \dots . \tag{37}$$

It is usual to call this eigenvalue problem Stekloff problem.

We consider also the mixed Stekloff problem, which is the following one

$$\begin{aligned} \Delta u &= 0 \text{ in } D, \\ \frac{\partial u}{\partial n} &= \sigma u \text{ on } C_1, \\ u &= 0 \text{ on } C_2, \end{aligned} \tag{38}$$

where $\partial D = C = C_1 \cup C_2$. There is also an infinity of eigenvalues with finite multiplicity

$$0 < \sigma_1 \leq \sigma_2 \leq \sigma_3 \leq \dots .$$

3.1. *Stekloff problem*

Transplanting the eigenvalue problem (36) conformally in the unit disk and using the Neumann function of the unit disk and the well-known fact that the Stekloff eigenfunctions are eigenfunctions of the Neumann function on the boundary, it follows a formula for the sum of all squares of reciprocals.

Theorem 14.(Dittmar 1998) If f is the univalent conformal mapping of the interior of the unit disk onto the domain D with $f(0) = f'(0)-1 = 0$, then for the eigenvalues of problem (36) with $p \equiv 1$ we get

$$\sum_{i=2}^{\infty} \frac{1}{\nu_i^2} = \int_0^{2\pi} \int_0^{2\pi} N_D^2(f(\xi), f(\eta)) |f'(\xi)||f'(\eta)| ds_\xi ds_\eta, \tag{39}$$

where N_D is Neumann's function of the domain D.

It follows once more[4]

Corollary 3. If D is a simply connected sufficiently smoothly bounded domain. Then

$$\sum_{j=2}^{\infty} \frac{1}{\nu_j^2} = \sum_{j=2}^{\infty} \frac{\int_U (\nabla b_j)^2 dA}{\nu_j^{(o)}}, \tag{40}$$

with a harmonic function b_j in the unit disk U which satisfied

$$\frac{\partial b_j}{\partial n} = u_j^{(o)} |f'(z)| - \frac{|f'(z)|}{L} \int_{\partial U} u_j^{(o)}(s) |f'(s)| ds, \text{ on } \partial U \tag{41}$$

where $u_j^{(o)}$ and $\nu_j^{(o)}$ are the eigenvalues of the unit disk.

It holds also here a lemma analogously to Lemma 1 and Lemma 4.

Lemma 5. If D is a simply connected smoothly bounded domain with $L = |\partial D|$ then it holds

$$\frac{L}{\pi} - \frac{1}{L} \Big(\int_{\partial U} u_{2j}^{(o)^2} |f'| ds \Big)^2 - \frac{1}{L} \Big(\int_{\partial U} {u_{2j+1}^{(o)}}^2 |f'| ds \Big)^2 \geq 2. \tag{42}$$

In the same way follows[4]
Theorem 15. Let D be a sufficiently smoothly bounded domain with the maximal conformal radius 1, then holds

$$\sum_2^{\infty} \frac{1}{\nu_j^2} \geq \sum_2^{\infty} \frac{1}{\nu_j^{(o)2}} = \frac{\pi^2}{3} \tag{43}$$

It is also possible to obtain isoperimetric results for the finite case. Using these and the Hardy, Littlewood and Pólya result it is easy to see that the inequalities are weaker than the result given by Hersch, Payne and Schiffer 1976.[13]

3.2. *Mixed Stekloff problem*

The eigenfunctions of the mixed Stekloff problem are the eigenfunctions of the Robin function. This is Green's function for the mixed problem. There are a lot of interesting results for the Robin function given by Duren and Schiffer.[10]

The situation in the mixed Stekloff problem is easier because the Robin function is invariant under conformal maps. We restrict ourselves to the case of the ring domain where C_2 is one of the components of the boundary. Hersch and Payne found out that in this case the conformal modulus of the domain D can be characterized by the mixed Stekloff eigenvalues.

It holds the following
Theorem 16.(Dittmar 1998) Let $f$map the annulus $A(1,R)$ onto the piece-wise analytically bounded ring domain D such that the circle $\{|z|=1\}$ corresponds to the boundary component C_1, then it holds

$$\int_{|z|=1}\int_{|z|=1} R^2(z,\zeta)|f'(z)||f'(\zeta)|ds_z ds_\zeta = \sum_{j=1}^{\infty} \frac{1}{\sigma_j^2}. \tag{44}$$

The Robin function of the annulus $A(1,R)$ was given by Duren and Schiffer in 1991:

$$R(z,\zeta) = \frac{\ln R}{2\pi} + \sum_{n=1}^{\infty} \frac{(R^n - R^{-n})}{\pi n(R^n + R^{-n})} \cos n(\vartheta - \varphi), z = Re^{i\varphi}, \zeta = Re^{i\vartheta}. \tag{45}$$

A simple calculation yields

$$R^2(z,\zeta) = \frac{B_o}{2} + \sum_{n=1}^{\infty} B_n \cos n(\vartheta - \varphi), \tag{46}$$

with

$$B_o = \frac{\ln^2 R}{2\pi^2} + \sum_{m=1}^{\infty} \frac{(R^m - R^{-m})^2}{\pi^2 m^2 (R^m + R^{-m})^2}, \tag{47}$$

and for $k = 1, 2, ...$

$$B_k = \frac{\ln R(R^k - R^{-k})}{\pi^2 k(R^k + R^{-k})} \cdots . \tag{48}$$

For this mixed Stekloff problem the following result holds

Corollary 4.(Dittmar 1998) Of all doubly connected conformally equivalent domains with a piece-wise analytic Jordan curve C_1 without outward-pointing cusps and a given length L of the boundary curve C_2 the annulus has the lowest amount of $\sum \sigma_j^{(-2)}$. In other words

$$\sum_{j=1}^{\infty} \frac{1}{\sigma_j^2} \geq \frac{L^2}{4\pi} B_o \tag{49}$$

and equality holds for the annulus.

It follows from the general procedure above

Theorem 17.(Dittmar, Hantke 2006) For the eigenvalues of problem (38) in Ω it holds for any n

$$\sum_{j=1}^{n} \frac{1}{\lambda_j} \geq \frac{\alpha_0}{2} \sum_{j=1}^{n} \frac{1}{\lambda_j^{(0)}}, \alpha_o = \frac{|B^{(o)}|}{\pi R}.$$

Equality holds if Ω is an annulus.

Example

Let the domain D be the unit disk U_1 furnished with a straight slit from $x_o, 0 < x_o < 1$, to 1. This simple connected domain is closely related to the Grötzsch ring domain. Let C_1 be the unit circle (without $z = 1$) and C_2 the straight slit. In[9] the eigenfunctions has been constructed and a formula for the eigenvalues was given. We get

$$\sum_{j=1}^{\infty} \frac{1}{\mu_j^2(x_o)} = \frac{1}{\pi^2} \int_0^{2\pi} \int_0^{2\pi} \frac{(1 - x_o^2)}{|1 + x_o z|^2} \frac{(1 - x_o^2)}{|1 + x_o \zeta|^2} \ln^2 \left| \frac{1 - \sqrt{\zeta \bar{z}}}{1 - \sqrt{\zeta z}} \right| =$$

$$= \frac{2}{3}\pi^2 + 8x_o + \frac{17}{2} x_o^2 + O(x_o^3).$$

References

1. C. Bandle, *Isoperimetric Inequalities and Applications.* Pitman Publ., London 1980.
2. B. Dittmar, *Isoperimetric inequalities for the sums of reciprocal eigenvalues.* In: Progress in partial differential equations. Pont-á-Mousson 1997. Volume 1 (eds. H. Amann, C. Bandle, M. Chipot, F. Conrad and I. Shafrir), Pitman Research Notes in Mathematics Series **383**, 78 - 87, Addison Wesley Longman Ltd., Harlow, Essex 1998.
3. B. Dittmar, *Sums of Reciprocal Eigenvalues of the Laplacian.* Math. Nachr. **237** (2002), 45 - 61.
4. B. Dittmar, *Sums of Reciprocal Stekloff Eigenvalues.* Math. Nachr. **268** (2004(, 44 - 49.
5. B. Dittmar, *Sums of free membrane eigenvalues.* J. d'Anal. Math. **95** (2005), 323 - 332.
6. B. Dittmar, *Sums of free membrane eigenvalues II.* (to appear)
7. B. Dittmar, *Eigenvalue problems and conformal mapping.*In: Handbook of Complex Analysis: Geometric Function Theory, Vol. 2,(ed. R. Kühnau) Amsterdam etc., Elsevier 2004.

8. B. Dittmar, M. Hantke,*Robin function and eigenvalue problems.* Preprint Martin-Luther-Universität Halle-Wittenberg 2006.
9. B. Dittmar, R. Kühnau, *Zur Konstruktion der Eigenfunktionen Stekloffscher Eigenwertaufgaben.* Z. angew. Math. Phys. **51** (2000), 806 - 819.
10. P. Duren, *Robin capacity.* In: Proceedings of the Third CMFT Conference (eds. N. Papamichael et al.) Singapore, New Jersey, Hong Kong: World Scientific 1997, 177-190.
11. M. Hantke, *Summen reziproker Eigenwerte.* Dissertation Martin-Luther-Universität Halle-Wittenberg 2006.
12. A. Henrot, *Extremum problems for eigenvalues of elliptic operators.* Birkhäuser Verlag 2006.
13. J. Hersch, L. E. Payne and M. M. Schiffer, *Some inequalities for Stekloff eigenvalues.* Arch. Rat. Mech. Anal. **57** (1974), 99 - 114.
14. P. Kröger, *Upper bounds for the Neumann eigenvalues on a bounded domain in euclidean space.* J. Funct. Anal. **106** (1992), 353 - 357.
15. P. Kröger, *On upper bounds for high order Neumann eigenvalues of convex domains in euclidean space.* Proc. Amer. Math. Soc. **127**(6) (1999), 1665 - 1669.
16. R. S. Laugesen, *Eigenvalues of Laplacians with mixed boundary conditions, under conformal mapping.* Ill. J. Math. **42** (1998), 19-39.
17. R. S. Laugesen, *Eigenvalues of the Laplacian on inhomogeneous membranes.* Amer. J. Math. **120** (1998), 305-344.
18. R. S. Laugesen, C. Marpurgo, *Extremals for eigenvalues of Laplacians under conformal mapping.* J. of Functional Analysis **155** (1998), 64-108.
19. Ü. Lumiste, J. Peetre, *Edgar Krahn 1894-1961 a Centenary Volume,* IOS Press Amsterdam, Oxford, Washington DC, Tokyo 1994.
20. N. Nadirashvili, *Conformal maps and isoperimetric inequalities for eigenvalues of the Neumann problem.* Israel Mathematical Conference Proceedings **11** (1997), 197-201.
21. G. Pólya, *Patterns of Plausible Inference.* Princeton, University Press, Princeton New Jersey 1954.
22. G. Pólya, *On the characteristic frequencies of a symmetric membrane.* Math. Zeitschr. **63** (1955), 331 - 337.
23. G. Pólya, *On the eigenvalues of vibrating membranes.* Proc. London Math. Soc. **11** (1961), 419 - 433.
24. G. Pólya and G. Szegö, *Isoperimetric Inequalities in Mathematical Physics.* Princeton University Press 1951.
25. G. Pólya and M. Schiffer, *Convexity of functionals by transplantation.* J. d'Anal. Math. **3** (1954), 245–345.
26. G. Szegö, *Inequalities for certain eigenvalues of a membrane of given area.* J. Rational Mech. Anal. **3** (1954), 343 - 356.
27. H. F. Weinberger, *An isoperimetric inequality for the N-dimensional free membrane problem.* J. Rat. Mech. Anal. **5** (1956), 633-636.
28. H. Weyl, *Das asymptotische Verteilungsgesetz der Eigenwerte linearer partieller Differentialgleichungen (mit einer Anwendung auf die Theorie der Hohlraumstrahlung).* Math. Ann. **71** (1912), 441-479.

GEOMETRY OF THE GENERAL BELTRAMI EQUATIONS

B. BOJARSKI

Warszawa

Although the quasiconformal mappings theory—QC-theory for short—originated in the context of geometric methods of pure complex analysis and holomorphic 2-d mappings ([28], [29], [32]) from the beginning it was connected with elliptic partial differential equations. These are the Beltrami equation (0.1) and the general Beltrami equation (0.2) which we write in the complex form

$$w_{\bar{z}} - \mu(z)w_z = 0 \tag{0.1}$$

and

$$w_{\bar{z}} - \mu(z)w_z - \nu\overline{w}_z = 0 \tag{0.2}$$

where

$$w_z = \tfrac{1}{2}(w_x - iw_y), \quad w_{\bar{z}} = \tfrac{1}{2}(w_x + iw_y).$$

Systems (0.1) and (0.2) are elliptic (at the point z) if

$$|\mu(z)| < 1 \quad \text{or} \quad |\mu(z)| + |\nu(z)| < 1. \tag{0.3}$$

The ellipticity is uniform in a subdomain D of the complex plane $\mathbb{C}$ if for some constant k

$$|\mu(z)| + |\nu(z)| \leq k < 1 \tag{0.4}$$

for all $z \in D$ ($\nu \equiv 0$ for the equation (0.1)).

For convenience of the reader not used to the complex operators $\frac{\partial}{\partial z}$ and $\frac{\partial}{\partial \bar{z}}$ we recall that the Beltrami equations (0.1) and (0.2) have the real form, for $w \equiv u + iv$,

$$\begin{aligned} v_y &= \alpha u_x + \beta u_y + au + bv + e, \\ -v_x &= \gamma u_x + \delta u_y + cu + dv + f \end{aligned} \tag{0.5}$$

with $a = b = c = d = e = f \equiv 0$.

The ellipticity condition takes the form

$$4\alpha\delta - (\beta + \gamma)^2 > 0. \tag{0.6}$$

System (0.5) is uniformly elliptic if

$$4\alpha\delta - (\beta + \gamma)^2 \geq K_0 > 0 \tag{0.7}$$

and the coefficients $\alpha, \beta, \gamma, \delta$ in (0.5) are uniformly bounded.

The discussed theory is also directly connected with second order linear equations for one unknown function $\phi(x, y)$:

$$A\phi_{xx} + 2B\phi_{xy} + C\phi_{yy} + E\phi_x + F\phi_y + G\phi = H, \quad A > 0, \tag{0.8}$$

and the non-symmetric divergent form equations

$$\frac{\partial}{\partial x}(\alpha u_x + \beta u_y) + \frac{\partial}{\partial y}(\gamma u_x + \delta u_y) = 0. \tag{0.9}$$

The proper Beltrami equation (0.1) ($\nu \equiv 0$) is obtained if $\beta \equiv \gamma$ in (0.5) and moreover

$$\alpha\delta - \beta^2 \equiv 1 \quad (\alpha, \beta, \gamma \text{ uniformly bounded}).$$

The general real system (0.5) can be written in the complex form

$$w_{\bar{z}} - \mu w_z - \nu \overline{w}_z = Aw + B\overline{w} + C \tag{0.10}$$

which is a linear differential equation over the field of reals.

Locally homeomorphic "solution" of the equation (0.1), for some μ satisfying the inequality (0.4), is by definition a quasiconformal mapping, QC-mapping for short or, more precisely, K-QC-mapping, with K defined by

$$k = \frac{K-1}{K+1}. \tag{0.11}$$

K is then called the dilatation of the QC-map and the coefficient μ in the Beltrami equation (0.1) satisfied by the mapping function $w(z)$ is traditionally termed the complex dilatation of the mapping.

Also coefficients (μ, ν) in the equation (0.2) will be called complex dilatations—complex characteristics—of the mapping.

So far we used the term "solutions" of (0.1) and (0.2) in a rather loose way. To be more precise we state now that in the following we restrict our considerations to equations (0.1) and (0.2) with measurable coefficients μ, ν compactly supported on the complex plane $\mathbb{C}$. This means that when considered on the whole complex plane $\mathbb{C}$ the equations (0.1) and (0.2) reduce to the Cauchy–Riemann equations near ∞ and the considered solutions are holomorphic functions of z for $|z|$ big enough.

"Solutions" of (0.1) and (0.2) are understood then in the generalized sense of Sobolev classes $W^{1,2}_{\text{loc}}(\mathbb{C})$; thus a priori discontinuous solutions are admitted. It is a fundamental fact, first established in 1954 and published in [13], that, under the uniform ellipticity assumptions (0.4), the generalized or weak solutions of Beltrami equations (0.1) and (0.2) are actually in $W^{1,p}_{\text{loc}}$, for some $p > 2$, depending on k only, and thus, by Sobolev imbedding theorem, continuous, more precisely, even Hölder continuous with exponent $\alpha = 1 - \frac{2}{p}$.

We recall the formula

$$\mathcal{J}w(z) = |w_z|^2 - |w_{\bar{z}}|^2$$

for the Jacobian of the mapping

$$w = w(z) \tag{0.12}$$

usually considered on an open subset D of the complex plane $\mathbb{C}$.

In the following we shall restrict our considerations to the orientation or sense preserving mappings meaning that at almost every differentiability point of the function $w = w(z)$ in the domain D the Jacobian $\mathcal{J}w(z)$ is positive

$$\mathcal{J}w(z) > 0 \quad \text{or} \quad |w_z| > 0 \text{ a.e. in } D. \tag{0.13}$$

1. Generating the Beltrami equations

Let us remark now that the elliptic Beltrami equations (0.1) and (0.2) naturally arise in the context of general theory of sense preserving transformations.

Indeed, if the mapping $w = w(z)$ satisfies $\mathcal{J}w(z) > 0$ a.e. then the differential identity

$$w_{\bar{z}} = \frac{w_{\bar{z}}}{w_z} \cdot w_z \quad \text{with} \quad \mu_w(z) = \frac{w_{\bar{z}}}{w_z} \tag{1.1}$$

is meaningful a.e. with some measurable function $\mu \equiv \mu_w$ satisfying the ellipticity condition (0.3) and produces the Beltrami equation (0.1) for $w = w(z)$ such that

$$\mathcal{J}w = |w_z|^2(1 - |\mu|^2) > 0.$$

Assume now that we have a pair $w = w(z)$ and $v = v(z)$ of sense preserving mappings, admitting the generalized derivatives $w_z, w_{\bar{z}}, v_z, v_{\bar{z}}$ a.e., satisfying moreover the condition

$$\Delta = 2i \operatorname{Im}(w_z \overline{v}_z) \neq 0 \text{ a.e.} \tag{1.2}$$

Again the differential identity

$$\Delta w_{\bar{z}} = (w_{\bar{z}}\overline{v}_z - v_{\bar{z}}\overline{w}_z)w_z - (w_{\bar{z}}v_z - v_{\bar{z}}w_z)\overline{w}_z \equiv \mu\Delta w_z + \nu\Delta\overline{w}_z \tag{1.3}$$

with μ and ν defined by the formulas

$$\begin{aligned} \mu\Delta &\equiv w_{\bar{z}}\overline{v}_z - v_{\bar{z}}\overline{w}_z \\ \nu\Delta &\equiv w_{\bar{z}}v_z - v_{\bar{z}}w_z \end{aligned} \tag{1.4}$$

implies the differential equation (0.2) for the functions $w = w(z)$ and $v = v(z)$.

Relations (1.3) and (1.4) can be considered as defining equations of the Beltrami equation (0.2). The natural question arises: what are the conditions on the pair $w(z)$, $v(z)$ assuring the ellipticity of the system (1.3)? The following propositions give the answer.

Proposition 1. *The complex equation* (1.4) *is uniformly elliptic* (k*-elliptic*) *if and only if the linear family*

$$w^{\alpha,\beta} = \alpha w + \beta v, \quad \alpha, \beta \text{ real}, \tag{1.5}$$

is a family of K-QC-mappings, $k = \frac{K-1}{K+1}$. *In other words, the k-ellipticity of system* (1.4) *is equivalent with the inequality*

$$\left|w_{\bar z}^{\alpha,\beta}\right| \le k\left|w_z^{\alpha,\beta}\right| \tag{1.6}$$

on a subset of full measure independent of α, β.

For the proof see Lemma 12.1 in [27].

Analogously holds

Proposition 2. *The equation* (1.4) *is elliptic* ($|\mu(z)| + |\nu(z)| < 1$) *iff all the mappings* $w^{\alpha,\beta}$ *are orientation preserving (i.e.* $\left|w_{\bar z}^{\alpha,\beta}\right| < \left|w_z^{\alpha,\beta}\right|$).

Actually the proof in [27] works for linear families of sense preserving mappings.

Our Theorem 3.1 below (see also [19]) states that any uniformly elliptic general Beltrami equation (0.2) admits a primary pair of solutions. Theorem 3.1 gives a positive answer to Conjecture 1 in [27].

Is this true for not necessary uniformly elliptic systems is an interesting delicate question, presumably not easy.

It is to be expected that no analogues of linear k-dimensional families of orientation preserving QC-mappings exist in dimensions $n \ge 3$, $k > 1$.

2. Principal homeomorphisms[1] of the Beltrami equations

These are global, i.e. defined on the whole complex plane $\mathbb{C}$, solutions of the equations (0.1) or (0.2) with compactly supported coefficients μ and ν

$$\mu \equiv \nu \equiv 0 \quad \text{for } |z| > R, \tag{2.1}$$

properly normalized at ∞.

Usually we shall consider the case

$$w(z) \sim az \quad \text{at } z \to \infty \tag{2.2}$$

with some complex constant $a \ne 0$.

Strictly following the analytical ideas and formulas developed in [13], [43], [44], we represent the considered homeomorphic solutions of (0.2) or the K-quasiconformal mappings, in the form

$$w(z) \equiv az - \frac{1}{\pi}\iint\limits_{\mathbb{C}} \frac{\omega(t)}{t-z}\, d\sigma_t \equiv az + T\omega(z) \tag{2.3}$$

for some complex constant a and $\omega \in L^p(\mathbb{C})$. Here $d\sigma_t$ is the Lebesgue measure on $\mathbb{C}$ or

$$d\sigma_t = -\frac{1}{2i}\, dt \wedge d\bar t.$$

[1] I. Vekua [44] calls them "basic" homeomorphisms ($\boldsymbol{a = 1}$).

The density $\omega(t)$ in (2.3) is assumed to be in L^p_{loc}, for some $p > 1$. In general this means that we are considering the solutions in the Sobolev spaces $W^{1,p}_{\text{loc}}(\mathbb{C})$. Then (2.3) may be differentiated and gives the formulas

$$\frac{\partial w}{\partial \bar{z}}(z) = \omega(z), \qquad \frac{\partial w}{\partial z} = a + \frac{\partial T\omega}{\partial z} \tag{2.4}$$

with

$$\frac{\partial T\omega}{\partial z} = -\frac{1}{\pi} \iint \frac{\omega(t)}{(t-z)^2}\, d\sigma_t \equiv S\omega(z). \tag{2.5}$$

Here $S\omega$ may be understood as principal value singular integral operator in the plane. It is classically meaningful for Hölder continuous densities ω, but makes sense also for $\omega \in L^p(\mathbb{C})$ for all $p > 1$, as a bounded operator $S : L^p \to L^p$

$$\|S\omega\|_{L^p} \leq A_p \cdot \|\omega\|_{L^p}. \tag{2.6}$$

Moreover for $p = 2$ $S\omega$ is an isometry $\|S\omega\|_{L^2} = \|\omega\|_{L^2}$, $A_2 = 1$ and

$$k \cdot A_p < 1 \quad \text{for } |p - 2| < \varepsilon \tag{2.7}$$

for some sufficiently small positive ε, depending on the uniform ellipticity constant in (0.4).

Then the system (0.2) for $w(z)$ reduces to the system of linear singular equations for the density ω

$$\omega - \mu S\omega - \nu \overline{S\omega} = h, \qquad h = a\mu + \bar{a}\nu \tag{2.8}$$

in the space $L^p(\mathbb{C})$ for p satisfying (2.7).

The equation (2.8) acts as an invertible operator in $L^p(\mathbb{C})$ for the range of p defined by (2.7). Most important is the case $p > 2$. Then the function $w(z)$ in (2.3) realizes a continuous, in view of the Sobolev imbedding theorems, even Hölder continuous (with exponent $\alpha = 1 - \frac{2}{p} > 0$) mapping of the complex plane. In fact the following proposition of fundamental importance for the theory of two-dimensional quasiconformal mappings (QC-maps) and the theory of elliptic equations in the complex plane holds.

Proposition 3. *The formulas* (2.3) *with the density* ω *satisfying the integral equation* (2.8) *define a* K*-quasiconformal homeomorphism of the complex plane. It is a generalized solution of the Beltrami equation* (0.1) *of the Sobolev class* $W^{1,p}_{\text{loc}}(\mathbb{C})$. *The mapping* $w = w(z)$ *is differentiable a.e. and its Jacobian* $\mathcal{J}w(z) = |w_z|^2 - |w_{\bar{z}}|^2$ *is positive a.e.*

Moreover, the L^p *norm of* ω *has an estimate* $\|\omega\|_{L^p} \leq C$ *with the constant* C *depending on* k, p *and the* L^p *norm of* h *in* (2.8) *only.*

The inverse mapping $z = z(w)$ also may be represented by the integral formulas of the type (2.3) in the complex plane of variable w and satisfies the corresponding Beltrami equations with coefficients $\widetilde{\mu}, \widetilde{\nu}$ satisfying inequalities (0.4) with the same k.

The coefficients $\widetilde{\mu}, \widetilde{\nu}$—the complex dilatations of the inverse map—in the complex plane w are expressed by simple formulas which may be found in [15] and [44] and in numerous references afterwards.

For the most important simplest case of the classical Beltrami equation (0.1), i.e., when the coefficient $\nu \equiv 0$,

$$w_{\bar{z}} - \mu(z) w_z = 0, \tag{2.9}$$

the equation (2.8) reduces to

$$\omega - \mu S \omega = h \tag{2.10}$$

which is linear over the complex field $\mathbb{C}$. The equation (2.8) has to be considered over the real field $\mathbb{R}$ only. However in both cases the ellipticity conditions (0.4) ensure the unique solvability in the form of convergent Neumann series of the non-homogeneous equations (2.8) and (2.10). All these facts have been discussed and used in [13], [14], [15] and are fundamental in the literature on quasiconformal mappings and planar elliptic partial differential equations.

We stress explicitly that the remarkable geometric properties of the principal solutions like infinitesimal quasiconformality, positive Jacobian and global (homeomorphism) properties of the mappings (2.3) are the result of the interplay of the form of (2.3), the asymptotics (2.2) and the equations (2.8), (0.2). If the integral equation (2.8) is removed, the representation (2.3) holds for any solutions of the non-homogeneous Cauchy–Riemann equations

$$\frac{\partial w}{\partial \bar{z}} = \omega, \quad \omega \in L^p_{\mathrm{loc}}(\mathbb{C}) \tag{2.11}$$

and the asymptotics (2.2) and these do not imply none of the mentioned geometric properties of $w(z)$ whatsoever. The representation (2.3) and the formulas (2.4)–(2.8) in the context of Sobolev spaces $W^{1,p}_{\mathrm{loc}}(\mathbb{C})$, $p > 2$, and Proposition 3 with the numerous other analytical and geometrical consequences first appeared in the papers of I. N. Vekua and the author [13], [14], [43]. L. Ahlfors in the paper [2], published somewhat later in 1955, independently came to the formulas (2.3)–(2.5) and (2.8) (for the classical complex Beltrami equation (0.1), with $\nu \equiv 0$), but, since he was working only in L^2 or $W^{1,2}_{\mathrm{loc}}$, he finally had to stop the construction of his global homeomorphisms of the form (2.3) at the level of Hölder classes H^α, $\mu, \omega \in H^\alpha(\mathbb{C})$, $0 < \alpha < 1$, $w \in C^{1+\alpha}$, though, actually almost one year earlier the Bojarski–Vekua theory as described above was available.

In the following years the formulas of type (2.3)–(2.8) and their numerous consequences have been often used in the immense and permanently growing literature on topics of QC-maps, univalent functions with QC-extensions, the study of dependence on parameters of families of QC-mappings and conformal structures, Teichmüller spaces, complex dynamics, harmonic analysis etc. and somehow dispersed in kind of an unidentified folklore or have been, rather occasionally, identified depending on the intentions of the referent and the context.

The recalled above amazingly wide range of applications and consequences of the representation formula (2.3) gives it some features of universality in the local and global complex analysis. The vision of this universality of (2.3) has been independently and in full parallel understood by I. N. Vekua and L. V. Ahlfors in their famous papers of 1955 [43], [2].

In the following years both L. V. Ahlfors and I. N. Vekua propagated the use and consequences of the formula in general complex analysis. Therefore in 2007, the year of their parallel centenary, as a due tribute to these great men of science, it is proper and highly justified to relate the names of these two remarkable mathematicians with the formula (2.3) and call it henceforth the Ahlfors–Vekua representation formula.

The basic representation formula (2.3) arises as a solution of the non-homogeneous Cauchy–Riemann equation

$$\frac{\partial w}{\partial \overline{z}} = \omega \tag{2.12}$$

with the asymptotic condition (2.2) at infinity.

The operator $S(\omega)$ appears then as the transformation

$$\frac{\partial w}{\partial z} = S\left(\frac{\partial w}{\partial \overline{z}}\right) \tag{2.13}$$

and the resulting obvious formula

$$\frac{\partial S\omega}{\partial \overline{z}} = \frac{\partial^2 w}{\partial z \partial \overline{z}} = \frac{\partial \omega}{\partial z} \tag{2.14}$$

implies the crucial isometry property of the operator $S : L^2 \to L^2$.

Analogous facts hold if instead of the whole plane we consider the unit disc K or the upper half plane H: $\operatorname{Im} z > 0$ with the boundary conditions

$$\operatorname{Re} w = 0 \quad \text{on the boundary } \partial K,\ |z| = 1 \tag{2.15}$$

$$\operatorname{Im} w = 0 \quad \text{on the boundary } \partial H\text{: real line } z = \overline{z}. \tag{2.16}$$

For the problem (2.12), (2.15) the representation formula reads

$$T_1(\omega) = -\frac{1}{\pi} \iint\limits_K \left(\frac{\omega(t)}{t - z} + \frac{z\overline{\omega(t)}}{1 - z\overline{t}} \right) d\sigma_t \tag{2.17}$$

and the singular integral operator

$$S_1(\omega) = \frac{\partial T_1(\omega)}{\partial z} = -\frac{1}{\pi} \iint\limits_K \left(\frac{\omega(t)}{(t - z)^2} + \frac{\overline{\omega(t)}}{(1 - z\overline{t})^2} \right) d\sigma_t \tag{2.18}$$

satisfies (2.14) which implies, together with the boundary condition (2.15), the L^2-isometry of S_1.

As in the case of the principal solutions, based on the Ahlfors–Vekua representation (2.3), if in the formula (2.17) we use as densities $\omega(t)$ the solutions of the corresponding singular integral equations of type (2.8), we obtain solutions of the boundary value problem (2.15) with the normalizing condition $w(1) = 0$, for non-homogeneous Beltrami equations (0.2)

$$\frac{\partial w}{\partial \bar{z}} - \mu \frac{\partial w}{\partial z} - \nu \overline{\frac{\partial w}{\partial z}} = h, \quad h \in L^p(K) \tag{2.19}$$

in the Sobolev spaces $W^{1,p}(K)$ with the precise estimates for the $W^{1,p}(K)$ norms of the solutions in terms of the ellipticity constant k, and the L^p norm of h. Because of (2.7) we obtain also global Hölder exponents for the solutions of th boundary problems (2.15) and (2.16).

We can also use the formula

$$w(z) = f(z) + \widetilde{T}_1(\omega), \quad f(z) = \frac{1-z}{1+z}, \tag{2.20}$$

to represent solutions of the Beltrami equation (0.1) satisfying the boundary conditions (2.15) and the conditions

$$w(0) = 1, \quad \operatorname{Im} w(1) = 0 \quad (w(1) = 0) \tag{2.21}$$

as QC-mappings of the disc K on the half-planes. The operator $\widetilde{T}_1$ in (2.20) is obtained from T_1 by adding obvious linear functionals, so that (2.21) holds. Moreover,

$$\frac{\partial \widetilde{T}_1}{\partial \bar{z}} = \frac{\partial T_1}{\partial \bar{z}} \equiv \omega, \quad S_1(\omega) \equiv \frac{\partial \widetilde{T}_1}{\partial z}.$$

Thus, if the density $\omega(z)$ in (2.20) satisfies the corresponding singular integral equations of type (2.8), $w(z)$ in (2.19) will be the solution of the Beltrami system (0.1), (0.2) mapping the disc K onto the half-plane $\operatorname{Re} w > 0$. Again we immediately obtain also the $W^{1,p}$ estimates for the homeomorphic solutions and consequently the Hölder estimates.

3. Structure theorem for general Beltrami equations

The existence theorem of a principal homeomorphism of the Beltrami equation (0.1) immediately implies the following structural theorem.

Theorem 1. *Let $\chi = \chi(z)$ be a homeomorphic solution of the Beltrami equation* (0.1) *in a planar domain G. Then any other generalized solution $w = w(z)$ of the equation* (0.1) *in G may be represented in the form*

$$w(z) = f(\chi(z)) \tag{3.1}$$

where $f(\chi)$ is holomorphic in the image domain $\chi(G)$.

Conversely, given a solution $\chi = \chi(z)$ of (0.1)*, not necessarily homeomorphic, and an arbitrary holomorphic function $f(\chi)$ in $\chi(G)$ the formula* (3.1) *gives a solution of* (0.1).

Corollary. Given two arbitrary Jordan regions G_1 and G_2 of the complex plane there always exists a solution $w = w(t)$ of the Beltrami equation (0.1) homeomorphically mapping G_1 onto G_2. Such a solution may be extended to the closed regions $\overline{G}_1 \to \overline{G}_2$ as a homeomorphism.

It is uniquely determined e.g. by the assignment of three arbitrary points on the boundary ∂G_1 or in some other way precisely as in the theory of conformal mappings.

This is the famous Riemann mapping theorem for quasiconformal mappings with the given (fixed) complex dilatation $\mu(z)$, satisfying the uniform ellipticity condition (0.4). Since our only assumption on the dilatation $\mu(z)$ is its measurability, it is called the *measurable Riemann mapping theorem.*

The homeomorphism $\chi = \chi(z)$ in (3.1) may be chosen in a variety of ways. If it is the principal homeomorphism of the Beltrami equation (0.1) then it has a univalent conformal extension to the complex plane over the boundary of G and then all the singularities of the behavior of the solution $w(z)$ are transferred to the boundary behavior of the holomorphic component $f(\chi)$ near the boundary of the image domain $\chi(G)$. If G is e.g. the unit disc D and $\chi = \chi_0(z)$ is the quasiconformal mapping of D onto D, normalized by fixing the origin $\chi(0) = 0$ and $\chi(1) = 1$, then the following a priori estimate holds ([14], [15]).

Proposition 4. *With the normalization above the $W^{1,p}(D)$ norms of $\chi(z)$ are estimated*

$$|\chi_z|_{L^p(D)},\ \|\chi_{\bar z}\|_{L^p(D)} \leq C \tag{3.2}$$

with the constant C depending on the ellipticity constant k in (0.4) only (for any admissible p, $p > 2$, in agreement with (2.7)).

In particular the homeomorphism $\chi = \chi(z)$ and the inverse homeomorphism $\chi^{-1}(\xi)$ satisfy uniform Hölder condition with exponent $\alpha = 1 - \frac{2}{p}$, depending on the ellipticity constant.

These a priori estimates are crucial for the extension of the measurable Riemann mapping theorems to general Beltrami equations (0.2).

The representation formula (3.1) holds also for the solutions of the general Beltrami equation (0.2) and is a direct consequence of Theorem 1 since we admit measurable dilatations: indeed, we can write

$$0 = w_{\bar z} - \mu w_z - \nu \overline{w}_z \equiv w_{\bar z} - \widetilde{\mu} w_z$$

with

$$\widetilde{\mu} = \mu + \nu \frac{\overline{w}_z}{w_z} \tag{3.3}$$

which is pointwise legitimate in the class of measurable coefficients μ only. Nevertheless, the crucial estimate (0.4)

$$|\widetilde{\mu}(z)| \leq |\mu| + |\nu| \leq k < 1 \text{ a.e.}$$

holds with the same constant k.

However, because of (3.3), the homeomorphism $\chi = \chi(z)$ depends on the represented solution $w = w(z)$, whereas for the Beltrami equation the component $\chi(z)$ was possible to be taken universal, depending only on the coefficient μ. It is of fundamental importance that the a priori estimates (3.2) hold with the same constants.

This fact and the estimates of Proposition 3 are crucial in the quite not trivial proof of the general measurable Riemann mapping theorem. Also the uniqueness questions in the mapping theorem for equation (0.2) are much more difficult than for the Beltrami equation (0.1).

Assume now that we have a fixed k-elliptic real Beltrami equation (0.2).

Theorem 2. *Let $F(\chi)$ be a given, in general meromorphic, function in the unit disc D. Then there exists a QC-homeomorphism $\chi = \chi(z)$ of the (closed) disc $\chi : D \to D$ normalized by $\chi(0) = 0$, $\chi(1) = 1$, such that the function*

$$w(z) = F(\chi(z)) \tag{3.4}$$

is the solution of the given general Beltrami equation.

The QC-homeomorphism $\chi(z)$ may be normalized in some other standard way used in the theory of holomorphic conformal mappings.

Theorem 3.2 is in some sense inverse to the structure Theorem 1 for measurable real Beltrami equations. The transition from the function $F(\chi)$ to $w(z)$ is a highly non-linear operation. Theorem 3.2 was first published in this generality in [14] in 1955, based on the tools elaborated in 1953–54 by I. N. Vekua and B. Bojarski and shortly described above.

As a corollary we get

Theorem 3. *For any uniformly elliptic real Beltrami equation* (0.2) *in the unit disc D, any fixed oriented triple of boundary points of ∂D and any given Jordan domain G with a fixed equally oriented triple of boundary points on ∂G, there exists a QC-homeomorphic solution of the equation mapping $(\overline{D}, z_1, z_2, z_3)$ onto (G, w_1, w_2, w_3).*

Proof. Take a conformal univalent map $f(\chi)$ mapping $(\overline{D}, z_1, z_2, z_3)$ onto (G, w_1, w_2, w_3) and "change" the variables in the disc D by Theorem 3.2.

The described procedure clearly applies to the construction of solutions of uniformly elliptic real Beltrami equations mapping simply connected domains more general than Jordan domains.

The first direct self-contained proof of the measurable Riemann mapping theorem was obtained in the framework of analytical methods and concepts elaborated by I. N. Vekua and the author in the early fifties of the last century.

Earlier work, going back even as far as to F. Gauss in the nineteenth century, was done with various smoothment assumptions on the coefficients and as a rule locally with classical methods of potential theory. The highlight here were the beautiful and deep works of L. Lichtenstein [36] from the time he worked in Kraków, Poland, in the A. Zaremba group, and published in 1916 in the Bulletin of the Cracow Academy of Sciences. Lichtenstein's work stopped at the level of Hölder continuous coefficients of complex Beltrami equations (0.1) in the real symmetric form (0.5). Though later often rather formally referred to (C. B. Morrey, L. Bers et al.) the Lichtenstein theory does not seem to have been incorporated into a modern, self-contained presentation of the topic. Important issue of distinction between local and global character of these results wait for a clear and concise presentation. References in loose style in various places to the general Koebe-Poincaré uniformization theory are rather unsatisfactory. It may be also that the precise conformally invariant estimates for measurable real and complex Beltrami equations as presented above in Propositions 3 and 4 open the shortest and most natural theory to the general planar uniformization theory of Poincaré–Koebe–R. Nevanlinna. This is certainly a challenging problem of the planar geometric function theory (though rather expository, taking into account all the a priori estimates and other strong tools available at the present moment in the general analytical theory of QC-maps).

The paper [39] of C. B. Morrey (1938) presented new important methods to discuss the local integral type weighted energy estimates for the general planar uniformly elliptic differential equations (0.5) with measurable bounded coefficients. These estimates imply the universal Hölder estimates for continuous homeomorphic solutions of (0.5), (0.2) in the Sobolev spaces $W^{1,2}_{\text{loc}}$ in terms of the ellipticity constants and L^2 bounds of the coefficients. Relying on these C. B. Morrey performed the limiting process in the Riemann mapping and boundary value problems for smooth, Hölder continuous coefficients theory of (0.5) or (0.9) and thus extended the L. Lichtenstein existence results to the general measurable Beltrami equations.

L. Bers, in several of his papers from late fifties and in Bers–Nirenberg [12] heavily relies on the results of L. Lichtenstein [36], M. Lavrentiev [32] and C. B. Morrey [39]. It is to be regretted that the evidently sketchy presentation in [12] has never been followed by a detailed paper, promised in [12] and elsewhere.

Looking at the literature of the topic from the perspective of the later years, and maybe even present, it is rather clear that the methods used by C. B. Morrey and the concepts and methods propagated by L. Bers were rather far from the universality and flexibility of the methods and analytical concepts of I. N.Vekua and his school from the early fifties of the 20th century, partially described above. Also the highly publicized paper [4] on Riemann mapping theorem "for variable metrics" in fact in a rather direct and explicit way relies on the analytic apparatus and estimates used several years earlier in the I. N. Vekua school.

It is to be regretted that in a part of mainly American origin literature the role of I. N. Vekua's school in the analytic theory of general measurable Beltrami equation is, gently speaking, at least slightly, underestimated by being referred to as an "interesting observation".

Most of the topics discussed in the present paper are directly related with the results and concepts of Vekua's school from the years 1954–1957, see [43], [44], [13], [14], [15], [16].

As a contribution to this point of view let us mention also the new elements in the present paper: 1) the formula (2.20) gives a direct explicit expression for the general QC-mapping of the unit disc K onto the half-plane $\operatorname{Re} w > 0$ for arbitrary measurable complex dilatations (μ, ν) of the equations (0.2) in complete analogy to the generality of the principal QC-homeomorphisms for general Beltrami equations. 2) The discussed below existence theorem for primary pairs, solving Conjecture 1 in [27].

4. Primary solutions of the general Beltrami equations

Primary solutions of the general Beltrami equation (0.2) allow to recover the dilatations μ and ν. They are thus crucial in discussing various structural properties and inverse type problems for equations (0.2). The important role of primary solutions was recognized in [27], [20], where the following definition was introduced.

Definition. A pair (φ, ψ) of global homeomorphic solutions (i.e., realizing the homeomorphic orientation preserving mapping of the complex plane $\varphi : \mathbb{C} \to \mathbb{C}$, $\varphi(\infty) = \infty$) of (0.2) is called *a primary pair* if the condition (1.2) is satisfied.

The fundamental problem was the existence of primary pairs for the general equation (0.2). It was stated in [27] as Conjecture 1 and proved for equations (0.2) satisfying the condition

$$k < \tfrac{1}{2} \quad \text{or equivalently} \quad K < 3 \tag{4.1}$$

when formulated in terms of associated K-QC-mappings.

An interesting special class of Beltrami equations (0.2) arises when

$$\nu + \mu \equiv 0, \quad \nu \equiv -\mu$$

and (0.2) takes the form

$$w_{\bar{z}} - \mu(w_z - \overline{w}_z) = 0. \tag{4.2}$$

The ellipticity condition (0.3) then reads $|\mu| < \frac{k}{2} < \frac{1}{2}$.

A characteristic feature of (4.2) is that the identity mapping $w(z) \equiv z$ is a principal homeomorphism of (4.2) for arbitrary dilatation μ. Then the condition (1.2) for a primary pair $(z, \psi(z))$ takes the form

$$\operatorname{Im}(\psi_z) \neq 0 \quad \text{a.e.} \tag{4.3}$$

Theorem 4. *The Beltrami equation*

$$w_{\bar{z}} - q(z)(w_z - \overline{w}_z) = 0$$

with the compactly supported measurable dilatation ($q \equiv 0$ in the neighborhood of ∞) satisfying the ellipticity condition (0.4) admits a global solution $\psi : \mathbb{C} \to \mathbb{C}$ satisfying condition (4.3). Then the pair $(z, \psi(z))$ is a primary pair for the Beltrami equation (4.2).

The main idea of the proof is the "factorization" of solutions of (4.2) through the solutions of a Beltrami equation

$$\frac{\partial w_0}{\partial \bar{z}} - q_1(z)\,\frac{\partial w_0}{\partial z} = 0 \tag{4.4}$$

and the "conjugate" or inverse Beltrami equation

$$\frac{\partial w}{\partial \bar{\zeta}} + \widetilde{q}(\zeta)\,\overline{\frac{\partial w}{\partial \zeta}} = 0 \tag{4.5}$$

with a special choice of the auxiliary dilatations $q_1(z)$ and $\widetilde{q}(\zeta)$ in (4.4) and (4.5). The function $q_1(z)$ in (4.4) is determined by $q(z)$ in (4.2) by the formula

$$q(z) = \frac{q_1(z)}{1 + |q_1(z)|^2}\,.$$

The uniform ellipticity condition (0.4) ensures that (4.4) is uniformly elliptic also. The choice of $q_1(z)$ is "canonical": it depends on $q(z)$ in (4.2) only.

The choice of $\widetilde{q}(\zeta)$ for the "conjugate" Beltrami equation depends on the selected global homeomorphic solution of (4.4). If w_0 is a normalised—"principal"—homeomorphism of (4.4) (i.e. it is determined by putting $a = 1$ in (2.3)) then any other global homeomorphism $\zeta(z)$ is determined by the formula

$$\zeta(z) = L(w_0(z)) \tag{4.6}$$

for some linear $L(w_0) = bw_0 + c$.

Assuming a $\zeta(z)$ fixed we determine $\widetilde{q}(\zeta)$ in the equation (4.5) by

$$\widetilde{q}(\zeta) \equiv q_1\left(z^{-1}(\zeta)\right) \quad \text{or} \quad \widetilde{q}(\zeta(z)) \equiv q_1(z). \tag{4.7}$$

For further details of the proof see [19].

Since the primary pair (ϕ, ψ) for (0.2) allows to recover the coefficients μ and ν of the general Beltrami equation (0.2),

$$\mu(z) = \frac{\phi_{\bar{z}}\overline{\psi_z} - \psi_{\bar{z}}\overline{\phi_z}}{2i\,\mathrm{Im}(\phi_z\overline{\psi_z})}, \qquad \nu(z) = -\frac{\phi_{\bar{z}}\psi_z - \psi_{\bar{z}}\phi_z}{2i\,\mathrm{Im}(\phi_z\overline{\psi_z})},$$

it plays such a fundamental role in describing the totality (infinite dimensional) of all solutions of (0.2), in particular in discussing the G-convergence phenomenon and the G-closed classes of elliptic operators in the complex plane (see [27]).

As is well known (see e.g. Vekua's monograph [44]), the auxiliary equations (4.4) and (4.5) admit global solutions without necessarily assuming that $|\mu| + |\nu|$ is compactly supported. The factorization procedure used in the proof of our theorem then can be extended to the general Beltrami equation (0.1) in the whole complex plane $\mathbb{C}$.

There are also many other interesting analytical and geometrical problems related with the important class of primary solutions of Beltrami equations (0.2). Some of these will be hopefully discussed in a subsequent publication [25].

5. Lavrentiev fields and quasiconformal mappings

The infinitesimal geometric meaning of a differentiable transformation $w = w(z)$ at a point z_0 is defined by the linear tangent map

$$Dw(z)(\xi) = w_z(z_0)\xi + w_{\bar{z}}(z_0)\overline{\xi} \tag{5.1}$$

It transforms invariantly ellipses in the tangent plane at z_0 into ellipses in the tangent plane at the image point $w(z_0)$.

Ellipses centered at z are defined up to a similarity transformation by the ratio $p \geq 1$ of their semiaxes and, if $p > 1$, the angle θ mod π between majoraxis and the positive z-axis, and denoted by $\mathcal{E}(p, \theta; z)$ or $\mathcal{E}_h(p, \theta; z)$ where h is the length of the same minoraxis. The pair (p, θ) is called the characteristic of the infinitesimal ellipse, and the family $\mathcal{E}_h(p, \theta; z)$, $h > 0$, $z \in G$, is a field of infinitesimal ellipses (Lavrentiev field). A homeomorphism $w = w(z)$ is said to map the infinitesimal ellipse $\mathcal{E}(p, \theta; z)$ onto the infinitesimal ellipse $\mathcal{E}(p_1, \theta_1; w(z))$ if the tangent map $Dw(z)$ transforms $\mathcal{E}(p, \theta; z)$ onto $\mathcal{E}(p_1, \theta_1; w(z))$. In the Euclidean case the infinitesimal ellipse $\mathcal{E}(p, \theta; z)$ at z is identified with geometric curve—ellipse—in the mapped domain G, for h sufficiently small and then it is meaningful to consider the image

$$w\big(\mathcal{E}_h(p, \theta; z)\big) \equiv \gamma_h$$

as a geometric curve γ_h in the w-plane. If the image curve γ_h lies in the ring domain between the curves $\mathcal{E}_k(p_1, \theta_1; w(z))$ and $\mathcal{E}_l(p_1, \theta_1; w(z))$, $k \leq l$, such that

$$\lim_{h \to 0} \frac{l}{k} = 1 \tag{5.2}$$

then the homeomorphism $w = w(z)$ is said to map the infinitesimal ellipse $\mathcal{E}(p, \theta; z)$ at z onto the infinitesimal ellipse $\mathcal{E}(p_1, \theta_1; w(z))$ at $w = w(z)$. The idea of Lavrentiev was to consider the infinitesimal fields of ellipses as conformal structures and the quasiconformal mappings as mappings compatible with the assigned conformal structures.

Proposition 5. *The general Beltrami equation*

$$w_{\bar{z}} - \mu(z)w_z - \nu(z)\overline{w}_z = 0$$

in a domain G of complex plane determines two conformal structures $\mathcal{E}(p, \theta; z)$ and $\mathcal{E}(p_1, \theta_1; z)$[2] *on G such that the QC-mapping determined by the solution $w = w(z)$ of the equation maps the infinitesimal field $\mathcal{E}(p, \theta; z)$ onto $\mathcal{E}(p_1, \theta_1; z)$.*

---

[2]We freely follow here, strictly speaking, the uncorrect tradition of identifying the notation $\mathcal{E}(p_1, \theta_1; z)$ with the notation $\mathcal{E}(p_1(w(z)), \theta_1(w(z)); w(z))$; it is legitimate only when the "source" z and the image $w = w(z)$ are described in the same Euclidean coordinate system. Correctness would require to call $\mathcal{E}(p_1, \theta_1, z)$ the pull-back by the map $w = w(z)$ of a Lavrentiev field on the image.

The fields are determined by the coefficients or complex dilatations of the Beltrami equations (0.2) by the (invertible) formulas

$$\mu(z) = -\frac{p - p^{-1}}{p + p^{-1} + p_1 + p_1^{-1}} e^{2i\theta}, \quad \nu(z) = \frac{p_1 - p_1^{-1}}{p + p^{-1} + p_1 + p_1^{-1}} e^{2i\theta_1}. \tag{5.3}$$

For details of the corresponding calculations see [41] or [45].

The conformal structure in the sense of Lavrentiev on the complex plane—one-dimensional complex manifold—corresponds to Lavrentiev fields with characteristics $p_1 \equiv 1$, p—uniformly bounded. They correspond to the complex (proper) Beltrami equations of the type (0.1).

We should stress that the clu of the Lavrentiev idea is that the "source" characteristic (p, θ), is mapped into the "target" characteristic (p_1, θ_1) independently of the considered particular solution of the general Beltrami equation (0.2) as long as the relation source z—target $w = w(z)$ is preserved. The formulae (5.3) describe then a pair of "distinguished" or canonical Lavrentiev fields intrinsic for the Beltrami system considered and a selected, pointwise correspondence $w = w(z)$. Of course any chosen solution $w = w(z)$ at every differentiability point transforms an arbitrary Lavrentiev field of infinitesimal ellipses into "some" Lavrentiev field whose characteristics p_1, θ_1 at the image point $w(z)$ depend on the behavior of the map at neighboring points, i.e., on the values of the derivatives $w_z, w_{\bar{z}}$ at z.

Lavrentiev in his seminal paper of 1935 [32] defined QC-mappings as homeomorphic mappings of the unit disc D onto itself such that at every point $z \in D$ the infinitesimal ellipse $\mathcal{E}(p(z), \theta(z); z)$ is mapped onto an infinitesimal circle in the sense defined above. He also proved the existence theorem for such mappings by a direct geometric construction without referring to any classical solutions of boundary value problems, e.g. in the L. Lichtenstein paper [36]. Thus Lavrentiev is probably to be credited for the first direct self-contained proof of the global Riemann mapping theorem for a rather general class of complex Beltrami equations with continuous coefficients $\mu(z)$.

In the following years the Lavrentiev ideas, unfortunately, seem to have been considered and developed mostly in the Russian language mathematical literature. The only paper in English is probably the one of M. Näätänen [40]. P. P. Belinskiĭ and I. N. Pesin in a series of papers starting with [11] made substantial progress in the Lavrentiev theory, introducing the class of generalized QC-maps and discussing their various subtle properties.

Later P. Belinskiĭ in [10] has shown that Lavrentiev's proof from 1935 [32] may be extended to measurable characteristics μ thus giving another full direct proof of the measurable Riemann mapping theorem.

The ideas of Belinskiĭ and Pesin have many geometric features relating this theory with the deep investigations of D. E. Menshov [37], [38] on the monogeneity and the pointwise theory of solutions of Cauchy–Riemann equations in the setting of classical real analysis.

From the point of view of partial differential equations the state of the art at this moment is described by the following theorem obtained by the parallel efforts of P. P. Belinskiĭ, B. Bojarski, I. N. Pesin, see also [40], [16].

Theorem 5. *Orientation preserving Lavrentiev maps are the generalized homeomorphic solutions of the complex Beltrami equations with continuous coefficients.*

The generalized QC-mappings in the sense of I. N. Pesin [11], [10] are Sobolev space $W^{1,p}_{\mathrm{loc}}(\mathbb{C})$ $(p > 2)$ *solutions of the general, uniformly elliptic, Beltrami equations with measurable coefficients* μ, ν.

In this area there are interesting deep problems to be studied, in particular in the spirit of D. E. Menshov's geometric theory of Cauchy–Riemann equations discussed in the language of advanced subtle methods of Real Analysis in the early decades of the last century.

6. Uniqueness in the general measurable Riemann mapping theorem

Here fundamental is the following proposition.

Proposition 6. *Let* $f = f(z)$ *be the generalized solution of the equation*

$$f_{\bar{z}} - q(z)(f_z + \overline{f}_z) = 0 \quad |q(z)| \le k < \tfrac{1}{2} \tag{6.1}$$

mapping the disc $|z| \le 1$ *onto the disc* $|f| \le 1$. *Then* $f(z) \equiv z$ *if one of the following holds:*

1) $f(z) \equiv z$ *in three different points of boundary* $|z| = 1$,
2) $f(z) \equiv z$ *in one interior point of the disc and one boundary point,*
3) $f(z) \equiv z$ *in two different points of the open disc.*

The equations (6.1) and (4.2) are obviously equivalent. This reflects the fact that the reduction of the uniqueness question for real Beltrami systems (0.2) to the consideration of (6.1) is analogous to the reduction of the search of primary pair for (0.2) to the problem (4.2).

The proof of Proposition 6 is analytically rather complicated; in particular it requires to appeal to the properties of the general homogeneous elliptic systems of the form (0.10) (with $C \equiv 0$).

In infinitesimal geometric interpretation in terms of Lavrentiev fields the equation (6.1) means that the solutions of (6.1) just realize the parallel shift (or translation) ($\nu \equiv \mu$) of the corresponding Lavrentiev ellipses. If, additionally, the global curve: the boundary $|z| = 1$ is preserved and has three fixed points then the mapping does not move any point at all. Proposition 6 should be compared with the famous D. E. Menshov [37] theorem on monogeneity conditions expressed in terms of the geometric requirement to preserve almost all infinitesimal circles.

References

1. L. Ahlfors, *Zur Theorie der Überlagerungsflächen*, Acta Math. 65 (1935).
2. L. Ahlfors, *Conformality with respect to Riemannian metrics*, Ann. Acad. Sci. Fenn. Ser. A. I, 1955, no. 206, 22 pp.
3. L. Ahlfors, *Lectures on Quasiconformal Mappings*, Van Nostrand, Princeton 1966.
4. L. Ahlfors, L. Bers, *Riemann's mapping theorem for variable metrics*, Ann. of Math. (2) 72 (1960), 385–404.
5. C. Andreian Cazacu, *Foundations of quasiconformal mappings*, in: Handbook of Complex Analysis: Geometric Function Theory, vol. 2, Ed. R. Kühnau, Elsevier, 2005.
6. K. Astala, *Analytic aspects of quasiconformality*, in: Proceedings of the International Congress of Mathematicians (Berlin 1998), vol. II.
7. K. Astala, T. Iwaniec, G. Martin, *Elliptic equations and quasiconformal mappings in the plane*, to appear.
8. K. Astala, T. Iwaniec, E. Saksman, *Beltrami operators in the plane*, Duke Math. J. 107 (2001), 27–56.
9. K. Astala, V. Nesi, *Composites and quasiconformal mappings: new optimal bounds in two dimensions*, Calc. Var. Partial Differential Equations 18 (2003), 335–355.
10. P. P. Belinskiĭ, *General Properties of Quasiconformal Mappings*, Nauka, Novosibirsk 1974 (in Russian).
11. P. P. Belinskiĭ, I. N. Pesin, *On the closure of the class of differentiable quasiconformal mappings*, Dokl. Akad. Nauk SSSR 102 (1955), 865–866 (in Russian).
12. L. Bers, L. Nirenberg, *On a representation theorem for linear elliptic systems with discontinuous coefficients and its applications*, in: Atti del Convegno Internaz. sulle Equazioni alle Derivate Parziali, Trieste 1954, Ed. Cremonese, Roma 1955, 111–140.
13. B. Bojarski, *Homeomorphic solutions of the Beltrami systems*, Doklady Akad. Nauk SSSR (N.S.) 102 (1955), 661–664 (in Russian).
14. B. Bojarski, *On solutions of elliptic systems in the plane*, Doklady Akad. Nauk SSSR (N.S.) 102 (1955), 871–874 (in Russian).
15. B. Bojarski, *Generalized solutions of a system of differential equations of first order and of elliptic type with discontinuous coefficients*, Mat. Sb. N.S. 43(85) (1957), 451–503 (in Russian).
16. B. Bojarski, *General properties of solutions of elliptic systems in the plane*, in: Issled. sovr. probl. teor. funktsiĭ kompl. peremen., ed. I. A. Markushevich, Gosfizmatizdat, Moscow 1960, 461–483 (in Russian). Transl. in: Fonctions d'une variable complexe, Probl. contemporaires, Paris 1962, 254–268.
17. B. Bojarski, *Quasiconformal mappings and general structural properties of systems of non-linear equations elliptic in the sense of Lavrentiev*, in: Symposia Mathematica XVIII (Convegno sulle Trasformazioni Quasiconformi e Questioni Connesse, INDAM, Rome 1974), Academic Press, London 1976, 485–499.
18. B. Bojarski, *Old and new on Beltrami equation*, in: Functional Analytic Methods in Complex Analysis and Applications to Partial Differential Equations (Trieste 1988), World Scientific, River Edge 1990.
19. B. Bojarski, *Primary solutions of general Beltrami equations*, to appear.
20. B. Bojarski, L. D'Onofrio, T. Iwaniec, C. Sbordone, *G-closed classes of elliptic operators in the complex plane*, to appear.
21. B. Bojarski, V. Ya. Gutlyanskii, *On the Beltrami equation*, in: Conference Proceedings and Lecture Notes on Analysis I (Tianjin 1992), ed. Zhong Li, International Press, Cambridge, MA 1994, 8–33.
22. B. Bojarski, T. Iwaniec, *Quasiconformal mappings and non-linear elliptic equations* I, Bull. Pol. Acad. Sci. 27 (1974), 473–478.

23. B. Bojarski, T. Iwaniec, *Quasiconformal mappings and non-linear elliptic equations* II, Bull. Pol. Acad. Sci. 27 (1974), 479–484.
24. B. Bojarski, T. Iwaniec, *Analytical foundations of the theory of quasiconformal mappings in* R^n, Ann. Acad. Sci. Fenn. Ser. A. I Math. 8 (1983), 257–324.
25. B. Bojarski, T. Iwaniec, *Primary solutions of planar elliptic systems*, in preparation.
26. F. W. Gehring, *Definitions for a class of plane quasiconformal mappings*, Nagoya Math. J. 29 (1967), 175–184.
27. F. Giannetti, T. Iwaniec, L. Kovalev, G. Moscariello, C. Sbordone, *On g-compactness of the Beltrami operators*, in: Nonlinear Homogenization and its Applications to Composites, Polycristals and Smart Materials, Kluwer, Dordrecht 2004, 107–138.
28. H. Grötzsch, *Über einige Extremalprobleme der konformen Abbildung* I, II, Ber. Verh. Sächs. Akad. Wiss. Math.-Nat. Kl. Leipzig 80 (1928), 367–376, 497–502.
29. H. Grötzsch, *Über die Verzerrung bei schlichten nichtkonformen Abbildungen und über eine damit zusammenhängende Erweiterung des Picardschen Satzes*, Ber. Verh. Sächs. Akad. Wiss. Math.-Nat. Kl. Leipzig 80 (1928), 503–507.
30. V. V. Jikov, S. M. Kozlov, O. A. Oleĭnik, *Homogenization of differential operators and integral functionals*, translated from the Russian by G. A. Yosifian, Springer, Berlin 1994.
31. R. Kühnau, *Herbert Grötzsch zum Gedächtnis*, Jahresber. Deutsch. Math.-Verein. 94 (1992), 141–169.
32. M. A. Lavrentiev, *Sur une classe de représentations continues*, Rec. Math. Moscou 42 (1935).
33. M. A. Lavrentiev, *A general problem of the theory of quasiconformal representation of plane regions*, Mat. Sb. N.S. 21 (1947) (in Russian).
34. M. A. Lavrentiev, *A fundamental theorem of the theory of quasiconformal mapping of plane regions*, Izv. Akad. Nauk SSSR 12 (1948) (in Russian).
35. O. Lehto, K. I. Virtanen, *Quasikonforme Abbildungen*, Grundlehren Math. Wiss. 126, Springer, Berlin 1965; 2nd ed.: *Quasiconformal Mappings in the Plane*, Springer, Berlin 1973.
36. L. Lichtenstein, *Zur Theorie der konformen Abbildung: Konforme Abbildung nichtanalytischer, singularitätenfreier Flächenstücke auf ebene Gebiete*, Bull. Acad. Sci. Cracovie, 1916, 192–217.
37. D. E. Menshov, *Les conditions de monogénéite*, Act. Sci. et Ind. 1936, vol. 329, 1–52.
38. D. E. Menshov, *On the asymptotic monogeneity*, Mat. Sb. 43 (1936), no. 2 (in Russian).
39. C. B. Morrey, Jr., *On the solutions of quasi-linear elliptic partial differential equations*, Trans. Amer. Math. Soc. 43 (1938), 126–166.
40. M. Näätänen, *Maps with continuous characteristic as a subclass of quasiconformal maps*, Ann. Acad. Sci. Fenn. Ser. A. I, 1967, no. 410, 28 pp.
41. H. Renelt, *Quasikonforme Abbildungen und elliptische Systeme*, Teubner Texte z. Mathematik 46, Leipzig 1982.
42. S. Stoïlow, *Leçons sur les principes topologiques de la théorie des fonctions analytiques*, Collection Borel, Gauthier–Villars, Paris 1938; 2nd ed. 1956; Russian ed.: Nauka, Moscow 1964.
43. I. N. Vekua, *The problem of reducing differential forms of elliptic type to canonical form and the generalized Cauchy–Riemann system*, Doklady Akad. Nauk SSSR 100 (1955), 197–200 (in Russian).
44. I. N. Vekua, *Obobshchennye analiticheskie funktsii*, Fizmatgiz, Moscow 1959.
45. L. I. Volkovysky, *Quasiconformal mappings*, Lvov 1954 (in Russian).

A PARTICULAR POLYHARMONIC DIRICHLET PROBLEM

H. BEGEHR

I. Math. Inst., FU Berlin,
Arnimallee 3,D-14195, Germany;
begehr@math.fu-berlin.de

A certain Dirichlet boundary value problem for the polyharmonic operator of arbitrary order n is explicitly solved in case of the unit disc of the complex plane. Bassisc tools are a related Cauchy-Pompeiu representation formula modified with a proper polyharmonic Green function. A similar but asymmetric Dirichlet problem is recently treated with the same method.

Keywords: Polyharmonic operator, Dirichlet problem, unit disc, Cauchy-Pompeiu representation, Green function

1. Biharmonic boundary value problems

For the biharmonic differential equation

$$(\partial_z \partial_{\bar{z}})^2 w = f \tag{1}$$

in some regular plane domain D of the complex plane $\mathbb{C}$ two boundary conditions may be described determining the solution. Preferring the Dirichlet condition two kinds of boundary value problems are possible.

Dirichlet-Dirichlet problem Find a solution to (1) in D satisfying

$$w = \gamma_0, \partial_z \partial_{\bar{z}} w = \gamma_2 \quad \text{on} \quad \partial D \tag{2}$$

Dirichlet-Neumann problem Find a solution to (1) in D satisfying

$$w = \gamma_0, \partial_\nu \partial_z \partial_{\bar{z}} w = \gamma_3 \quad \text{on} \quad \partial D \tag{3}$$

Here $_\nu$ denotes the outward normal derivative on D.

Similarly with respect to the Neumann condition there are two more boundary value problems.

Neumann-Dirichlet problem Find a solution to (1) in D satisfying

$$\partial_\nu w = \gamma_1, \partial_z \partial_{\bar{z}} w = \gamma_2 \quad \text{on} \quad \partial D \tag{4}$$

Neumann-Neumann problem Find a solution to (1) in D satisfying

$$\partial_\nu w = \gamma_1, \partial_\nu \partial_z \partial_{\bar{z}} w = \gamma_3 \quad \text{on} \quad \partial D \tag{5}$$

With proper biharmonic Green and Neumann functions problems (2) and (5) respectively can be solved. Problems (3) and (4) are solvable using a hybrid Green-Neumann function, see.[6,7] Of course wherever a Neumann condition is involved as well some solvability condition as a normalization condition appears.

However there is another biharmonic boundary value problem different from the ones stated before.

Dirichlet problem Find a solution to (1) in D satisfying

$$w = \gamma_0, \partial_\nu w = \gamma_1 \quad \text{on} \quad \partial D \tag{6}$$

While problems (2) to (5) are corresponding to the decomposition of equation (1) into the system of two Poisson equations

$$\partial_z \partial_{\bar{z}} w = \omega, \partial_z \partial_{\bar{z}} \omega = f$$

and their solutions can be attained by iterating the solutions to these equations, problem (6) does not go along with this decomposition. Problem (6) requires another kind of biharmonic Green function.[2] Moreover it requires more smoothness assumptions on the boundary data.[4,5,14–16]

A problem related to (6) is

$$w = \gamma_0, \partial_{\bar{z}} w = \gamma_1 \quad \text{on} \quad \partial D \tag{7}$$

or equivalently

$$w = \gamma_0, \partial_z w = \gamma_1 \quad \text{on} \quad \partial D \tag{7'}$$

see[14–16] . Obviously the variety of boundary value problems for the polyharmonic operator grows with its order. All kinds of combinations of conditions from (2) to (7′) are possible and other conditions as e.g. of Robin type like

$$w + \alpha \partial_\nu w = \gamma \quad \text{on} \quad \partial D$$

can be added. In[14–16] Dirichlet problems of type (7) are treated for the polyharmonic equation in the unit disc and the upper half plane, respectively. Here the Dirichlet problem of type (6) is solved for the ployharmonic equation of arbitrary order.

2. A representation formula

In order to be explicit the domain is chosen to be the unit disc Δ of the complex plane $\mathbb{C}$.

Dirichlet problem Find a solution to

$$(\partial_z \partial_{\bar{z}})^\nu w = f \quad \text{in} \quad \mathbb{D} \tag{8}$$

satisfying on $\partial \mathbb{D}$

$$\begin{aligned} (\partial_z \partial_{\bar{z}})^\mu \, w &= \gamma_\mu, \, 0 \le 2\mu \le n-1, \\ \partial_\nu (\partial_z \partial_{\bar{z}})^\mu \, w &= \hat{\gamma}_\mu, \, 0 \le 2\mu \le n-2. \end{aligned} \tag{9}$$

The solution will be described with the polyharmonic Green-Almansi function[2]

$$(n-1)!^2 G_n(z,\zeta) = |\zeta - z|^{2(n-1)} \log \left| \frac{1 - z\bar{\zeta}}{\zeta - z} \right|^2 + \sum_{\nu=1}^{n-1} \frac{(-1)^\nu}{\nu} |\zeta - z|^{2(n-1-\nu)} (1 - |z|^2)^\nu (1 - |\zeta|^2)^\nu. \tag{10}$$

It has the properties

- $G_n(z,\zeta)$ is polyharmonic for $z \in \mathbb{D} \setminus \{\zeta\}$ for $\zeta \in \mathbb{D}$,
- $G_n(z,\zeta) + \frac{|\zeta - z|^{2(n-1)}}{(n-1)!^2} \log |\zeta - z|^2$ is polyharmonic for $z \in \mathbb{D}, \zeta \in \mathbb{D}$,
- $(\partial_z \partial_{\bar{z}})^\mu G_n(z,\zeta) = 0$ for $0 \le 2\rho \le n-1$ for $z \in \partial\mathbb{D}, \zeta \in \mathbb{D}$,
- $\partial_\nu (\partial_z \partial_{\bar{z}})^\mu G_n(z,\zeta) = 0$ for $0 \le 2\rho \le n-2$ for $z \in \mathbb{D}, \zeta \in \mathbb{D}$,
- $G_n(z,\zeta) = G_n(\zeta, z)$ for $z, \zeta \in \mathbb{D}, z \neq \zeta$.

Basic for treating boundary value problems are integral representation formulas of Cauchy-Pompeiu type.[1–16] For the polyharmonic operator this is, see[10] ,

$$\begin{aligned} w(z) =& \frac{1}{2\pi i} \int_{\partial\mathbb{D}} \frac{w(\zeta)}{\zeta - z} d\zeta \\ &+ \sum_{\nu=1}^{n-1} \frac{1}{2\pi i} \int_{\partial\mathbb{D}} \frac{(\zeta - z)^{\nu-1} (\overline{\zeta - z})^\nu}{(\nu-1)!\nu!} \left[\log |\zeta - z|^2 - \sum_{\rho=1}^{\nu-1} \frac{1}{\rho} - \sum_{\sigma=1}^{\nu} \frac{1}{\sigma} \right] \\ &\times (\partial_\zeta \partial_{\bar{\zeta}})^\nu w(\zeta) d\zeta \\ &+ \sum_{\nu=0}^{n-1} \frac{1}{2\pi i} \int_{\partial\mathbb{D}} \frac{|\zeta - z|^{2\nu}}{\nu!^2} \left[\log |\zeta - z|^2 - 2 \sum_{\rho=1}^{\nu} \frac{1}{\rho} \right] (\partial_\zeta \partial_{\bar{\zeta}})^\nu \partial_{\bar{\zeta}} w(\zeta) d\bar{\zeta} \\ &+ \frac{1}{\pi} \int_D \frac{|\zeta - z|^{2(n-1)}}{(n-1)!^2} \left[\log |\zeta - z|^2 - 2 \sum_{\rho=1}^{n-1} \frac{1}{\rho} \right] (\partial_\zeta \partial_{\bar{\zeta}})^n w(\zeta) d\xi d\eta. \end{aligned} \tag{11}$$

By combining this formula with its complex conjugate counterpart the representation becomes symmetric in z and $\bar{z}$, namely

$$
\begin{aligned}
w(z) &= \frac{1}{4\pi i}\int_{\partial\mathbb{D}}\left(\frac{\zeta}{\zeta-z}+\frac{\bar{\zeta}}{\overline{\zeta-z}}\right)w(\zeta)\frac{d\zeta}{\zeta} \\
&+\sum_{\mu=1}^{n-1}\frac{1}{4\pi i}\int_{\partial\mathbb{D}}\frac{|\zeta-z|^{2\mu}}{(\mu-1)!\mu!}\left(\frac{\zeta}{\zeta-z}+\frac{\bar{\zeta}}{\overline{\zeta-z}}\right) \\
&\times\left[\log|\zeta-z|^2-\sum_{\rho=1}^{\mu-1}\frac{1}{\rho}-\sum_{\sigma=1}^{\mu}\frac{1}{\sigma}\right](\partial_\zeta\partial_{\bar{\zeta}})^\mu w(\zeta)\frac{d\zeta}{\zeta} \\
&-\sum_{\mu=1}^{n-1}\frac{1}{4\pi i}\int_{\partial\mathbb{D}}\frac{|\zeta-z|^{2\mu}}{\mu!^2}\left[\log|\zeta-z|^2-2\sum_{\rho=1}^{\mu}\frac{1}{\rho}\right]\partial_\nu(\partial_\zeta\partial_{\bar{\zeta}})^\mu w(\zeta)\frac{d\zeta}{\zeta} \\
&+\frac{1}{\pi}\int_{\mathbb{D}}\frac{|\zeta-z|^{2(n-1)}}{(n-1)!^2}\left[\log|\zeta-z|^2-2\sum_{\rho=1}^{n-1}\frac{1}{\rho}\right](\partial_\zeta\partial_{\bar{\zeta}})^n w(\zeta)d\xi d\eta
\end{aligned}
\tag{12}
$$

Denoting the sum of the kernel function in the area integral of (12) and $G_n(z,\zeta)$ by A_n then as in[14]

$$
\begin{aligned}
(n-1)!^2 A_n &= |\zeta-z|^{2(n-1)}\left[\log|1-z\bar{\zeta}|^2-2\sum_{\rho=1}^{n-1}\frac{1}{\rho}\right] \\
&+\sum_{\nu=1}^{n-1}\frac{(-1)^\nu}{\nu}|\zeta-z|^{2(n-1-\nu)}(1-|z|^2)^\nu(1-|\zeta|^2)^\nu.
\end{aligned}
\tag{13}
$$

Lemma 1 *For* $2 \le n$

$$
\begin{aligned}
\partial_{\nu_\zeta} A_n &= \frac{|\zeta-z|^{2(n-1)}}{(n-2)!(n-1)!}\left[\frac{\zeta}{\zeta-z}+\frac{\bar{\zeta}}{\overline{\zeta-z}}\right] \\
&\times\left[\log|1-z\bar{\zeta}|^2-\sum_{\rho=1}^{n-1}\frac{1}{\rho}-\sum_{\sigma=1}^{n-2}\frac{1}{\sigma}\right] \\
&+\frac{1}{(n-1)!^2}\sum_{\mu=1}^{n-2}\frac{(-1)^\mu}{\mu}|\zeta-z|^{2(n-1-\mu)} \\
&\times\left[\frac{\zeta}{\zeta-z}+\frac{\bar{\zeta}}{\overline{\zeta-z}}\right](1-|z|^2)^\mu(1-|\zeta|^2)^\mu \\
&+\frac{(-1)^n}{(n-1)!^2}\left[\frac{\bar{z}\zeta}{1-\bar{z}\zeta}+\frac{z\bar{\zeta}}{1-z\bar{\zeta}}\right](1-|z|^2)^{n-1}(1-|\zeta|^2)^{n-1},
\end{aligned}
\tag{14}
$$

$$
\partial_\zeta\partial_{\bar{\zeta}} A_n = A_{n-1} + a_{n-1} \tag{15}
$$

where for $1 \le \nu \le n-1$

$$a_\nu = \frac{(-1)^\nu}{(\nu-1)!\nu!}(1-|z|^2)^\nu(1-|\zeta|^2)^{\nu-1}g_1(z,\zeta)$$
$$= -\frac{1}{(\nu-1)!\nu!}(1-|z|^2)^\nu \partial_\zeta \partial_{\bar\zeta} \tilde G_{\nu+1}(z,\zeta)$$

with, see[2] ,

$$\partial_\zeta \partial_{\bar\zeta} \tilde G_{\nu+1}(z,\zeta) = (|\zeta|^2-1)^{\nu-1} g_1(z,\zeta). \tag{16}$$

Proof Differentiating and summing up geometric series shows

$$(n-1)!^2 \partial_\zeta A_n = (n-1)$$
$$\times \Bigg[(\zeta-z)^{n-2}(\overline{\zeta-z})^{n-1}\Bigg[\log|1-z\tilde\zeta|^2 - \sum_{\rho=1}^{n-1}\frac{1}{\rho} - \sum_{\sigma=1}^{n-2}\frac{1}{\sigma}\Bigg]$$
$$+\sum_{\nu=1}^{n-2}\frac{(-1)^\nu}{\nu}(\zeta-z)^{n-2-\nu}(\overline{\zeta-z})^{n-1-\nu}(1-|z|^2)^\nu(1-|\zeta|^2)^\nu\Bigg]$$
$$-|\zeta-z|^{2(n-2)}\frac{\overline{\zeta-z}}{1-\bar z\zeta}(1-|z|^2)$$
$$+\frac{|\zeta-z|^{2(n-2)}-(-1)^{n-2}(1-|z|^2)^{n-2}(1-|\zeta|^2)^{n-2}}{|1-z\bar\zeta|^2}$$
$$\times(\overline{\zeta-z})(1-|z|^2)(1-|\zeta|^2)$$
$$+\frac{|\zeta-z|^{2(n-1)}-(-1)^{n-1}(1-|z|^2)^{n-1}(1-|\zeta|^2)^{n-1}}{|1-z\bar\zeta|^2}\bar\zeta(1-|z|^2)$$

which leads to (14) and

$$(n-1)!^2\partial_\zeta\partial_{\bar\zeta}A_n = (n-1)^2\Bigg[|\zeta-z|^{2(n-2)}\Bigg[\log|1-z\bar\zeta|^2 - 2\sum_{\rho=1}^{n-2}\frac{1}{\rho}\Bigg]$$
$$+\sum_{\nu=1}^{n-2}\frac{(-1)^\nu}{\nu}|\zeta-z|^{2(n-2-\nu)}((1-|z|^2)^\nu(1-|\zeta|^2)^\nu\Bigg]$$
$$-(n-1)|\zeta-z|^{2(n-2)}\frac{1-|z|^2}{1-z\bar\zeta}$$
$$+(n-1)\frac{|\zeta-z|^{2(n-2)}-(-1)^{n-2}(1-|z|^2)^{n-2}(1-|\zeta|^2)^{n-2}}{|1-z\bar\zeta|^2}$$

$$\times(1-|z|^2)(1-|\zeta|^2)$$

$$+(n-1)\frac{|\zeta-z|^{2(n-2)}-(-1)^{n-2}(1-|z|^2)^{n-2}(1-|\zeta|^2)^{n-2}}{|1-z\bar{\zeta}|^2}$$

$$\times(|\zeta|^2-\bar{\zeta}\zeta)(1-|z|^2)$$

$$+(-1)^{n-1}(n-1)\bar{z}\zeta\frac{(1-|z|^2)^{n-1}(1-|\zeta|^2)^{n-2}}{1-\bar{z}\zeta}$$

which sums up to (15).

Theorem 1 *Any* $w \in C^{2n}(D;C) \cap C^{2n-1}(\bar{D};\mathbb{C})$ *is representable as*

$$w(z) = \frac{1}{2\pi i}\int_{\partial\mathbb{D}} g_1(z,\zeta)w(\zeta)\frac{d\zeta}{\zeta} + \sum_{\mu=1}^{n-1}\frac{1}{\pi}\int_{\mathbb{D}}\frac{(-1)^\mu}{(\mu-1)!\mu!}$$

$$\times(1-|z|^2)^\mu(1-|\zeta|^2)^{\mu-1}g_1(z,\zeta)(\partial_\zeta\partial_{\bar{\zeta}})^\mu w(\zeta)d\xi d\eta \tag{17}$$

$$-\frac{1}{\pi}\int_{\Delta} G_n(z,\zeta)(\partial_\zeta\partial_{\bar{\zeta}})^n w(\zeta)d\xi d\eta.$$

Proof Applying the Gauss theorem, see,[1] in the forms

$$\frac{1}{2\pi i}\int_{\partial\mathbb{D}} w(\zeta)d\zeta = \frac{1}{\pi}\int_{D} w_{\bar{\zeta}}(\zeta)d\xi d\eta \tag{18}$$

and

$$\frac{1}{2\pi i}\int_{\partial\mathbb{D}} w(\zeta)d\bar{\zeta} = -\frac{1}{\pi}\int_{D} w_{\zeta}(\zeta)d\xi d\eta \tag{18'}$$

shows

$$\frac{1}{\pi}\int_{\mathbb{D}} A_n(\partial_\zeta\partial_{\bar{\zeta}})^n w\, d\xi d\eta$$

$$= \frac{1}{2\pi}\int_{\mathbb{D}}\Big\{\partial_\zeta\left[A_n\partial_\zeta^{n-1}\partial_{\bar{\zeta}}^n w\right]$$

$$+\partial_{\bar{\zeta}}[A_n\partial_\zeta^n\partial_{\bar{\zeta}}^{n-1}w] - \partial_\zeta A_n\partial_\zeta^{n-1}\partial_{\bar{\zeta}}^n w - \partial_{\bar{\zeta}}A_n\partial_\zeta^n\partial_{\bar{\zeta}}^{n-1}w\Big\}\,d\xi d\eta$$

$$= \frac{1}{4\pi i}\int_{\partial\mathbb{D}} A_n\partial_\nu(\partial_\zeta\partial_{\bar{\zeta}})^{n-1}w\frac{d\zeta}{\zeta}$$

$$-\frac{1}{2\pi}\int_{\mathbb{D}}\Big\{\partial_{\bar{\zeta}}\left[\partial_\zeta A_n(\partial_\zeta\partial_{\bar{\zeta}})^{n-1}w\right] + \partial_\zeta[\partial_{\bar{\zeta}}A_n(\partial_\zeta\partial_{\bar{\zeta}})^{n-1}w]$$

$$-\partial_\zeta\partial_{\bar{\zeta}}A_n(\partial_\zeta\partial_{\bar{\zeta}})^{n-1}w - 2\partial_\zeta\partial_{\bar{\zeta}}A_n(\partial_\zeta\partial_{\bar{\zeta}})^{n-1}w\Big\}d\xi d\eta$$

$$= \frac{1}{4\pi i} \int\limits_{\partial\mathbb{D}} \left\{ A_n \partial_\nu (\partial_\zeta \partial_{\bar\zeta})^{n-1} w - \partial_\nu A_n (\partial_\zeta \partial_{\bar\zeta})^{n-1} w \right\} \frac{d\zeta}{\zeta}$$

$$+ \frac{1}{\pi} \int\limits_{\mathbb{D}} (A_{n-1} + a_{n-1})(\partial_\zeta \partial_{\bar\zeta})^{n-1} w d\xi d\eta.$$

In the same manner induction gives for $0 \le k \le n-1$

$$\frac{1}{\pi} \int\limits_{\mathbb{D}} A_n (\partial_\zeta \partial_{\bar\zeta})^n w d\xi d\eta = \frac{1}{\pi} \int\limits_{\mathbb{D}} A_{n-k} (\partial_\zeta \partial_{\bar\zeta})^{n-k} w d\xi d\eta$$

$$+ \sum_{\mu=1}^{k} \frac{1}{\pi} \int\limits_{\mathbb{D}} a_{n-\mu} (\partial_\zeta \partial_{\bar\zeta})^{n-\mu} w d\xi d\eta$$

$$+ \sum_{\mu=0}^{k-1} \frac{1}{4\pi i} \int\limits_{\partial\mathbb{D}} \left\{ A_{n-\mu} \partial_\nu (\partial_\zeta \partial_{\bar\zeta})^{n-\mu-1} w - \partial_\nu A_{n-\mu} \partial_\zeta \partial_{\bar\zeta}^{n-\mu-1} w \right\} \frac{d\zeta}{\zeta}$$

$$= \frac{1}{\pi} \int\limits_{\mathbb{D}} A_1 (\partial_\zeta \partial_{\bar\zeta}) w d\xi d\eta + \sum_{\mu=1}^{n-1} \frac{1}{\pi} \int\limits_{\mathbb{D}} a_{n-\mu} (\partial_\zeta \partial_{\bar\zeta})^{n-\mu} w d\xi d\eta$$

$$+ \sum_{\nu=0}^{n-2} \frac{1}{4\pi i} \int\limits_{\partial\mathbb{D}} \left\{ A_{n-\mu} \partial_\nu (\partial_\zeta \partial_{\bar\zeta})^{n-\mu-1} w - \partial_\nu A_{n-\mu} (\partial_\zeta \partial_{\bar\zeta})^{n-\mu-1} w \right\} \frac{d\zeta}{\zeta}.$$

As

$$\frac{1}{\pi} \int\limits_{\mathbb{D}} \log |1 - z\bar\zeta|^2 \partial_\zeta \partial_{\bar\zeta} w d\xi d\eta$$

$$= \frac{1}{4\pi i} \int\limits_{\partial\mathbb{D}} \left\{ \log |\zeta - z|^2 \partial_\nu w + \left[\frac{\bar z \zeta}{1 - \bar z \zeta} + \frac{z\bar\zeta}{1 - z\bar\zeta} \right] w \right\} \frac{d\zeta}{\zeta}$$

the last relation becomes

$$\frac{1}{\pi} \int\limits_{D} A_n (\partial_\zeta \partial_{\bar\zeta})^n w d\xi d\eta = \sum_{\mu=1}^{n-1} \frac{1}{\pi} \int\limits_{\mathbb{D}} a_{n-\mu} (\partial_\zeta \partial_{\bar\zeta})^{n-\mu} w d\xi d\eta$$

$$+ \sum_{\nu=0}^{n-1} \frac{1}{4\pi i} \int\limits_{\partial\mathbb{D}} \left\{ A_{n-\mu} \partial_\nu (\partial_\zeta \partial_{\bar\zeta})^{n-\mu-1} w +_\nu A_{n-\mu} (\partial_\zeta \partial_{\bar\zeta})^{n-\mu-1} w \right\} \frac{d\zeta}{\zeta}$$

$$= \sum_{\mu=1}^{n-1} \frac{1}{\pi} \int\limits_{\mathbb{D}} a_\mu (\partial_\zeta \partial_{\bar\zeta})^\mu w d\xi d\eta \tag{19}$$

$$+ \sum_{\mu=0}^{n-1} \frac{1}{4\pi i}{}_{\Delta} \left\{ A_{\mu+1} \partial_\nu (\partial_\zeta \partial_{\bar\zeta})^\mu w - \partial_\nu A_{\mu+1} (\partial_\zeta \partial_{\bar\zeta})^\mu w d\zeta \right\}.$$

Rewriting (12) as

$$w(z) = \frac{1}{2\pi i}\int\limits_{\partial\Delta} g_1 w \frac{d\zeta}{\zeta} + \sum_{\mu=0}^{n-1}\frac{1}{4\pi i}\int\limits_{\partial\mathbb{D}}\Big\{\partial_\nu A_{\mu+1}(\partial_\zeta\partial_{\bar\zeta})^\mu w - A_{\mu+1}\partial_\nu(\partial_\zeta\partial_{\bar\zeta})^\mu w\Big\}\frac{d\zeta}{\zeta}$$
$$+\frac{1}{\pi}_{\mathbb{D}}(A_n - G_n)(\partial_\zeta\partial_{\bar\zeta})^n w d\xi d\eta$$

and substituting (19) proves (17).

Theorem 2 *Any $w \in C^{2n}(\mathbb{D};\mathbb{C}) \cap C^{2n-1}(\overline{\mathbb{D}};\mathbb{C})$ is representable as*

$$w(z) = \frac{1}{4\pi i}\int\limits_{\partial}\Big\{\Big[2g_1(z,\zeta) + \sum_{\mu=1}^{n-1}[g_{\mu+1}(z,\zeta) - g_\mu(z,\zeta)](1-|z|^2)^\mu\Big]w(\zeta)$$
$$-\sum_{\mu=1}^{n-1}\frac{1}{\mu}g_\mu(z,\zeta)(1-|z|^2)^\mu_\nu w(\zeta)$$
$$+\sum_{\kappa=1}^{[\frac{n-1}{2}]}\sum_{\mu=2\kappa}^{n-1}\sum_{\lambda=0}^{\mu-2\kappa}\frac{1}{\kappa!(\kappa+1)!}\frac{\binom{\mu-\kappa-1-\lambda}{\kappa-1}}{\binom{\mu}{\kappa+1}}$$
$$\times\Big[2\kappa g_{\lambda+1}(z,\zeta) + \lambda(g_{\lambda+1}(z,\zeta) - g_\lambda(z,\zeta))\Big](1-|z|^2)^\mu(\partial_\zeta\partial_{\bar\zeta})^\kappa w(\zeta)$$
$$-\sum_{\kappa=1}^{[\frac{n}{2}]-1}\sum_{\mu=2\kappa+1}^{n-1}\sum_{\lambda=1}^{\mu-2\kappa}\frac{1}{\kappa!(\kappa+1)!}\frac{\binom{\mu-\kappa-1-\lambda}{\kappa-1}}{\binom{\mu}{\kappa+1}} \tag{20}$$
$$\times\, g_\lambda(z,\zeta)(1-|z|^2)^\mu\partial_\nu(\partial_\zeta\partial_{\bar\zeta})^\kappa w(\zeta)\Big\}\frac{d\zeta}{\zeta} - \frac{1}{\pi}\int\limits_{\mathbb{D}} G_n(z,\zeta)(\partial_\zeta\partial_{\bar\zeta})^n w(\zeta)d\xi d\eta.$$

Proof The integral under the sum in (1) is by repeated application of the Gauß theorems (18) and (18′) and using (16)

$$\frac{1}{\pi}\int\limits_{\mathbb{D}}(-1)^\mu(1-|\zeta|^2)^{\mu-1}g_1(z,\zeta)(\partial_\zeta\partial_{\bar\zeta})^\mu w(\zeta)d\xi d\eta$$
$$= -\frac{1}{\pi}\int\limits_{\mathbb{D}}\partial_\zeta\partial_{\bar\zeta}\tilde G_{\mu+1}(z,\zeta)(\partial_\zeta\partial_{\bar\zeta})^\mu w(\zeta)d\xi d\eta$$
$$= \sum_{\kappa=1}^{\mu}\frac{1}{2\pi i}\int\limits_{\partial\mathbb{D}}\Big\{\partial_\nu(\partial_\zeta\partial_{\bar\zeta})^\kappa\tilde G_{\mu+1}(z,\zeta)(\partial_\zeta\partial_{\bar\zeta})^{\mu-\kappa}w(\zeta)$$
$$-(\partial_\zeta\partial_{\bar\zeta})^\kappa\tilde G_{\mu+1}(z,\zeta)\partial_\nu(\partial_\zeta\partial_{\bar\zeta})^{\mu-\kappa}w(\zeta)\Big\}\frac{d\zeta}{\zeta}$$
$$-\frac{1}{\pi}\int\limits_{D}(\partial_\zeta\partial_{\bar\zeta})^{\mu+1}\tilde G_{\mu+1}(z,\zeta)w(\zeta)d\xi d\eta.$$

On the basis of[2], Lemma 2, see[3], for $1 \le \kappa \le \mu - 1$

$$\frac{1}{(\mu-1)!\mu!}(\partial_\zeta \partial_{\bar\zeta})^\kappa \tilde{G}_{\mu+1}(z,\zeta) = \frac{(\kappa-1)!}{\mu!(\mu-\kappa)!} \sum_{\tau=0}^{\min\{\kappa-1,\mu-\kappa\}} \binom{\mu-\kappa}{\tau}$$

$$\times (|\zeta|^2 - 1)^{\mu-\kappa-\tau} \sum_{\lambda=0}^{\kappa-1-\tau} \binom{\mu-\kappa-1+\lambda}{\lambda} g_{\kappa-\tau-\lambda}(z,\zeta),$$

$$\frac{1}{(\mu-1)!\mu!}\partial_\nu (\partial_\zeta \partial_{\bar\zeta})^\kappa \tilde{G}_{\mu+1}(z,\zeta)$$

$$= \frac{2(\kappa-1)!}{\mu!(\mu-\kappa)!} \sum_{\tau=0}^{\min\{\kappa-1,\mu-\kappa-1\}} \frac{(\mu-\kappa)!}{\tau!(\mu-\kappa-\tau-1)!}$$

$$\times (|\zeta|^2 - 1)^{\mu-\kappa-\tau-1} |\zeta|^2 \sum_{\lambda=0}^{\kappa-1-\tau} \binom{\mu-\kappa-1+\lambda}{\lambda} g_{\kappa-\tau-\lambda}(z,\zeta)$$

$$+ \frac{(\kappa-1)!}{\mu!} \sum_{\tau=0}^{\min\{\kappa-1,\mu-\kappa\}} \frac{1}{\tau!(\mu-\kappa-\tau)!} (|\zeta|^2 - 1)^{\mu-\kappa-\tau}$$

$$\times \sum_{\lambda=0}^{\kappa-1-\tau} \frac{(\mu-\kappa-1+\lambda)!}{\lambda!(\mu-\kappa-1)!} (\kappa-\tau-\lambda)[g_{\kappa-\tau-\lambda+1}(z,\zeta) - g_{\kappa-\tau-\lambda}(z,\zeta)]$$

and

$$\frac{1}{(\mu-1)!\mu!}(\partial_\zeta \partial_{\bar\zeta})^\mu \tilde{G}_{\mu+1}(z,\zeta) = \frac{1}{\mu} g_\mu(z,\zeta),$$

$$\frac{1}{(\mu-1)!\mu!}\partial_\nu (\partial_\zeta \partial_\zeta)^\mu \tilde{G}_{\mu+1}(z,\zeta) = g_{\mu+1}(z,\zeta) - g_\mu(z,\zeta),$$

$$\frac{1}{(\mu-1)!\mu!}(\partial_\zeta \partial_{\bar\zeta})^{\mu+1} \tilde{G}_{\mu+1}(z,\zeta) = 0,$$

so that on $|\zeta| = 1$ for $1 \le \kappa \le \mu - 1$

$$\tfrac{1}{(\mu-1)!\mu!}(\partial_\zeta \partial_{\bar\zeta})^\kappa \tilde{G}_{\mu+1}(z,\zeta)$$

$$= \begin{cases} 0\,, \ 2\kappa < \mu+1\,, \\ \frac{(\kappa-1)!}{\mu!(\mu-\kappa)!} \sum\limits_{\lambda=0}^{2\kappa-1-\mu} \binom{\mu-1-\kappa+\lambda}{\lambda} \ g_{2\kappa-\mu-\lambda}(z,\zeta), \mu+1 \le 2\kappa, \end{cases}$$

$$\frac{1}{(\mu-1)!\mu!}\,\partial_\zeta(\partial_{\zeta\bar\zeta})^\kappa \tilde G_{\mu+1}(z,\zeta)$$

$$= \begin{cases} 0\ ,\ 2\kappa<\mu, \\ \frac{2}{(2\mu)!}g_1(z,\zeta),\ 2\kappa=\mu, \\ \frac{2(\kappa-1)!}{\mu!(\mu-\kappa-1)!}\sum\limits_{\lambda=0}^{2\kappa-\mu}\binom{\mu-\kappa-1+\lambda}{\lambda}g_{2\kappa-\mu+1-\lambda}(z,\zeta) \\ +\frac{(\kappa-1)!}{\mu!(\mu-\kappa)!}\sum\limits_{\lambda=0}^{2\kappa-\mu-1}\binom{\mu-\kappa-1+\lambda}{\lambda}(2\kappa-\mu-\lambda) \\ \times[g_{2\kappa-\mu-\lambda+1}(z,\zeta)-g_{2\kappa-\mu-\lambda}(z,\zeta)],\ \mu<2\kappa. \end{cases}$$

Thus

$$\sum_{\mu=1}^{n-1}\frac{1}{\pi}\int_{\mathbb{D}}\frac{(-1)^\mu}{(\mu-1)!\mu!}(1-|z|^2)^\mu(1-|\zeta|^2)^{\mu-1}g_1(z,\zeta)(\partial_\zeta\partial_{\bar\zeta})^\mu w(\zeta)d\xi d\eta$$

$$=\sum_{\mu=1}^{n-1}\frac{(1-|z|^2)^\mu}{4\pi i}\int_{\partial\mathbb{D}}\Big\{[g_{\mu+1}(z,\zeta)-g_\mu(z,\zeta)]w(\zeta)-\frac{1}{\mu}g_\mu(z,\zeta)_\nu w(\zeta)\Big\}$$

$$+\sum_{\mu=1}^{n-1}\sum_{\mu=1}^{\mu-1}\frac{(1-|z|^2)^\mu}{4\pi i}$$

$$\times\int_{\partial\mathbb{D}}\Big\{\frac{1}{(\mu-1)!\mu!}\partial_\nu(\partial_\zeta\partial_{\bar\zeta})^{\mu-\kappa}\tilde G_{\mu+1}(z,\zeta)(\partial_\zeta\partial)^\kappa w(\zeta)$$

$$-\frac{1}{(\mu-1)!\mu!}{}_\zeta(\partial_\zeta\partial_{\bar\zeta})^{\mu-\kappa}\tilde G_{\mu+1}(z,\zeta)\partial_\nu(\partial_\zeta\partial_{\bar\zeta})^\kappa w(\zeta)\Big\}\frac{d\zeta}{\zeta}.$$

Here the last sum becomes

$$-\sum_{\kappa=1}^{[\frac{n}{2}]-1}\sum_{\mu=2\kappa+1}^{n-1}\frac{(1-|z|^2)^\mu}{4\pi i}\int_{\partial\mathbb{D}}\frac{(\mu-\kappa-1)!}{\mu!\kappa!}$$

$$\times\sum_{\lambda=0}^{\mu-2\kappa-1}\binom{\mu+\lambda-1}{\kappa-1}g_{\mu-2\kappa-\lambda}(z,\zeta)\partial_\nu(\partial_{\zeta\bar\zeta})^\kappa w(\zeta)\frac{d\zeta}{\zeta}$$

$$+\sum_{\kappa=1}^{[\frac{n-1}{2}]}\sum_{\mu=2\kappa}^{n-1}\frac{(1-|z|^2)^\mu}{4\pi i}$$

$$\times \int_{\partial\mathbb{D}} \Big\{ 2\frac{(\mu-\kappa-1)!}{\mu!(\kappa-1)!} \sum_{\lambda=0}^{\mu-2\kappa} \binom{\kappa+\lambda-1}{\kappa-1} g_{\mu-2\kappa+1-\lambda}(z,\zeta)$$

$$+\frac{(\mu-\kappa-1)!}{\mu!\kappa!} \sum_{\lambda=0}^{\mu-2\kappa-1} \binom{\kappa+\lambda-1}{\kappa-1} (\mu-2\kappa-\lambda)$$

$$\times [g_{\mu-2\kappa-\lambda+1}(z,\zeta) - g_{\mu-2\kappa-\lambda}(z,\zeta)] \Big\} (\partial_\zeta \partial_{\bar\zeta})^\kappa w(\zeta) \frac{d\zeta}{\zeta}.$$

Thus (20) is attained.

3. A polyharmonic Dirichlet problem

As is known e.g. from the Cauchy formula not every integral representation formula provides a solution to a related boundary value problem. The problem may be overdetermined so that it is sovable only under some solvability conditions, see e.g.[4,5] The representation formula (20) gives rise to the solution of a Dirichlet problem for the polyharmonic equation.

Theorem 3 *The unique solution to the Dirichlet problem*

$$(\partial_z \partial_{\bar z})^n w(z) = f(z) \text{ in } \mathbb{D},$$
$$(\partial_z \partial_{\bar z})^\mu w(z) = \gamma_\mu(z) \text{ on } \partial\mathbb{D} \text{ for } 0 \le 2\mu \le n-1,$$
$$\partial_z^\mu \partial_{\bar z}^{\mu+1} w(z) = \hat\gamma_\mu(z) \text{ on } \partial\mathbb{D} \text{ for } 0 \le 2\mu \le n-2,$$

for $f \in L_1(\mathbb{D};\mathbb{C}), \gamma_\mu \in C^{n-2\mu}(\mathbb{D};\mathbb{C}), \hat\gamma_\mu \in C^{n-1-2\mu}(\mathbb{D};\mathbb{C})$ *in the distributional sense is given as*

$$w(z) = \frac{1}{2\pi i} \int_{\partial\mathbb{D}} \Big\{ \Big[g_1(z,\zeta) + \frac{1}{2} \sum_{\mu=1}^{n-1} [g_{\mu+1}(z,\zeta) - g_\mu(z,\zeta)](1-|z|^2)^\mu \Big] \gamma_0(\zeta)$$

$$-\sum_{\mu=1}^{n-1} \frac{1}{2\mu} g_\mu(z,\zeta)(1-|z|^2)^\mu \hat\gamma_0(\zeta)$$

$$+\sum_{\kappa=1}^{[\frac{n-1}{2}]} \sum_{\mu=2\kappa}^{n-1} \sum_{\lambda=0}^{\mu-2\kappa} \frac{1}{\kappa!(\kappa+1)!} \frac{\binom{\mu-\kappa-1-\lambda}{\kappa-1}}{\binom{\mu}{\kappa+1}}$$

$$\times \Big[\kappa g_{\lambda+1}(z,\zeta) + \frac{\lambda}{2}(g_{\lambda+1}(z,\zeta) - g_\lambda(z,\zeta)) \Big] (1-|z|^2)^\mu \gamma_\kappa(\zeta)$$

$$-\sum_{\kappa=1}^{[\frac{n}{2}]-1} \sum_{\mu=2\kappa+1}^{n-1} \sum_{\lambda=1}^{\mu-2\kappa} \frac{1}{2\kappa!(\kappa+1)!} \frac{\binom{\mu-\kappa-1-\lambda}{\kappa-1}}{\binom{\mu}{\kappa+1}} \tag{21}$$

$$\times\, g_\lambda(z,\zeta)(1-|z|^2)^\mu \hat\gamma_\kappa(\zeta) \Big\} \frac{d\zeta}{\zeta} - \frac{1}{\pi} \int_{\mathbb{D}} G_n(z,\zeta) f(\zeta) d\xi d\eta.$$

Proof The uniqueness of the solution follows from the representation formula (20). If there is a function satisfying the inhomogeneous polyharmonic equation

and the above Dirichlet conditions then it is representable via (21). Thus the related homogeneous problem is only trivially solvable. The existence of a solution can be proved by verifying (21) to be a solution.
For this purpose the properties

$$(\partial_z\partial_{\bar z})^\mu G_n(z,\zeta) = 0 \text{ on } \Delta \text{ for } 0 \le 2\mu \le n-1,$$
$$\partial_z^\mu\partial_{\bar z}^{\mu+1} G_n(z,\zeta) = 0 \text{ on } \Delta \text{ for } 0 \le 2\mu \le n-2,$$

of the polyharmonic Green function $G_n(z,\zeta)$ are important, see,[2] which together with

$$(\partial_z\partial_{\bar z})^n\Big\{-\frac{1}{\pi}\int\limits_{\mathbb{D}} G_n(z,\zeta)f(\zeta)d\xi d\eta\Big\} = f(z)$$

in the weak (distributional) sense shows that this area integral in (21) provides a solution to the inhomogeneous polyharmonic equation satisfying the above homogeneous Dirichlet data. As the $g_\nu(z,\zeta)$ are harmonic in z (and in ζ) the boundary integral in (21) obviously is a polyharmonic function of order n. Hence it only remains to prove the proper boundary behaviour of this boundary integral. For this aim the respective derivatives have to be calculated. The following rule will be used.
Lemma A *For $f,g \in C^{2\rho}(D;\mathbb{C}), 1 \le \rho, D \subset \mathbb{C}$ open, and g harmonic, i.e. $\partial_z\partial_{\bar z}g = 0$, in D*

$$(\partial_z\partial_{\bar z})^\rho(fg) = g(\partial_z\partial_{\bar z})^\rho f + \sum_{\tau=0}^{\rho-1}\binom{\rho}{\tau}\Big\{\partial_z^\tau\partial_{\bar z}^\rho f\partial_z^{\rho-\tau}g + \partial_z^\rho\partial_{\bar z}^\tau f\partial_{\bar z}^{\rho-\tau}g\Big\}.$$

Verifying the boundary conditions will be done on the basis of the next two lemmas.
Lemma B *Let $\gamma \in C(\partial\mathbb{D};\mathbb{C})$ and $1 \le k \le n,\ k,n \in \mathbb{N}$. Then for any $\zeta_0 \in \partial\mathbb{D}$*

$$\lim_{z\to\zeta_0} {}_\Delta\gamma(\zeta)\frac{(1-|z|^2)^n}{(1-\bar z\zeta)^k}d\zeta = 0,\ \lim_{z\to\zeta_0}\int\limits_{\partial\mathbb{D}}\gamma(\zeta)\frac{(1-|z|^2)^n}{(1-z\bar\zeta)^k}d\zeta = 0.$$

Proof Observing the property of the Poisson kernel function

$$\lim_{z\to\zeta_0}\frac{1}{2\pi i}\int\limits_{\partial\mathbb{D}}\gamma(\zeta)g_1(z,\zeta)\frac{d\zeta}{\zeta} = \gamma(\zeta_0),\ g_1(z,\zeta) = \frac{1-|z|^2}{|1-z\bar\zeta|^2}$$

the first part of the lemma follows from

$$\int\limits_{\partial\mathbb{D}}\gamma(\zeta)\frac{(1-|z|^2)^n}{(1-\bar z\zeta)^k}d\zeta$$

$$= \int\limits_{\partial\mathbb{D}}\gamma(\zeta)(\zeta-z)(1-|z|^2)^{n-k}\left(\frac{1-|z|^2}{1-\bar z\zeta}\right)^{k-1} g_1(z,\zeta)\frac{d\zeta}{\zeta}.$$

The second follows similarly or by complex conjugation.

Lemma C *Let* $\gamma \in C^m(\partial\mathbb{D};\mathbb{C}), m,n \in \mathbb{N}$. *Then for any* $\zeta_0 \in \partial\mathbb{D}$

$$\lim_{z\to\zeta_0} \int_{\partial\mathbb{D}} \gamma(\zeta)\frac{(1-|z|^2)^n}{(1-\bar{z}\zeta)^{m+n}}d\zeta = 0, \ \lim_{z\to\zeta_0} \int_{\partial\mathbb{D}} \gamma(\zeta)\frac{(1-|z|^2)^n}{(1-z\bar{\zeta})^{m+n}}d\zeta = 0.$$

Proof Arguing as before after m times integrating by parts shows

$$\int_{\partial\mathbb{D}} \gamma(\zeta)\frac{(1-|z|^2)^n}{(1-\bar{z}\zeta)^{m+n}}d\zeta = \frac{(n-1)!}{(m+n-1)!\bar{z}^m}{}_{\partial\mathbb{D}}\, \gamma(\zeta)\partial_\zeta^m\left(\frac{1-|z|^2}{1-\bar{z}\zeta}\right)^n d\zeta$$

$$= \frac{(-1)^m(n-1)!}{(m+n-1)!\bar{z}^m}\int_{\partial\mathbb{D}} \partial_\zeta^m\gamma(\zeta)\left(\frac{1-|z|^2}{1-\bar{z}\zeta}\right)^n d\zeta$$

and

$$\overline{\int_{\partial\mathbb{D}} \gamma(\zeta)\frac{(1-|z|^2)^n}{(1-z\bar{\zeta})^{m+n}}d\zeta} = -\int_{\partial\mathbb{D}} \frac{\overline{\gamma(\zeta)}}{\zeta^2}\,\frac{(1-|z|^2)^n}{(1-\bar{z}\zeta)^{m+n}}d\zeta$$

tend to zero with $z \in \mathbb{D}$ tending to $\zeta_0 \in \partial\mathbb{D}$.

(i) Differentiating

$$A = g_1(z,\zeta) + \sum_{\nu=1}^{n-1}\frac{\bar{z}\zeta}{(1-\bar{z}\zeta)^{\nu+1}}(1-|z|^2)^\nu \tag{22}$$

once with respect to gives

$$A_{\bar{z}} = n\zeta\frac{(1-|z|^2)^{n-1}}{(1-\bar{z}\zeta)^{n+1}},$$

$$\partial_{\nu_z}A = n[g_{n+1}(z,\zeta) - g_n(z,\zeta)](1-|z|^2)^{n-1} \tag{22'}$$

Similarly,

$$A_{z\bar{z}} = -(n-1)n\bar{z}\zeta\frac{(1-|z|^2)^{n-2}}{(1-\bar{z}\zeta)^{n-1}}, \tag{22''}$$

$$({}_z)^m A = \sum_{\tau=0}^{m-1}\sum_{\sigma=0}^{\tau}\sum_{\rho=0}^{\tau-\sigma}\sum_{k=0}^{m-\tau}(-1)^{k+\rho+\tau}(2m)!$$

$$\times\binom{m-1}{\tau}\binom{\tau}{\sigma}\binom{m}{\sigma}\binom{\tau-\sigma}{\rho}\binom{m-\tau}{\kappa}\frac{\binom{n+m-\tau-1}{2m-\sigma}}{\binom{2\nu}{\sigma}}\;\frac{(1-|z|^2)^{n-m-\rho-1}}{(1-\bar{z}\zeta)^{n+\kappa}} \tag{22'''}$$

for $2 \leq 2m \leq n-1$ and

$$\partial_z^m\partial_{\bar{z}}^{m+1}A = \sum_{\tau=0}^{m}\sum_{\sigma=0}^{\tau}\sum_{\rho=0}^{\tau-\sigma}\sum_{\kappa=0}^{m-\tau}(-1)^{\kappa+\rho+\tau}(2m+1)!$$

$$\times\binom{m}{\tau}\binom{\tau}{\sigma}\binom{m}{\sigma}\binom{\tau-\sigma}{\rho}\binom{m-\tau}{\kappa}\frac{\binom{n+m-\tau}{2m+1-\sigma}}{\binom{2m+1}{\sigma}}\ \frac{\zeta(1-|z|^2)^{n-m-\rho-1}}{(1-\bar{z}\zeta)^{n+\kappa+1}}$$

$$\nu_z(\partial_z\partial_{\bar{z}})^m A = \sum_{\tau=0}^{m}\sum_{\sigma=0}^{\tau}\sum_{\rho=0}^{\tau-\sigma}\sum_{\kappa=0}^{m-\tau}(-1)^{\kappa+\rho+\tau}(2m+1)!$$

$$\times\binom{m}{\tau}\binom{\tau}{\sigma}\binom{m}{\tau}\binom{\tau-\sigma}{\rho}\binom{m-\tau}{\kappa}\frac{\binom{n+m-\tau}{2m+1-\sigma}}{\binom{2m+1}{\sigma}} \tag{22^{IV}}$$

$$\times[g_{n+\kappa+1}(z,\zeta)-g_{n+\kappa}(z,\zeta)](1-|z|^2)^{n-m-\rho-1}$$

for $0 \leq 2m \leq n-2$.

(ii) Similarly for

$$B = \sum_{\nu=1}^{n-1}\frac{1}{\nu}g_\nu(z,\zeta)(1-|z|^2)^\nu \tag{23}$$

it is seen

$$B_{\bar{z}} = -zg_1(z,\zeta) + \zeta\frac{(1-|z|^2)^{n-1}}{(1-\bar{z}\zeta)^n}$$

$$+\sum_{\nu=1}^{n-2}\left[\frac{\zeta(1-|z|^2)^\nu}{(1-\bar{z}\zeta)^{\nu+1}} - zg_{\nu+1}(z,\zeta)(1-|z|^2)^\nu\right],$$

$$\partial_{\nu_z}B = -2g_1(z,\zeta)+\sum_{\mu=1}^{n-1}(1-|z|^2)^\mu[g_\mu(z,\zeta)-g_{\mu+1}(z,\zeta)]+2(1-|z|^2)^{n-1}g_n(z,\zeta) \tag{23$'$}$$

and

$$B_{z\bar{z}} = -(n-1)g_n(z,\zeta)(1-|z|^2)^{n-2}. \tag{23$''$}$$

Moreover,

$$(\partial_z\partial_{\bar{z}})^m B = \sum_{\tau=0}^{m-1}\sum_{\sigma=0}^{\tau}\sum_{\rho=0}^{\tau-\sigma}\sum_{\kappa=0}^{m-\tau-1}(-1)^{\kappa+\rho+\tau-1}(2m-1)!\frac{\binom{n+m-\tau-2}{2m-1-\sigma}}{\binom{2m-1}{\sigma}}$$

$$\times\binom{m-1}{\tau}\binom{\tau}{\sigma}\binom{m-1}{\sigma}\binom{\tau-\sigma}{\rho}\binom{m-\tau-1}{\kappa}g_{n+\kappa}(z,\zeta)(1-|z|^2)^{n-m-\rho-1} \tag{23$'''$}$$

for $2 \leq 2m \leq n-1$ and

$$\partial_z^m\partial_{\bar{z}}^{m+1}B = \sum_{\tau=0}^{m-1}\sum_{\sigma=0}^{\tau}\sum_{\rho=0}^{\tau-\sigma}\sum_{\kappa=0}^{m-\tau-1}(-1)^{\kappa+\rho+\tau-1}(2m-1)!\frac{\binom{n+m-\tau-2}{2m-1-\sigma}}{\binom{2m-1}{\sigma}}$$

$$\times\binom{m-1}{\tau}\binom{\tau}{\sigma}\binom{m-1}{\sigma}\binom{\tau-\sigma}{\rho}\binom{m-\tau-1}{\kappa}$$

$$\times\left[(n+\kappa)\zeta\frac{(1-|z|^2)^{n-m-\rho-1}}{(1-\bar{z}\zeta)^{n+\kappa+1}}-(n-m-\rho-1)zg_{n+\kappa}(z,\zeta)(1-|z|^2)^{n-m-\rho-2}\right],$$

$$\partial_{\nu_z}(\partial_z\partial_{\bar{z}})^m B=\sum_{\tau=0}^{m-1}\sum_{\sigma=0}^{\tau}\sum_{\rho=0}^{\tau-\sigma}\sum_{\kappa=0}^{m-\tau-1}(-1)^{\kappa+\rho+\tau-1}(2m-1)!$$

$$\times\frac{\binom{n+m-\tau-2}{2m-1-\sigma}}{\binom{2m-1}{\sigma}}\binom{m-1}{\tau}\binom{\tau}{\sigma}\binom{m-1}{\sigma}\binom{\tau-\sigma}{\rho}\binom{m-\tau-1}{\kappa}$$

$$\times\Bigg[(n+\kappa)[g_{n+\kappa+1}(z,\zeta)-g_{n+\kappa}(z,\zeta)](1-|z|^2)^{n-m-\rho-1}$$

$$-2(n-m-\rho-1)|z|^2 g_{n+\kappa}(z,\zeta)(1-|z|^2)^{n-m-\rho-2}\Bigg] \tag{23^{IV}}$$

for $2\le 2m\le n-2$.

(iii) Let

$$C=\sum_{\kappa=1}^{[\frac{n-1}{2}]}C_{\kappa\kappa}\gamma_\kappa \tag{24}$$

with $C_{\kappa\kappa}=C_{\kappa\kappa 1}+C_{\kappa\kappa 2}$ where

$$\begin{aligned}C_{\kappa\kappa 1}&=\sum_{\mu=2\kappa}^{n-1}\sum_{\lambda=0}^{\mu-2\kappa}\frac{1}{\kappa!(\kappa+1)!}\frac{\binom{\mu-\kappa-\lambda-1}{\kappa-1}}{\binom{\mu}{\kappa+1}}\mu g_{\lambda+1}(z,\zeta)(1-|z|^2)^{\mu}\\&=\sum_{\mu=0}^{n-1-2\kappa}\sum_{\lambda=0}^{\mu}\frac{1}{(\kappa-1)!(\kappa+1)!}\frac{\binom{\mu+\kappa-\lambda-1}{\kappa-1}}{\binom{\mu+2\kappa}{\kappa+1}}g_{\lambda+1}(z,\zeta)(1-|z|^2)^{\mu+2\kappa}\end{aligned}$$

and

$$\begin{aligned}C_{\kappa\kappa 2}&=\sum_{\mu=2\kappa}^{n-1}\sum_{\lambda=0}^{\mu-2\kappa}\frac{\lambda}{2\kappa!(\kappa+1)!}\frac{\binom{\mu-\kappa-\lambda-1}{\kappa-1}}{\binom{\mu}{\kappa+1}}[g_{\lambda+1}(z,\zeta)-g_\lambda(z,\zeta)](1-|z|^2)^{\mu}\\&=\sum_{\mu=0}^{n-1-2\kappa}\sum_{\lambda=0}^{\mu}\frac{\lambda}{2\kappa!(\kappa+1)!}\frac{\binom{\mu+\kappa-\lambda-1}{\kappa-1}}{\binom{\mu+2\kappa}{\kappa+1}}[g_{\lambda+1}(z,\zeta)-g_\lambda(z,\zeta)](1-|z|^2)^{\mu+2\kappa}.\end{aligned}$$

Lemma 2 *For* $0\le\rho\le\kappa-2$

$$\partial_z\partial_{\bar{z}}C_{\kappa\kappa-\rho\ 1}=C_{\kappa\kappa-\rho-1\ 1}-\hat{C}_{\kappa\kappa-\rho-1\ 1}(1-|z|^2)^{n-2\rho-2}, \tag{25}$$

where

$$C_{\kappa\kappa-\rho\ 1} = \sum_{\mu=0}^{n-1-2\kappa} \sum_{\lambda=0}^{\mu} \frac{1}{(\kappa-\rho-1)!(\kappa-\rho+1)!}$$

$$\times \frac{\binom{\mu+\kappa-\lambda-\rho-1}{\kappa-\rho-1}}{\binom{\mu+2\kappa-2\rho}{\kappa+1-\rho}} g_{\lambda+1}(z,\zeta)(1-|z|^2)^{\mu+2\kappa-2\rho}$$

and

$$\hat{C}_{\kappa\kappa-\rho-1\ 1} = \sum_{\lambda=0}^{n-2\kappa} \left[\frac{1}{(\kappa-\rho-1)!^2} \frac{\binom{n-1-\kappa-\rho-\lambda}{\kappa-\rho-1}}{\binom{n-2}{\kappa-\rho-1}} \right.$$

$$\left. - \frac{1}{(\kappa-\rho-2)!(\kappa-\rho)!} \frac{\binom{n-2-\kappa-\rho-\lambda}{\kappa-\rho-2}}{\binom{n-2}{\kappa-\rho}} \right] g_{\lambda+1}(z,\zeta).$$

Lemma 2′ *For*

$$C_{\kappa 1\ 1} = \sum_{\mu=0}^{n-1-2\kappa} \sum_{\lambda=0}^{\mu} \frac{1}{(\mu+1)(\mu+2)} g_{\lambda+1}(z,\zeta)(1-|z|^2)^{\mu+2}$$

holds

$$\partial_z\partial_{\bar{z}} C_{\kappa 1\ 1} = g_1(z,\zeta) - \sum_{\lambda=0}^{n-2\kappa} g_{\lambda+1}(z,\zeta)(1-|z|^2)^{n-2\kappa}.$$

Lemma 3 *For* $0 \le \rho \le \kappa - 2$

$$\partial_z\partial_{\bar{z}} C_{\kappa\kappa-\rho\ 2} = C_{\kappa\kappa-\rho-1\ 2} + \hat{C}_{\kappa\kappa-\rho-1\ 2}(1-|z|^2)^{n-2\rho-2} \tag{26}$$

where

$$C_{\kappa\kappa-\rho\ 2} = \sum_{\mu=0}^{n-1-2\kappa} \sum_{\lambda=0}^{\mu} \frac{\lambda}{2(\kappa-\rho)!(\kappa-\rho+1)!}$$

$$\times \frac{\binom{\mu+\kappa-\lambda-\rho-1}{\kappa-\rho-1}}{\binom{\mu+2\kappa-2\rho}{\kappa+1-\rho}} [g_{\lambda+1}(z,\zeta) - g_\lambda(z,\zeta)](1-|z|^2)^{\mu+2\kappa-2\rho}$$

and

$$\hat{C}_{\kappa\kappa-\rho-1\ 2} = \sum_{\lambda=0}^{n-2\kappa} \frac{\lambda}{(\kappa-\rho-1)!(\kappa-\rho)!}$$

$$\times \left[\frac{\binom{n-1-\kappa-\rho-\lambda}{\kappa-\rho-1}}{\binom{n-2}{\kappa-\rho-1}} + \frac{\binom{n-2-\kappa-\rho-\lambda}{\kappa-\rho-2}}{\binom{n-2}{\kappa-\rho}} \right] [g_{\lambda+1}(z,\zeta) - g_\lambda(z,\zeta)].$$

Lemma 3′ *For*

$$C_{\kappa 1\ 2} = \sum_{\mu=0}^{n-1-2\kappa} \sum_{\lambda=0}^{\mu} \frac{\lambda}{2(\mu+1)(\mu+2)} [g_{\lambda+1}(z,\zeta) - g_\lambda(z,\zeta)](1-|z|^2)^{\mu+2}$$

holds

$$\partial_z\partial_{\bar z}C_{\kappa 1\ 2} = \frac{1}{2}{}_{\mu=1}^{n-2\kappa}\,[g_{\mu+1}(z,\zeta) - g_\mu(z,\zeta)](1-|z|^2)^\mu$$
$$-\frac{1}{2}\sum_{\lambda=1}^{n-2\kappa}\lambda\big[g_{\lambda+1}(z,\zeta) - g_\lambda(z,\zeta\big](1-|z|^2)^{n-2\kappa}.$$

Corollary 1 *For $C_{\kappa 1} = C_{\kappa 1\ 1} + C_{\kappa 1\ 2}$ follows*

$$\partial_z\partial_{\bar z}C_{\kappa 1} = C_{\kappa 0} - \sum_{\lambda=0}^{n-2\kappa}\left[g_{\lambda+1}(z,\zeta) + \frac{\lambda}{2}[g_{\lambda+1}(z,\zeta) - g_\lambda(z,\zeta)]\right](1-|z|^2)^{n-2\kappa},$$

where

$$C_{\kappa 0} = g_1(z,\zeta) + \frac{1}{2}\sum_{\mu=1}^{n-2\kappa}\big[g_{\mu+1}(z,\zeta) - g_\mu(z,\zeta)\big](1-|z|^2)^\mu,$$

and

$$\begin{aligned}
&\partial_{\nu_z}C_{\kappa 0} = \frac{1}{2}g_2(z,\zeta)(1-|z|^2)\\
&+\frac{1}{2}\sum_{\mu=2}^{n-2\kappa}\mu\big[g_{\mu+1}(z,\zeta) - g_\mu(z,\zeta)\big]\big[(1-|z|^2)^\mu - (1-|z|^2)^{\mu-1}\big]\\
&+\frac{n-2\kappa+1}{2}\big[g_{n-2\kappa+2}(z,\zeta) - g_{n-2\kappa+1}(z,\zeta)\big](1-|z|^2)^{n-2\kappa},\\
&\partial_z\partial_{\bar z}C_{\kappa 0} = -\frac{1}{2}(n-2\kappa)(n-2\kappa+1)\big[g_{n-2\kappa+2}(z,\zeta) - g_{n-2\kappa+1}(z,\zeta)\big]\\
&\times(1-|z|^2)^{n-2\kappa-1}.
\end{aligned}$$

Also

$$\begin{aligned}
-\hat C_{\kappa 0} &= \sum_{\lambda=0}^{n-2\kappa}\left[g_{\lambda+1}(z,\zeta) + \frac{\lambda}{2}[g_{\lambda+1}(z,\zeta) - g_\lambda(z,\zeta)]\right]\\
&= \frac{n-2\kappa+2}{2}g_{n-2\kappa+1}(z,\zeta) + \frac{1}{2}\sum_{\lambda=1}^{n-2\mu}g_\lambda(z,\zeta).
\end{aligned}$$

Lemma 4 *For $0 \le \sigma \le \kappa - 1$*

$$(\partial_z\partial_{\bar z})^\sigma C_{\kappa\kappa} = C_{\kappa\kappa-\sigma} +_{\tau=1}^{\sigma}(\partial_z\partial_{\bar z})^{\sigma-\tau}\big[\hat C_{\kappa\kappa-\tau}(1-|z|^2)^{n-2\tau}\big],$$

where $C_{\kappa\kappa-\sigma} = C_{\kappa\kappa-\sigma 1} + C_{\kappa\kappa-\sigma 2}, \hat C_{\kappa\kappa-\sigma} = \hat C_{\kappa\kappa-\sigma 1} + \hat C_{\kappa\kappa-\sigma 2}$ and $(\partial_z\partial_{\bar z})^{\sigma-\tau}$ $\big[\hat C_{\kappa\kappa-\tau}(1-|z|^2)^{n-2\tau}\big]$ according to the definitions in lemmas 2 and 3 is a proper *linear combination of*

$$(\partial_z\partial_{\bar z})^{\sigma-\tau}\left[g_{\lambda+1}(z,\zeta)(1-|z|^2)^{n-2\tau}\right]$$

$$=\sum_{\nu=0}^{\sigma-\tau}\sum_{\rho=0}^{\sigma-\tau-\nu}\sum_{\kappa=0}^{\nu}\sum_{\iota=0}^{\sigma-\tau-\nu-\rho}(-1)^{\kappa+\sigma-\tau-\nu-\iota}(\sigma-\tau)!^2$$

$$\times\binom{n-2\tau}{\sigma-\tau}\binom{n-\tau-\sigma}{\sigma-\tau-\nu-\rho}\binom{\nu}{\kappa}\binom{\lambda+\nu}{\lambda}\binom{\sigma-\tau}{\rho}\binom{\sigma-\tau-\nu-\rho}{\iota}$$

$$\times g_{\lambda+\kappa+1}(z,\zeta)(1-|z|^2)^{n-\sigma-\tau-\iota}$$

for $0\le\lambda\le n-2\kappa$.

Lemma 4′ *For* $0\le\sigma\le\kappa$

$$\partial_{\nu_z}(\partial_z\partial_{\bar z})^{\sigma}C_{\kappa}=\partial_{\nu_z}C_{\kappa\kappa-\sigma}+\sum_{\tau=1}^{\sigma}\partial_{\nu_z}(\partial_z\partial_{\bar z})^{\sigma-\tau}\left[\hat C_{\kappa\kappa-\tau}(1-|z|^2)^{n-2\tau}\right]$$

is some proper linear combination similarly as in the preceding lemma of

$$\partial_{\nu_z}(\partial_z\partial_{\bar z})^{\sigma-\tau}\left[g_{\lambda+1}(z,\zeta)(1-|z|^2)^{n-2\tau}\right]$$

$$=\sum_{\nu=0}^{\sigma-\tau}\sum_{\rho=0}^{\sigma-\tau-\nu}\sum_{\kappa=0}^{\nu}\sum_{\iota=0}^{\sigma-\tau-\nu-\rho}(-1)^{\kappa+\sigma-\tau-\nu-\iota}(\sigma-\tau)!^2$$

$$\times\binom{n-2\tau}{\sigma-\tau}\binom{n-\tau-\sigma}{\sigma-\tau-\nu-\rho}\binom{\nu}{\kappa}\binom{\lambda+\nu}{\lambda}\binom{\sigma-\tau}{\rho}\binom{\sigma-\tau-\nu-\rho}{\iota}$$

$$\times\Big[(\lambda+\kappa+1)\zeta\left[g_{\lambda+\kappa+2}(z,\zeta)-g_{\lambda+\kappa+1}(z,\zeta)\right](1-|z|^2)^{n-\sigma-\tau-\iota}$$

$$+2(n-\sigma-\tau-\iota)g_{\lambda+\kappa+1}(z,\zeta)\left[(1-|z|^2)^{n-\sigma-\tau-\iota}-(1-|z|^2)^{n-\sigma-\tau-\iota-1}\right]$$

for $0\le\lambda\le n-2\kappa$.

Lemma 5

$$(\partial_z\partial_{\bar z})^{\kappa}C_{\kappa\kappa}=\partial_z\partial_{\bar z}C_{\kappa1}+\sum_{\tau=1}^{\kappa-1}(\partial_z\partial_{\bar z})^{\kappa-\tau}\left[\hat C_{\kappa\kappa-\tau}(1-|z|^2)^{n-2\tau}\right],$$

where $\partial_z\partial_{\bar z}C_{\kappa1}$ *is given in Corollary 1 and the terms* $(\partial_z\partial_{\bar z})^{\kappa-\tau}\left[\hat C_{\kappa\kappa-\tau}(1-|z|^2)^{n-2\tau}\right]$ *are expressible as in Lemma 4 with* $\sigma=\kappa$.

Lemma 5′

$$\partial\nu_z(\partial_z\partial_{\bar z})^{\kappa}C_{\kappa\kappa}=\partial_{\nu_z}C_{\kappa0}+\sum_{\tau=1}^{\kappa}\partial_{\nu_z}(\partial_z\partial_{\bar z})^{\kappa-\tau}\left[\hat C_{\kappa-\tau}(1-|z|^2)^{n-2\tau}\right],$$

where $\partial_{\nu_z}C_{\kappa0}$ *is given in Corollary 1 and the last sum is expressible according to Lemma 4′ with* $\sigma=\kappa$.

Lemma 6 *For* $1\le\sigma,2(\kappa+\sigma)\le n-1$

$$(\partial_z\partial_{\bar z})^{\kappa+\sigma}C_{\kappa\kappa}=(\partial_{z\bar z})^{\sigma}C_{\kappa0}+\textstyle\sum_{\tau=1}^{\kappa-1}(\partial_z\partial_{\bar z})^{\kappa+\sigma-\tau}\left[\hat C_{\kappa-\tau}(1-|z|^2)^{n-2\tau}\right].$$

Here $\partial_z\partial_{\bar z}C_{\kappa 0}$ is given in Corollary 1 and $(\partial_z\partial_{\bar z})^\sigma C_{\kappa 0}$ is linearly composed by terms as

$$(\partial_z\partial_{\bar z})^{\sigma-1}g_{n-2\kappa+2}(z,\zeta)(1-|z|^2)^{n-2\kappa-1}$$

$$=\sum_{\nu=0}^{\sigma-1}\sum_{\tau=0}^{\nu}\sum_{\lambda=0}^{\nu-\tau}\sum_{\iota=0}^{\sigma-1-\nu}(-1)^{\nu+\lambda+\iota}(\sigma-1)!^2\binom{n-2\kappa-\sigma}{\nu-\tau}$$

$$\times\binom{n-2\kappa-1}{\sigma-1}\binom{n-2\kappa+\sigma-\nu}{\sigma-1-\nu}\binom{\sigma-1}{\tau}\binom{\sigma-1-\nu}{\iota}\binom{\nu-\tau}{\lambda}$$

$$\times g_{n-2\kappa+\iota+2}(z,\zeta)(1-|z|^2)^{n-2\kappa-\sigma-\lambda},$$

and another one like this with n formally replaced by $n-1$. The other terms are expressible as in Lemma 4 with $\kappa+\sigma$ instead of σ.

Lemma 6′ *For $1\le\sigma, 2(\kappa+\sigma)\le n-2$*

$$\partial_{\nu_z}(\partial_z\partial_{\bar z})^{\kappa+\sigma}C_{\kappa\kappa}=\partial_{\nu_z}(\partial_z\partial_{\bar z})^{\sigma}C_{\kappa 0}+\sum_{\tau=1}^{\kappa}\partial_{\nu_z}(\partial_z\partial_{\bar z})^{\kappa+\sigma-\tau}\left[\hat C_{\mu-\tau}(1-|z|^2)^{n-2\tau}\right].$$

The first term is linearly expressible by terms like

$$\partial_{\nu_z}(\partial_z\partial_{\bar z})^{\sigma-1}g_{n-2\kappa+2}(z,\zeta)(1-|z|^2)^{n-2\kappa-1}$$

$$=\sum_{\nu=0}^{\sigma-1}\sum_{\tau=0}^{\nu}\sum_{\lambda=0}^{\nu-\tau}\sum_{\iota=0}^{\sigma-1-\nu}(-1)^{\nu+\lambda+\iota}(\sigma-1)!^2\binom{n-2\kappa-\sigma}{\nu-\tau}\binom{n-2\kappa-1}{\sigma-1}$$

$$\times\binom{n-2\kappa+\sigma-\nu}{\sigma-1-\nu}\binom{\sigma-1}{\tau}\binom{\sigma-1-\nu}{\iota}\binom{\nu-\tau}{\lambda}$$

$$\times\Big[(n-2\kappa+\iota+2)\zeta\left[g_{n-2\kappa+\iota+3}(z,\zeta)-g_{n-2\kappa+\iota+2}(z,\zeta)\right](1-|z|^2)^{n-2\kappa-\sigma-\lambda}$$

$$-2(n-2\kappa-\sigma-\lambda)g_{n-2\kappa+\iota+2}(z,\zeta)$$

$$\times\left[(1-|z|^2)^{n-2\kappa-\sigma-\lambda-1}-(1-|z|^2)^{n-2\kappa-\sigma-\lambda}\right],$$

and the same expression with n formally replaced by $n-1$. The remaining terms can be expressed as in Lemma 4′ with $\mu+\sigma$ instead of σ.

(iv) Let

$$D=\sum_{\kappa=1}^{[\frac{n}{2}]-1}D_{\kappa\kappa}\hat\gamma_\kappa$$

with

$$D_{\kappa\kappa} = \sum_{\mu=2\kappa+1}^{n-1} \sum_{\lambda=1}^{\mu-2\kappa} \frac{1}{\kappa!(\kappa+1)!} \frac{\binom{\mu-\kappa-\lambda-1}{\kappa-1}}{\binom{\mu}{\kappa+1}} g_\lambda(z,\zeta)(1-|z|^2)^\mu$$

$$= \sum_{\mu=1}^{n-1-2\kappa} \sum_{\lambda=1}^{\mu} \frac{1}{\kappa!(\kappa+1)!} \frac{\binom{\mu+\kappa-\lambda-1}{\kappa-1}}{\binom{\mu+2\kappa}{\kappa+1}} g_\lambda(z,\zeta)(1-|z|^2)^{\mu+2\kappa}.$$

Lemma 7 *For* $0 \le \rho \le \kappa - 2$

$$\partial_z \partial_{\bar z} D_{\kappa\kappa-\rho} = D_{\kappa\kappa-\rho-1} + \widehat{D}_{\kappa\kappa-\rho-1}(1-|z|^2)^{n-2\rho-2}, \tag{27}$$

where

$$D_{\kappa\kappa-\rho} = \sum_{\mu=1}^{n-2\kappa-1} \sum_{\lambda=1}^{\mu} \frac{1}{(\kappa-\rho)!(\kappa-\rho+1)!} \frac{\binom{\mu+\kappa-\lambda-\rho-1}{\kappa-\rho-1}}{\binom{\mu+2\kappa-2\rho}{\kappa-\rho+1}}$$

$$\times g_\lambda(z,\zeta)(1-|z|^2)^{\mu+2\kappa-2\rho}$$

and

$$\hat{D}_{\kappa\kappa-\rho-1} = \sum_{\lambda=1}^{n-2\kappa} \frac{1}{(\kappa-\rho-1)!(\kappa-\rho)!}$$

$$\times \left[\frac{\binom{n-\kappa-\lambda-\rho-2}{\kappa-\rho-2}}{\binom{n-2}{\kappa-\rho}} - \frac{\binom{n-\kappa-\lambda-\rho-1}{\kappa-\rho-1}}{\binom{n-2}{\kappa-\rho-1}} \right] g_\lambda(z,\zeta).$$

Lemma 7′ *For*

$$D_{\kappa 1} = \sum_{\mu=1}^{n-1-2\kappa} \sum_{\lambda=1}^{\mu} \frac{1}{(\mu+1)(\mu+2)} g_\lambda(z,\zeta)(1-|z|^2)^{\mu+2}$$

holds

$$\partial_z \partial_{\bar z} D_{\kappa 1} = D_{\kappa 0} + \widehat{D}_{\kappa 0}(1-|z|^2)^{n-2\kappa} \tag{28}$$

with

$$D_{\kappa 0} = \sum_{\mu=1}^{n-1-2\kappa} \frac{1}{\mu} g_\mu(z,\zeta)(1-|z|^2)^\mu,$$

$$\widehat{D}_{\kappa 0} = \frac{1}{n-2\kappa} g_{n-2\kappa}(z,\zeta) - \sum_{\lambda=1}^{n-2\kappa} g_\lambda(z,\zeta),$$

$$\partial_{\nu_z} D_{\kappa 0} = -2g_1(z,\zeta) - \sum_{\mu=2}^{n-1-2\kappa} g_\mu(z,\zeta)(1-|z|^2)^{\mu-1} \tag{29}$$

$$+ \sum_{\mu=1}^{n-1-2\kappa} g_\mu(z,\zeta)(1-|z|^2)^\mu + g_{n-2\kappa}(z,\zeta)(1-|z|^2)^{n-2\kappa-1},$$

and

$$\partial_z \partial_{\bar{z}} D_{\kappa 0} = -(n-2\kappa-1) g_{n-2\kappa}(z,\zeta)(1-|z|^2)^{n-2-2\kappa}, \tag{30}$$

$$\begin{aligned} \partial_z \partial_{\bar{z}} \left[\widehat{D}_{\kappa 0}(1-|z|^2)^{n-2\kappa} \right] &= (n-2\kappa-1) g_{n-2\kappa}(z,\zeta)(1-|z|^2)^{n-2\kappa-2} \\ +(n-2\kappa-1)(n-2\kappa) &\Bigg[\sum_{\lambda=1}^{n-2\kappa+1} g_\lambda(z,\zeta)(1-|z|^2)^{n-2\kappa-1} \\ &\quad - \sum_{\lambda=1}^{n-2\kappa} g_\lambda(z,\zeta)(1-|z|^2)^{n-2\kappa-2} \Bigg]. \end{aligned} \tag{31}$$

Lemma 8 *For* $0 \le \sigma \le \kappa - 1$

$$(\partial_z \partial_{\bar{z}})^\sigma D_{\kappa\kappa} = D_{\kappa\kappa-\sigma} + \sum_{\tau=1}^{\sigma} (\partial_z \partial_{\bar{z}})^{\sigma-\tau} \left[\hat{D}_{\kappa\kappa-\tau}(1-|z|^2)^{n-2\tau} \right],$$

where $(\partial_z \partial_{\bar{z}})^{\sigma-\tau} \left[\hat{D}_{\kappa\kappa-\tau}(1-|z|^2)^{n-2\tau} \right]$ *is according to the definition in Lemma 7 a certain linear combination of*

$$\begin{aligned} &(\partial_z \partial_{\bar{z}})^{\sigma-\tau} \left[g_\lambda(z,\zeta)(1-|z|^2)^{n-2\tau} \right] \\ &= \sum_{\nu=0}^{\sigma-\tau} \sum_{\rho=0}^{\sigma-\tau-\nu} \sum_{\mu=0}^{\nu} \sum_{\iota=0}^{\sigma-\tau-\nu-\rho} (-1)^{\mu+\sigma-\tau-\nu-\iota} (\sigma-\tau)!^2 \binom{n-2\tau}{\sigma-\tau} \\ &\times \binom{n-\tau-\sigma}{\sigma-\tau-\nu-\rho} \binom{\nu}{\mu} \binom{\lambda+\nu-1}{\nu} \binom{\sigma-\tau}{\rho} \binom{\sigma-\tau-\nu-\rho}{\iota} \\ &\times g_{\lambda+\mu}(z,\zeta)(1-|z|^2)^{n-\sigma-\tau-\iota} \end{aligned}$$

for $1 \le \lambda \le n - 2\kappa$.

Lemma 8′ *For* $0 \le \sigma \le \kappa - 1$

$$\partial_{\nu_z} (\partial_z \partial_{\bar{z}})^\sigma D_{\kappa\kappa} = \partial_{\nu_z} D_{\kappa\kappa-\sigma} + \sum_{\tau=0}^{\sigma} \partial_{\nu_z} (\partial_z \partial_{\bar{z}})^{\sigma-\tau} \left[\hat{D}_{\kappa\kappa-\tau}(1-|z|^2)^{n-2\tau} \right]$$

is a certain linear combination of

$$\partial_{\nu_z}(\partial_z\partial_{\bar z})^{\sigma-\tau}\left[g_\lambda(z,\zeta)(1-|z|^2)^{n-2\tau}\right]$$

$$=\sum_{\nu=0}^{\sigma-\tau}\sum_{\rho=0}^{\sigma-\tau-\nu}\sum_{\mu=0}^{\nu}\sum_{\iota=0}^{\sigma-\tau-\nu-\rho}(-1)^{\mu+\sigma-\tau-\nu-\iota}(\sigma-\tau)!^2$$

$$\times\binom{n-2\tau}{\sigma-\tau}\binom{n-\tau-\sigma}{\sigma-\tau-\nu-\rho}\binom{\nu}{\mu}\binom{\lambda+\nu-1}{\nu}\binom{\sigma-\tau}{\rho}$$

$$\times\binom{\sigma-\tau-\nu-\rho}{\iota}\Big[(\lambda+\mu)\left[g_{\lambda+\mu+1}(z,\zeta)-g_{\lambda+\mu}(z,\zeta)\right](1-|z|^2)^{n-\sigma-\tau-\iota}$$

$$-2(n-\sigma-\tau-\iota)g_{\lambda+\mu}(z,\zeta)\left[(1-|z|^2)^{n-\sigma-\tau-\iota-1}-(1-|z|^2)^{n-\sigma-\tau-\iota}\right]$$

for $1\le\lambda\le n-2\kappa$.

Lemma 9

$$(\partial_z\partial_{\bar z})^{\kappa}D_{\kappa\kappa}=D_{\kappa 0}+\hat D_{\kappa 0}(1-|z|^2)^{n-2\kappa}+\sum_{\tau=1}^{\kappa}(\partial_z\partial_{\bar z})^{\kappa-\tau}\left[\hat D_{\kappa\kappa-\tau}(1-|z|^2)^{n-2\tau}\right]$$

where $D_0,\hat D_0$ *are defined in Lemma 7′ and the terms* $(\partial_z\partial_{\bar z})^{\kappa-\tau}\hat D_{\kappa\kappa-\tau}(1-|z|^2)^{n-2\tau}$ *are expressible as in Lemma 8 with* $\sigma=\kappa$.

Lemma 9′

$$\partial_{\nu_z}(\partial_z\partial_{\bar z})^{\kappa}D_{\kappa\kappa}=\partial_{\nu_z}D_{\kappa 0}+\left[g_{n-2\kappa+1}(z,\zeta)-g_{n-2\kappa}(z,\zeta)\right](1-|z|^2)^{n-2\kappa}$$

$$-\sum_{\lambda=1}^{n-2\kappa}\lambda\left[g_{\lambda+1}(z,\zeta)-g_\lambda(z,\zeta)\right](1-|z|^2)^{n-2\kappa}$$

$$-2g_{n-2\kappa}(z,\zeta)\left[(1-|z|^2)^{n-2\kappa-1}-(1-|z|^2)^{n-2\kappa}\right]$$

$$+\sum_{\lambda=1}^{n-2\kappa}2(n-2\kappa)g_\lambda(z,\zeta)\left[(1-|z|^2)^{n-2\kappa-1}-(1-|z|^2)^{n-2\kappa}\right]$$

$$+\sum_{\tau=0}^{\kappa}\partial_{\nu_z}(\partial_z\partial_{\bar z})^{\kappa-\tau}\left[\hat D_{\kappa-\tau}(1-|z|^2)^{n-2\tau}\right],$$

where the last sums are linear combinations of terms as in Lemma 8′ with $\sigma=\kappa$ *and* $\partial_{\bar z}D_{\kappa 0}$ *is given in* (29).

Lemma 10

$$
\begin{aligned}
(\partial_z\partial_{\bar z})^{\kappa+1} D_{\kappa\kappa} &= (n-2\kappa-1)(n-2\kappa) \\
&\quad \times \Bigg[(1-|z|^2)^{n-2\kappa-1}\sum_{\lambda=1}^{n-2\kappa+1} g_\lambda(z,\zeta) \\
&\qquad -(1-|z|^2)^{n-2\kappa-2}\,{}^{n-2\kappa}_{\lambda=1}\, g_\lambda(z,\zeta)\Bigg] \\
&\quad + \sum_{\tau=1}^{\kappa}(\partial_z\partial_{\bar z})^{\kappa-\tau+1}\big[\hat D_{\kappa\kappa-\tau}(1-|z|^2)^{n-2\tau}\big],
\end{aligned}
$$

where the last sum is a certain linear combination of terms as in Lemma 8 with $\sigma = \kappa+1$.

Lemma 11 *For* $1 \le \sigma, 2(\kappa+\sigma) \le n-1$

$$
\begin{aligned}
&(\partial_z\partial_{\bar z})^{\kappa+\sigma} D_{\kappa\kappa} \\
&= \sum_{\lambda=1}^{n-2\kappa+1}(n-2\kappa-1)(n-2\kappa)(\partial_z\partial_{\bar z})^{\sigma-1}\big[g_\lambda(z,\zeta)(1-|z|^2)^{n-2\kappa-1}\big] \\
&- \sum_{\lambda=1}^{n-2\kappa}(n-2\kappa-1)(n-2\kappa)(\partial_z\partial_{\bar z})^{\sigma-1}\big[g_\lambda(z,\zeta)(1-|z|^2)^{n-2\kappa-2}\big] \\
&+ \sum_{\tau=1}^{\kappa}(\partial_z\partial_{\bar z})^{\mu+\sigma-\tau}\big[\hat D_{\kappa\kappa-\tau}(1-|z|^2)^{n-2\tau}\big].
\end{aligned}
$$

Here the terms in the first two sums are

$$
\begin{aligned}
&(\partial_z\partial_{\bar z})^{\sigma-1}\big[g_\lambda(z,\zeta)(1-|z|^2)^{m-2\kappa}\big] \\
&= \sum_{\rho=0}^{\sigma-1}\sum_{\tau=0}^{\rho}\sum_{\mu=0}^{\sigma-1-\rho}\sum_{\iota=0}^{\rho-\tau}(-1)^{\rho+\mu+\iota}(\sigma-1)!^2\binom{m-2\kappa}{\sigma-1} \\
&\times\binom{m+1-2\kappa-\sigma}{\rho-\tau}\binom{\sigma-1}{\tau}\binom{\sigma-1-\rho}{\mu}\binom{\sigma-\tau}{\iota}\binom{\lambda-2+\sigma-\rho}{\lambda-1} \\
&\times g_{\lambda+\mu}(z,\zeta)(1-|z|^2)^{m+1-2\kappa-\iota}
\end{aligned}
$$

for either $m = n-2$ *or* $m = n-1$ *and the last sum is a linear combination of terms as in Lemma 8 with* σ *replaced by* $\mu+\sigma$. *Here* $1 \le \lambda \le n-2\kappa$.

Lemma 11′ *For $1 \leq \sigma, 2(\kappa+\sigma) \leq n-2$*

$$\partial_{\nu_z}(\partial_z\partial_{\bar z})^{\kappa+\sigma} D_{\kappa\kappa}$$

$$= \sum_{\lambda=1}^{n-2\kappa+1} (n-2\mu-1)(n-2\mu)\partial_{\nu_z}(\partial_z\partial_{\bar z})^{\sigma-1}\left[g_\lambda(z,\zeta)^2(1-|z|^2)^{n-2\kappa-1}\right]$$

$$-\sum_{\lambda=1}^{n-2\kappa}(n-2\kappa-1)(n-2\kappa)\partial_{\nu_z}(\partial_z\partial_{\bar z})^{\sigma-1}\left[g_\lambda(z,\zeta)(1-|z|^2)^{n-\kappa-2}\right]$$

$$+\sum_{\tau=1}^{\kappa}\partial_{\nu_z}(\partial_z\partial_{\bar z})^{\kappa+\sigma-\tau}\left[\hat D_{\kappa\kappa-\tau}(1-|z|^2)^{n-2\kappa}\right]$$

where the terms in the first two sums are expressible as

$$\partial_{\nu_z}(\partial_z\partial_{\bar z})^{\sigma-1}\left[g_\lambda(z,\zeta)(1-|z|^2)^{m-2\kappa}\right]$$

$$=\sum_{\rho=0}^{\sigma-1}\sum_{\tau=0}^{\rho}\sum_{\mu=0}^{\sigma-1-\rho}\sum_{\iota=0}^{\rho-\tau}(-1)^{\rho+\mu+\iota}(\sigma-1)!^2\binom{m-2\kappa}{\sigma-1}$$

$$\times\binom{m+1-2\kappa-\sigma}{\rho-\tau}\binom{\sigma-1}{\tau}\binom{\sigma-1-\rho}{\mu}\binom{\sigma-\tau}{\iota}\binom{\lambda-2+\sigma-\rho}{\lambda-1}$$

$$\times\Big[(\lambda+\mu)\left[g_{\lambda+\mu+1}(z,\zeta)-g_{\lambda+\mu}(z,\zeta)\right](1-|z|^2)^{m+1-2\kappa-\iota}$$

$$-2(m+1-2\kappa-\iota)g_{\lambda+\kappa}(z,\zeta)\left[(1-|z|^2)^{m-2\kappa-\iota}-(1-|z|^2)^{m-2\kappa-\iota+1}\right]\Big]$$

for $m=n-2$ and $m=n-1$ and the third sum is a linear combination of terms as in Lemma 8′ with σ replaced by $\kappa+\sigma$. Here $1\leq\lambda\leq n-2$.

The proof of some of these lemmas will be given in the Appendix.

From (22), the property of the Poisson kernel $g_1(z,\zeta)$, Lemma B and Lemma C

$$\lim_{z\to\zeta} w(z)=\gamma_0(\zeta) \text{ for } \zeta\in\partial\mathbb{D}$$

follows. As $\gamma_0\in C^n(\partial\mathbb{D};\mathbb{C})$ there is no contribution from A to the boundary values or the derivatives of w.

From (23′) and lemmata B and C similarly

$$\lim_{z\to\zeta}\partial_{\nu_z}w(z)=\hat\gamma_0(\zeta) \text{ for } \zeta\in\partial\mathbb{D}$$

is seen and (23″) together with (23‴), (23^{IV}) indicate that no contribution may be expected for the other higher order derivatives from this second summand because $\hat\gamma_0\in C^{n-1}(\partial\mathbb{D};\mathbb{C})$.

For the third summand Corollary 1 and lemmata 2 to 6′ for $\mu = 1$ show

$$\lim_{z\to\zeta} \partial_z\partial_{\bar z} w(z) = \gamma_1(\zeta) \text{ for } \zeta \in \partial\mathbb{D}$$

and also that no contribution comes from this part to the boundary behaviour of the other higher order derivatives involved as $\gamma_1 \in C^{n-2}(\partial\mathbb{D};\mathbb{C})$.
Proceeding inductively next the term $C_{\kappa\kappa}\gamma_\kappa$ is considered for $2 \le 2\kappa \le n-1$. Rewriting (25) and (26) as

$$\partial_z\partial_{\bar z} C_{\kappa\kappa-\rho} = C_{\kappa\kappa-\rho-1} + \hat{C}_{\kappa\kappa-\rho-1}(1-|z|^2)^{n-2\rho-2}$$

iteration gives for $1 \le \sigma \le \kappa - 1$

$$(\partial_z\partial_{\bar z})^\sigma C_{\kappa\kappa} = C_{\kappa\kappa-\sigma} + \sum_{\tau=1}^{\sigma} (\partial_z\partial_{\bar z})^{\sigma-\tau}\Big[\widehat{C}_{\kappa\kappa-\tau}(1-|z|^2)^{n-2\tau}\Big],$$

in particular

$$(\partial_z\partial_{\bar z})^{\kappa-1} C_{\kappa\kappa} = C_{\kappa 1} + \sum_{\tau=1}^{\kappa-1} (\partial_z\partial_{\bar z})^{\kappa-\tau-1}\Big[\widehat{C}_{\kappa\kappa-\tau}(1-|z|^2)^{n-2\tau}\Big].$$

On the basis of Corollary 1 and lemmata 2 to 6′ then

$$\lim_{z\to\zeta} (\partial_z\partial_{\bar z})^\kappa w = \gamma_\kappa \text{ for } \zeta \in \partial\mathbb{D}$$

is seen and also that this term does not contribute to the boundary behaviour of any other higher order derivative involved because $\gamma_\kappa \in C^{n-2\kappa}(\partial\mathbb{D};\mathbb{C})$.

At last the terms $D_{\kappa\kappa}\hat{\gamma}_\kappa, 2 \le 2\kappa \le n-2$, are considered. As for the $C_{\kappa\kappa}$ from (27) follows for $1 \le \sigma \le \kappa - 1$

$$(\partial_z\partial_{\bar z})^\sigma D_{\kappa\kappa} = D_{\kappa\kappa-\sigma} + \sum_{\tau=1}^{\sigma} (\partial_z\partial_{\bar z})^{\sigma-\tau}\Big[\widehat{D}_{\kappa\kappa-\tau}(1-|z|^2)^{n-2\tau}\Big],$$

in particular

$$(\partial_z\partial_{\bar z})^{\kappa-1} D_{\kappa\kappa} = D_{\kappa 1} + \sum_{\tau=1}^{\mu-1} (\partial_z\partial_{\bar z})^{\kappa-1-\tau}\Big[\widehat{D}_{\kappa\kappa-\tau}(1-|z|^2)^{n-2\tau}\Big]$$

and via (28) and (29)

$$(\partial_z\partial_{\bar z})^{\kappa} D_{\kappa\kappa} = D_{\kappa 0} + \sum_{\tau=1}^{\kappa} (\partial_z\partial_{\bar z})^{\kappa-\tau}\Big[\widehat{D}_{\kappa\kappa-\tau}(1-|z|^2)^{n-2\tau}\Big]$$

and

$$\begin{aligned}\partial_{\nu_z}(\partial_z\partial_{\bar z})^\kappa D_{\kappa\kappa} = {}& -2g_1(z,\zeta) - \sum_{\mu=2}^{n-2\kappa-1} g_\mu(z,\zeta)(1-|z|^2)^{\mu-1} \\ & + \sum_{\mu=1}^{n-2\kappa-1} g_\mu(z,\zeta)(1-|z|^2)^\mu + g_{n-2\kappa}(z,\zeta)(1-|z|^2)^{n-2\kappa-1} \\ & + \sum_{\tau=1}^{\kappa} \partial_{\nu_z}(\partial_z\partial_{\bar z})^{\kappa-\tau}\Big[\widehat{D}_{\kappa\kappa-\tau}(1-|z|^2)^{n-2\tau}\Big].\end{aligned}$$

Hence, observing lemmata 7 to 11′

$$\lim_{z\to\zeta} \partial_{\nu_z} (\partial_z \partial_{\bar z})^\kappa w(z) = \hat\gamma_\kappa(\zeta) \text{ for } \zeta \in \Delta.$$

Moreover, from (30) and Lemma 10

$$\begin{aligned}
&(\partial_z\partial_{\bar z})^{\kappa+1} D_{\kappa\kappa} = (n-2\kappa-1)(n-2\kappa) \\
&\times\Big[1-|z|^2)^{n-2\kappa-1} \sum_{\lambda=1}^{n-2\kappa-1} g_\lambda(z,\zeta) - (1-|z|^2)^{n-2\kappa-2} \sum_{\lambda=1}^{n-2\kappa} g_\lambda(z,\zeta)\Big] \\
&+\sum_{\tau=1}^{\kappa} (\partial_z\partial_{\bar z})^{\kappa-\tau+1}\Big[\widehat D_{\kappa\kappa-\tau}(1-|z|^2)^{n-2\tau}\Big]
\end{aligned}$$

is seen, showing that this term does not contribute to any other higher order derivative on the boundary. Here $\hat\gamma_\kappa \in C^{n-2\kappa-1}(\partial\mathbb{D};\mathbb{C})$ is needed.
That higher order terms vice versa do not contribute to lower order derivatives of w on the boundary follows from the size of powers of the factor $(1-|z|^2)$ in relation to the powers in the denominators, see Lemma B.

4. Appendix

Here the proofs of some lemmas in Section 3 are provided.

(i) **Proof of Lemma 2** For simplicity ρ is taken to be 0. From

$$\begin{aligned}
\partial_{\bar z}\Big[g_{\lambda+1}(z,\zeta)(1-|z|^2)^{\mu+2\kappa}\Big] &= \frac{(\lambda+1)\zeta}{(1-\zeta)^{\lambda+2}}(1-|z|^2)^{\mu+2\kappa} \\
&\quad -(\mu+2\kappa) z g_{\lambda+1}(z,\zeta)(1-|z|^2)^{\mu+2\kappa-1}
\end{aligned}$$

and

$$\begin{aligned}
&\partial_z\partial_{\bar z}\Big[g_{\lambda+1}(z,\zeta)(1-|z|^2)^{\mu+2\kappa}\Big] \\
&= (\mu+2\kappa)\Big[(\mu+2\kappa-1)|z|^2 g_{\lambda+1}(z,\zeta)(1-|z|^2)^{\mu+2\kappa-2} \\
&-g_{\lambda+1}(z,\zeta)(1-|z|^2)^{\mu+2\kappa-1} \\
&-(\lambda+1)\Big(\frac{\bar z\zeta}{(1-\bar z\zeta)^{\lambda+2}} + \frac{z}{(1-z\bar\zeta)^{\lambda+2}}\Big)(1-|z|^2)^{\mu+2\kappa-1}\Big]
\end{aligned}$$

it follows

$$\begin{aligned}
\partial_z\partial_{\bar z} C_{\kappa\kappa 1} &= \sum_{\mu=0}^{n-1-2\kappa} \sum_{\lambda=0}^{\mu} \frac{1}{(\kappa-1)!\kappa!} \frac{\binom{\mu+\kappa-\lambda-1}{\kappa-1}}{\binom{\mu+2\kappa-1}{\kappa}} \\
&\times\Big[(\mu+2\kappa-1) g_{\lambda+1}(z,\zeta)(1-|z|^2)^{\mu+2\kappa-2} \\
&\quad -(\mu+2\kappa-1) g_{\lambda+1}(z,\zeta)(1-|z|^2)^{\mu+2\kappa-1} \\
&\quad +(\lambda g_{\lambda+1}(z,\zeta) - (\lambda+1) g_{\lambda+2}(z,\zeta))(1-|z|^2)^{\mu+2\kappa-1}\Big].
\end{aligned}$$

Observing for $2 \le \kappa$

$$\sum_{\lambda=1}^{\mu+1} \lambda g_{\lambda+1}(z,\zeta)\binom{\mu+\kappa-\lambda}{\kappa-1} - \sum_{\lambda=0}^{\mu} \lambda g_{\lambda+1}(z,\zeta)\binom{\mu+\kappa-\lambda-1}{\kappa-1}$$

$$= \sum_{\lambda=1}^{\mu+1} \binom{\mu+\kappa-\lambda-1}{\kappa-2} g_{\lambda+1}(z,\zeta)$$

and

$$\binom{\mu+\kappa-\lambda-1}{\kappa-1}(\mu+2\kappa-1) + \binom{\mu+\kappa-\lambda-1}{\kappa-2}\lambda$$

$$= \binom{\mu+\kappa-\lambda-1}{\kappa-1}(\mu+\kappa) + \binom{\mu+\kappa-\lambda-1}{\kappa-2}(\mu+1)$$

shows

$$\partial_z \partial_{\bar{z}} C_{\kappa\kappa 1} = \frac{(1-|z|^2)^{2\kappa-2}}{(2\kappa-2)!} g_1(z,\zeta)$$

$$+ \sum_{\mu=0}^{n-2-2\kappa} \sum_{\lambda=0}^{\mu+1} \frac{(1-|z|^2)^{\mu+2\kappa-1}}{(\kappa-1)!\kappa!} \frac{\binom{\mu+\kappa-\lambda}{\kappa-1}}{\binom{\mu+2\kappa}{\kappa}} (\mu+2\kappa) g_{\lambda+1}(z,\zeta)$$

$$
-\sum_{\mu=0}^{n-1-2\kappa}\sum_{\lambda=0}^{\mu-1}\frac{(1-|z|^2)^{\mu+2\kappa-1}}{(\kappa-1)!\kappa!}\frac{1}{\binom{\mu+2\kappa-1}{\kappa}}
$$

$$
\times\left[\binom{\mu+\kappa-\lambda-1}{\kappa-1}(\mu+\kappa)\right.
$$

$$
\left.+\binom{\mu+\kappa-\lambda-1}{\kappa-2}(\mu+1)\right]g_{\lambda+1}(z,\zeta)
$$

$$
-\sum_{\mu=0}^{n-1-2\kappa}\frac{(1-|z|^2)^{\mu+2\kappa-1}}{(\kappa-1)!\kappa!}\frac{\mu+1}{\binom{\mu+2\kappa-1}{\kappa}}g_{\mu+2}(z,\zeta)
$$

$$
-\sum_{\mu=0}^{n-1-2\kappa}\frac{(1-|z|^2)^{\mu+2\kappa-1}}{(\kappa-1)!\kappa!}\frac{1}{\binom{\mu+2\kappa-1}{\kappa}}
$$

$$
\times[\mu+2\kappa-1+(\kappa-1)\mu]g_{\mu+1}(z,\zeta)
$$

$$
=\sum_{\mu=0}^{n-1-2\kappa}\sum_{\lambda=0}^{\mu}\frac{1}{(\kappa-2)!\kappa!}\frac{\binom{\mu+\kappa-\lambda-2}{\kappa-2}}{\binom{\mu+2\kappa-2}{\kappa}}g_{\lambda+1}(z,\zeta)(1-|z|^2)^{\mu+2\kappa-2}
$$

$$
-\sum_{\lambda=0}^{n-2\kappa}\left[\frac{1}{(\kappa-1)!^2}\frac{\binom{n-1-\kappa-\lambda}{\kappa-1}}{\binom{n-2}{\kappa-1}}\right.
$$

$$
\left.-\frac{1}{(\kappa-2)!\kappa!}\frac{\binom{n-2-\kappa-\lambda}{\kappa-2}}{\binom{n-2}{\kappa}}\right]g_{\lambda+1}(z,\zeta)(1-|z|^2)^{n-2}.
$$

This is (25).

(ii) **Proof of Lemma 3** As before without loss of generality $\rho = 0$. Rewriting $C_{\kappa\kappa 2} = ReC_{\kappa\kappa 3}$. With

$$
C_{\kappa\kappa 3}=\sum_{\mu=0}^{n-1-2\kappa}\sum_{\lambda=0}^{\mu}\frac{\lambda}{\kappa!(\kappa+1)!}\frac{\binom{\mu+\kappa-\lambda-1}{\kappa-1}}{\binom{\mu+2\kappa}{\kappa+1}}\frac{\bar{z}\zeta}{(1-\bar{z}\zeta)^{\lambda+1}}(1-|z|^2)^{\mu+2\kappa}
$$

then

$$
\partial_{\bar{z}}C_{\kappa\kappa 3}=\sum_{\mu=0}^{n-1-2\kappa}\sum_{\lambda=0}^{\mu}\frac{\lambda}{\kappa!(\kappa+1)!}\frac{\binom{\mu+\kappa-\lambda-1}{\kappa-1}}{\binom{\mu+2\kappa}{\kappa+1}}
$$

$$
\times\left[\frac{(\lambda+1)\zeta^2}{(1-\bar{z}\zeta)^{\lambda+2}}(1-|z|^2)^{\mu+2\kappa}+\frac{\zeta(1-|z|^2)^{\mu+2\kappa}}{(1-\bar{z}\zeta)^{\lambda+1}}\right.
$$

$$
\left.-\frac{(\mu+2\kappa)|z|^2\zeta}{(1-\bar{z}\zeta)^{\lambda}}(1-|z|^2)^{\mu+2\kappa-1}\right].
$$

After some rearrangements and using for $2 \le \mu$

$$\frac{\lambda-1}{\kappa+1}\frac{\binom{\mu+\kappa-\lambda-1}{\kappa-2}}{\binom{\mu+2\kappa}{\kappa+1}}+\frac{\binom{\mu+\kappa-\lambda-1}{\kappa-1}}{\binom{\mu+2\kappa-1}{\kappa}}-\frac{\binom{\mu+\kappa-\lambda}{\kappa-1}}{\binom{\mu+2\kappa}{\kappa}}-\frac{\binom{\mu+\kappa-\lambda-1}{\kappa-1}}{(\kappa+1)\binom{\mu+2\kappa}{\kappa+1}}$$

$$=-\frac{\kappa}{\kappa+1}\frac{\binom{\mu+\kappa-\lambda-1}{\kappa-2}}{\binom{\mu+2\kappa}{\kappa+1}}$$

it follows

$$\partial_{\bar z}C_{\kappa\kappa 3}=-\sum_{\mu=0}^{n-1-2\kappa}\sum_{\lambda=0}^{\mu+1}\frac{\lambda}{(\kappa-1)(\kappa+1)!}$$

$$\times\frac{\binom{\mu+\kappa-\lambda-1}{\kappa-2}}{\binom{\mu+2\kappa}{\kappa+1}}\frac{\zeta}{(1-\bar z\zeta)^{\lambda+1}}(1-|z|^2)^{\mu+2\kappa}$$

$$-\sum_{\lambda=0}^{n-2\kappa}\frac{\lambda}{\kappa!^2}\frac{\binom{n-1-\kappa-1}{\kappa-1}}{\binom{n-1}{\kappa}}\frac{\zeta}{(1-\bar z\zeta)^{\lambda+1}}(1-|z|^2)^{n-1}.$$

Differentiating again gives

$$\partial_z\partial_{\bar z}C_{\kappa\kappa 3}=\sum_{\mu=0}^{n-1-2\kappa}\sum_{\lambda=0}^{\mu+1}\frac{\lambda(\mu+2\kappa)}{(\kappa-1)!(\kappa+1)!}$$

$$\times\frac{\binom{\mu+\kappa-\lambda-1}{\kappa-2}}{\binom{\mu+2\kappa}{\kappa+1}}\frac{\zeta\bar z}{(1-\bar z\zeta)^{\lambda+1}}(1-|z|^2)^{\mu+2\kappa-1}$$

$$+\sum_{\lambda=0}^{n-2\kappa}\frac{\lambda(n-1)}{\kappa!^2}\frac{\binom{n-1-\kappa-\lambda}{\kappa-1}}{\binom{n-1}{\mu}}\frac{\zeta\bar z}{(1-\bar z\zeta)^{\lambda+1}}(1-|z|^2)^{n-2}$$

$$=\sum_{\mu=0}^{n-1-2\kappa}\sum_{\lambda=0}^{\mu}\frac{\lambda}{(\kappa-1)!\kappa!}\frac{\binom{\mu+\kappa-\lambda-2}{\kappa-2}}{\binom{\mu+2\kappa-2}{\kappa}}$$

$$\times\frac{\bar z\zeta}{(1-\bar z\zeta)^{\lambda+1}}(1-|z|^2)^{\mu+2\kappa-2}$$

$$+\sum_{\lambda=0}^{n-2\kappa}\frac{\lambda}{(\kappa-1)!\kappa!}\left[\frac{\binom{n-2-\kappa-\lambda}{\kappa-2}}{\binom{n-2}{\kappa}}+\frac{\binom{n-1-\kappa-\lambda}{\kappa-1}}{\binom{n-2}{\kappa-1}}\right]\frac{\bar z\zeta}{(1-\bar z\zeta)^{\lambda+1}}$$

$$\times(1-|z|^2)^{n-2}.$$

Thus

$$2\partial_z\partial_{\bar z}C_{\kappa\kappa 2} = \sum_{\mu=0}^{n-1-2\kappa}\sum_{\lambda=0}^{\mu}\frac{\lambda}{(\kappa-1)!\kappa!}$$

$$\times\frac{\binom{\mu+\kappa-\lambda-2}{\kappa-2}}{\binom{\mu+2\kappa-2}{\kappa}}\left[g_{\lambda+1}(z,\zeta)-g_\lambda(z,\zeta)\right](1-|z|^2)^{\mu+2\kappa-2}$$

$$+\sum_{\lambda=0}^{n-2\kappa}\frac{\lambda}{(\kappa-1)!\kappa!}$$

$$\times\left[\frac{\binom{n-2-\kappa-\lambda}{\kappa-2}}{\binom{n-2}{\kappa}}+\frac{\binom{n-1-\kappa-\lambda}{\kappa-1}}{\binom{n-2}{\kappa-1}}\right]\left[g_{\lambda+1}(z,\zeta)-g_\lambda(z,\zeta)\right](1-|z|^2)^{n-2}.$$

This is (26).

(iii) **Proof of Lemma 7** Again $\rho = 0$ is no restriction. It holds

$$\partial_{\bar z}D_{\kappa\kappa} = \sum_{\mu=1}^{n-1-2\kappa}\sum_{\lambda=1}^{\mu}\frac{1}{\kappa!(\kappa+1)!}\frac{\binom{\mu+\kappa-\lambda-1}{\kappa-1}}{\binom{\mu+2\kappa}{\kappa+1}}$$

$$\times\left[\frac{\lambda\zeta}{(1-\zeta)^{\lambda+1}}(1-|z|^2)^{\mu+2\kappa}-(\mu+2\kappa)zg_1(z,\zeta)(1-|z|^2)^{\mu+2\kappa-1}\right],$$

and

$$\partial_z\partial_{\bar z}D_{\kappa\kappa} = \sum_{\mu=1}^{n-1-2\kappa}\sum_{\lambda=1}^{\mu}\frac{\mu+2\kappa}{\kappa!(\kappa+1)!}\frac{\binom{\mu+\kappa-\lambda-1}{\kappa-1}}{\binom{\mu+2\kappa}{\kappa+1}}$$

$$\times\left[\frac{-\lambda\zeta}{(1-\zeta)^{\lambda+1}}(1-|z|^2)^{\mu+2\kappa-1}-g_\lambda(z,\zeta)(1-|z|^2)^{\mu+2\kappa-1}\right.$$

$$-\frac{\lambda z\bar\zeta}{(1-z)^{\lambda+1}}(1-|z|^2)^{\mu+2\kappa-1}$$

$$\left.+(\mu+2\kappa-1)|z|^2g_\lambda(z,\zeta)(1-|z|^2)^{\mu+2\kappa-2}\right].$$

Using for $2\le\mu$

$$\frac{1}{\binom{\mu+2\kappa-1}{\kappa-2}}-\frac{\nu}{\kappa\binom{\mu+2\kappa-1}{\kappa}}=\frac{1}{\binom{\mu+2\kappa-1}{\kappa}}$$

and

$$\frac{\binom{\mu+\kappa-\lambda}{\kappa-1}}{\binom{\mu+2\kappa-1}{\kappa-1}}-\frac{\binom{\mu+\kappa-\lambda-1}{\kappa-1}}{\binom{\mu+2\kappa-2}{\kappa-1}}-\frac{\lambda-1}{\kappa}\frac{\binom{\mu+\kappa-\lambda-1}{\kappa-2}}{\binom{\mu+2\kappa-1}{\kappa}}=\frac{\binom{\mu+\kappa-\lambda-1}{\kappa-2}}{\binom{\mu+2\kappa-1}{\kappa}}$$

then

$$\partial_z\partial_{\bar z} D_{\kappa\kappa} = \sum_{\mu=0}^{n-1-2\kappa}\sum_{\lambda=1}^{\mu+1}\frac{1}{(\kappa-1)!\kappa!}\frac{\binom{\mu+\kappa-\lambda-1}{\kappa-2}}{\binom{\mu+2\kappa-1}{\kappa}}g_\lambda(z,\zeta)(1-|z|^2)^{\mu+2\kappa-1}$$

$$-\sum_{\lambda=1}^{n-2\kappa}\frac{1}{(\kappa-1)!\kappa!}\frac{\binom{n-1-\kappa-\lambda}{\kappa-2}}{\binom{\mu-2}{\kappa-1}}g_\lambda(z,\zeta)(1-|z|^2)^{n-2}$$

$$=\sum_{\mu=1}^{n-1-2\kappa}\sum_{\lambda=1}^{\mu}\frac{1}{(\kappa-1)!\kappa!}\frac{\binom{\mu+\kappa-\lambda-2}{\kappa-2}}{\binom{\mu+2\kappa-2}{\kappa}}g_\lambda(z,\zeta)(1-|z|^2)^{\mu+2\kappa-2}$$

$$+\sum_{\lambda=1}^{n-2\kappa}\frac{1}{(\kappa-1)!\kappa!}$$

$$\times\left[\frac{\binom{n-\kappa-\lambda-2}{\kappa-2}}{\binom{n-2}{\kappa}}-\frac{\binom{n-1-\kappa-\lambda}{\kappa-1}}{\binom{n-2}{\kappa-1}}\right]g_\lambda(z,\zeta)(1-|z|^2)^{n-2}.$$

This is (27).

References

1. H. Begehr: Complex analytic methods for partial differential equations. An introductory text. World Scientific, Singapore, 1994.
2. H. Begehr: Orthogonal decompositions of the function space $L_2(\bar{D};\mathbb{C})$. J. Reine Angew. Math. 549(2002), 191–219.
3. H. Begehr: Combined integral representations. Advances in Analyis, Proc. 4th Intern. ISAAC Congress, Toronto, 2003, eds. H. Begehr et al., World Scientific, Singapore, 2005, 187–195.
4. H. Begehr: Boundary value problems in complex analysis. I, II, Bol. Asoc. Mat. Venezolana 12 (2005), 65–85; 217–250.
5. H. Begehr: The main theorem of calculus in complex analysis. Ann. EAS, 2005, 184–210.
6. H. Begehr: Six biharmonic Dirichlet problems in complex analysis. Preprint, FU Berlin, 2006.
7. H. Begehr: Biharmonic Green functions. Preprint, FU Berlin.
8. H. Begehr, G. Harutyunyan: Robin boundary value problem for the Cauchy-Riemann operator. Complex Var., Theory Appl. 50 (2005), 1125–1136.
9. H. Begehr, G. Harutyunyan: Robin boundary value problem for the Poisson equation. J. Anal. Appl., 4 (2006), 201–213.
10. H. Begehr, G.N. Hile: A hierarchy of integral operators. Rocky Mountain J. Math. 27(1997), 669–706.
11. H. Begehr, A. Kumar: Boundary value problems for the inhomogeneous polyanalytic equation. I, Analysis 25 (2005), 55–71; II, to appear.
12. H. Begehr, D. Schmersau: The Schwarz problem for polyanalytic functions. ZAA 24 (2005), 341–351.

13. H. Begehr, C.J. Vanegas: Iterated Neumann problem for the higher order Poisson equation. Math. Nachr. 279 (2006), 38–57.
14. H. Begehr, T.N.N. VU, Z.X. Zhang: Polyharmonic Dirichlet problems. Preprint, FU Berlin, 2005.
15. H. Begehr; E. Gaertner: A Dirichlet problem for the inhomogeneous polyharmonic equation in the upper half plane. Preprint, FU Berlin, 2006.
16. E. Gaertner: Basic complex boundary value problems in the upper half plane. Ph.D. thesis, FU Berlin, 2006; http://www.diss.fu-berlin.de/2006/320.

FINELY MEROMORPHIC FUNCTIONS IN CONTOUR-SOLID PROBLEMS

T. ALIYEV AZEROĞLU

Department of mathematics,
Gebze Institute of Technology, Gebze,
41410 Kocaeli, Turkey
aliyev@gyte.edu.tr

P.M. TAMRAZOV

Institute of mathematics of National
Academy of Sciences of Ukraine,
Tereshshenkivska str., 3, 01601,
Kiev, MSP, Ukraine
tamrazov@gyte.edu.tr

We establish contour-solid theorems for finely meromorphic functions taking into account zeroes and the multivalence of functions.

1. Introduction

In[1] the purely fine contour-solid theory for finely holomorphic and finely hypoharmonic functions was established. That theory contains refined, strengthened and extended theorems for the mentioned classes of functions in finely open sets of the complex plane with preservable majorants (from the maximal classes of such majorants for the mentioned function classes). On the other hand, in[2],[3],[4],[5],[6] certain contour-solid theorems for analytic functions from earlier authors' works were extended onto meromorphic functions and strengthened with taking into account zeroes and the multivalence of functions.

In the present work we extend and strengthen some results of[1] related to finely holomorphic functions. The generalization is related to considering finely meromorphic functions instead of finely holomorphic, and strengthening is connected with taking into account zeroes and the multivalence of functions.

We need to recall a number of definitions and notations from[1].

Let $\mathfrak{M}$ be the class of all functions $\mu : (0, +\infty) \to [0, +\infty)$ for each of which the set $I^\mu := \{x : \mu(x) > 0\}$ is connected and the restriction of the function $\log \mu(x)$ to I^μ is concave with respect to $\log x$. Let $\mathfrak{M}^*$ be the class of all $\mu \in \mathfrak{M}$ for which I^μ is non-empty.

For $\mu \in \mathfrak{M}^*$ let us denote by x_-^μ and x_+^μ the left and the right ends of the interval I^μ, respectively. Obviously, $0 \le x_-^\mu \le x_+^\mu \le +\infty$. When $x_-^\mu < x_+^\mu$, the mentioned concavity condition is equivalent to the combination of the following conditions: the function $\log \mu(x)$ is concave with respect to $\log x$ (and therefore continuous) in the interval (x_-^μ, x_+^μ) and lower semicontinuous on I^μ. For $\mu \in \mathfrak{M}$ the limits

$$\mu_0 := \lim_{x\to 0} \frac{\log\ \mu(x)}{\log\ x}, \quad \mu_\infty := \lim_{x\to+\infty} \frac{\log\ \mu(x)}{\log\ x}$$

exist, and we have

$$\mu_0 \ge \mu_\infty, \quad \mu_0 > -\infty, \quad \mu_\infty < +\infty.$$

In particular, if $x_-^\mu > 0$ (analogously, if $x_+^\mu < +\infty$), then $\mu_0 = +\infty$ ($\mu_\infty = -\infty$, respectively). When $\mu_0 < +\infty$, define the integer m_0 by the conditions $m_0 - 1 < \mu_0 \le m_0$, and when $\mu_\infty > -\infty$, define the integer m_∞ by the conditions $m_\infty \le \mu_\infty < m_\infty + 1$.

For every fixed $\alpha \in \mathbb{R}$, $\beta \in (0, +\infty)$, the function $\mu(x) := \beta x^\alpha$ belongs to $\mathfrak{M}^*$, and for it we have $\mu_0 = \mu_\infty = \alpha$. If, moreover, α is an integer, then $m_0 = m_\infty = \alpha$.

Let $\widehat{\mathbb{C}}$ be the compact Riemann sphere.

We refer to[7], [8], [9] concerning the fine topology and related notions such as thinnes, the fine boundary and the fine closure of a set, fine limits of functions, fine superior and fine inferior limits of functions, finely hypoharmonic, finely hyperharmonic, finely subharmonic, finely harmonic, finely meromorphic functions, the generalized harmonic measure, the Green's function for a fine domain and so on.

Let $E \subset \widehat{\mathbb{C}}$. Denote by $\widehat{E}$ the standard closure of E in $\widehat{\mathbb{C}}$, and by $\overline{E}$ the standard closure of a set $E \subset \mathbb{C}$ in $\mathbb{C}$. The set of all points $x \in \widehat{\mathbb{C}}$ in which E is not thin is called *the base of the set* E *in* $\widehat{\mathbb{C}}$ and is denoted by $b(E)$. The set $\tilde{E} := E \cup b(E)$ is called *the fine closure of the set E in* $\widehat{\mathbb{C}}$. Clearly, $\tilde{E} \subset \widehat{E}$. Denote by $\widehat{\partial_f} E$ the fine boundary of E in $\widehat{\mathbb{C}}$. Let $\partial_f E := \mathbb{C} \cap \widehat{\partial_f} E$, $(E)_i := E \setminus b(E)$, $(E)_r := E \setminus (E)_i$. Points $x \in (E)_r$ and $x \in (E)_i$ are called *regular* and *irregular* points, respectively, of the set E.

For a set $E \subset \widehat{\mathbb{C}}$ let us denote $\widehat{\mathbb{C}} \setminus E =: FE$, and for a set $E \subset \mathbb{C}$, denote also $\mathbb{C} \setminus E =: CE$.

Let $G \subset \widehat{\mathbb{C}}$ be a finely open set, the set FG is non-polar and $z \in G$. A set $Q \subset FG$ will be called *nearly negligible relative to* G if for every finely connected component T of G the set $Q \cap \partial_f T$ contains no compact subset K of the harmonic measure $\omega_z^T(K) > 0$ at some (and therefore at any) point $z \in T$. This requirement is equivalent to the following alternative: either FG is polar (and then Q is also polar), or for every finely connected component T of G both $\partial_f T$ is non-polar and Q is a set of inner harmonic measure zero relative to T and any point $z \in T$.

If a set $E \subset FG$ is such that for every finely connected component T of G it contains no compact subset $K \subset \partial_f T$ of logarithmic capacity $\operatorname{Cap} K > 0$, then E is nearly negligible relative to G.

In particular, any set $E \subset FG$ of inner logarithmic capacity zero is nearly negligible relative to G.

For any finely open set $G \subset \mathbb{C}$ with a non-polar complement $\mathbb{C} \setminus G$, and $w \in \widehat{\mathbb{C}}$, $\zeta \in G, w \neq \zeta$ there exists Green's function $g_G(w, \zeta)$.

Let D be a fine domain in $\widehat{\mathbb{C}}$, i. e. a finely open, finely connected set. Then $\tilde{D} = \widehat{D}$. In particular, then D is finely separable from a point $z \in \widehat{\mathbb{C}}$ if and only if it is separable from z in the standard topology. So in such a situation we may speak of separability not specifying in what sense.

Let G be a finely open set in $\widehat{\mathbb{C}}$, and $z \in \widehat{\partial_f G}$. Given any functions $u : G \to [-\infty, +\infty]$ and $h : G \to \widehat{\mathbb{C}}$ we introduce the following notations for fine superior limits of functions:

$$\text{fine lim sup}_{\zeta \to z,\ \zeta \in G}\ u(\zeta) =: \breve{u}_{G,f}(z),$$

$$\text{fine lim sup}_{\zeta \to z,\ \zeta \in G}\ |h(\zeta)| =: \bar{h}_{G,f}(z).$$

Let G be a finely open set in $\mathbb{C}$ and $a \in CG$ be a fixed point. For a function $h : G \to \widehat{\mathbb{C}}$ let us denote:

$$h_{a,G,f} := \begin{cases} \text{fine lim sup}_{\zeta \to a,\ \zeta \in G} \frac{\log|f(\zeta)|}{|\log|\zeta - z||} & \text{when} \quad a \in \partial_f G, \\ 0 & \text{when} \quad a \notin \partial_f G. \end{cases}$$

$$h_{\infty,G,f} :== \begin{cases} \text{fine lim sup}_{\zeta \to \infty,\ \zeta \in G} \frac{\log|f(\zeta)|}{\log|\zeta|} & \text{when} \quad \infty \in \widehat{\partial_f G}, \\ 0 & \text{when} \quad \infty \notin \widehat{\partial_f G}, \end{cases}$$

If the function f is finely meromorphic in G then we denote by $k(f, w)$ the multiplicity of its value $f(w)$ at the point $w \in G$.

Given functions $p : X \to \widehat{\mathbb{C}}$ and $q : X \to \widehat{\mathbb{C}}$ on a set $X \subset \widehat{\mathbb{C}}$ and a finely limit point w for X, we use the following notations. If there exist a finite number $l \geq 0$ and a fine neighbourhood U of w for which $|p(z)| \leq l\ |q(z)| \quad \forall\ z \in\ X \cap U$, then we write

$$p(z) = \text{fine } O(q(z)) \quad (z \to w, z \in X),$$

and if for every $\epsilon > 0$ there exists a fine neighbourhood U of w for which $|p(z)| \leq \epsilon\ |q(z)| \quad \forall\ z \in\ X \cap U$, then we write

$$p(z) = \text{fine } o(q(z)) \quad (z \to w, z \in X).$$

$$p(z) = \text{fine } o(q(z)) \quad (z \to w, z \in X).$$

Let $G \subset \mathbb{C}$ be a finely open set, $h : G \to \widehat{\mathbb{C}}$ be a finely meromorphic function, and $\mu \in \mathfrak{M}$. Consider the following conditions:

(A, ∞) $\quad \infty \in b(CG)$ and for every finely connected component T of G with $\infty \in b(T)$ there holds $h_{\infty,T,f} < +\infty$;

(B, ∞) $\quad \infty \notin b(CG), \quad \mu_\infty > -\infty$ and

$$h(\zeta) \; = \; \text{fine } o(|\zeta|^{m_\infty+1}) \; (\zeta \to \infty, \; \zeta \in G); \tag{1}$$

(B_0, ∞) $\quad \infty \notin b(CG), \quad \mu_\infty \geq 0$ and (1) is true.

If $z \in \mathbb{C}$ is a fixed point, then we consider also the following conditions:

(A, z) $\quad z \in b(CG)$ and for every finely connected component T of G with $z \in b(T)$ there holds $h_{z,T,f} < +\infty$;

(B, z) $\quad z \notin b(CG), \quad \mu_0 < +\infty$ and

$$h(\zeta) \; = \; \text{fine } o(|\zeta - z|^{m_0-1}) \; (\zeta \to z, \; \zeta \in G); \tag{2}$$

(B_1, z) $\quad z \notin b(CG), \quad \mu_0 \leq 1$ and (2) is true.

We will use the following commutative rule for possible indefinite expressions: $0 \cdot (\pm\infty) = 0, \; -\infty + \infty = -\infty$.

2. Main Results

We prove the following results.

Theorem 2.1. *Let $a \in \mathbb{C}$ be a fixed point; $G \subset \mathbb{C}\backslash\{a\}$ be a finely open set with non-polar fine boundary $\partial_f G$; $\mu \in \mathfrak{M}$; $h : G \to \widehat{\mathbb{C}}$ be a finely meromorphic function for which*

$$\bar{h}_{G,f}(z) \leqslant \mu(|\, z - a\,|) \qquad \forall z \in \partial_f G\backslash\{a\}.$$

Denote $z_1 := a, \; z_2 := \infty$ and suppose that for each $s = 1, 2$ (independently from each other) one of the conditions (A,z_s) or (B,z_s) is satisfied. Let $\mathfrak{P}$ be the set of all poles of h in G. Then

$$|\, h(\zeta)\,| \exp\left[-\sum_{p \in \mathfrak{P}} g_G(p,\zeta) \cdot k(h,p)\right] \leqslant$$

$$\mu(|\, \zeta - a\,|) \; \exp\left[-\sum_{w : h(w)=0} g_G(w,\zeta) \cdot k(h,w)\right] \quad \forall \zeta \in G\backslash\, \mathfrak{P}. \tag{3}$$

Theorem 2.2. *Let $a \in \mathbb{C}$ be a fixed point; $G \subset \mathbb{C}\backslash\{a\}$ be a finely open set with non-polar fine boundary $\partial_f G$; Q be a set contained in FG and containing the points a and ∞; $\mu \in \mathfrak{M}$; $h : G \to \widehat{\mathbb{C}}$ be a finely meromorphic function. Denote $z_1 := a, \; z_2 := \infty$ and suppose that for each $s = 1, 2$ (independently from each other) one of the conditions (A,z_s) or (B,z_s) is satisfied. Let $\mathfrak{P}$ be the set of all poles of h in G.*

Denote

$$\lambda(x) := \log \mu(x) \qquad \forall x > 0,$$

$$u(\zeta) := \log |h(\zeta)| \; - \sum_{p \in \mathfrak{P}} g_G(p,\zeta) \cdot k(h,p) \quad \zeta \in G.$$

Suppose that for every finely connected component T of G the following conditions are satisfied: Q is nearly negligible relative to T and

$$\breve{u}_{T,f}(z) < \infty \qquad \forall z \in \partial_f T \backslash \{a\};$$

$$\breve{u}_{T,f}(z) \leq \lambda(|z-a|) \qquad \forall z \in \partial_f T \backslash Q.$$

Then (3) holds.

Notice that in formulations of Theorems 2, 8.1, 8.2 of[1] the following obvious correction should be done: Q must be assumed to be contained in FG instead of $\widehat{\partial_f G}$.

Theorem 2.3. *Let $G \subset \mathbb{C}$ be a finely open set; $\mu \in \mathfrak{M}$; $h : \tilde{G} \cap \mathbb{C} \to \widehat{\mathbb{C}}$ be a function finely meromorphic in G and satisfying the condition*

$$| \, h(z) - h(\zeta) \, | \leqslant \mu(| \, z - \zeta \, |) \quad \forall z, \zeta \in \partial_f G, \; \zeta \neq z \tag{4}$$

Let $\mathfrak{P}$ be the set of all poles of h in $\tilde{G} \cap \mathbb{C}$. Let one of the conditions (A, ∞) or (B_0, ∞) be satisfied for the restriction of h onto G (instead of h). Suppose also that for every finely connected component T of G the restriction of h onto $(\tilde{T} \cap \mathbb{C})\backslash \mathfrak{P}$ is finely continuous.

Under these assumptions we have

$$| \, h(\zeta) - h(z) \, | \exp \left[- \sum_{p \in \mathfrak{P}} (g_G(p,\zeta) + g_G(p,z)) \cdot k(h,p) \right] \leq \mu(| \, \zeta - z \, |) \times$$

$$\exp \left[- \sum_{\substack{w \in G \\ h(w) = h(\zeta)}} g_G(w,z) \cdot k(h,w) \right] \quad \forall z \in (\partial_f G)_r \quad \forall \zeta \in (\tilde{G} \cap \mathbb{C}) \backslash \mathfrak{P}, \; \zeta \neq z. \tag{5}$$

Theorem 2.4. *Let G, μ, h satisfy all assumptions of Theorem 3. Additionally suppose that $z_0 \in (CG)_i \cup G$ is a fixed point, $\mu_0 < +\infty$ and*

$$| \, h(\zeta) - h(z_0) \, | = \text{ fine } o(|\zeta - z_0|^{m_0 - 1}) \qquad (\zeta \to z_0, \; \zeta \in G \backslash \mathfrak{P}).$$

Then

$$| \, h(\zeta) - h(z_0) \, | \exp \left[- \sum_{p \in \mathfrak{P}} (g_G(p,\zeta) + g_G(p,z_0)) \cdot k(h,p) \right] \leq \mu(| \, \zeta - z_0 \, |) \times$$

$$\exp \left[- \sum_{\substack{w \in G \\ h(w) = h(\zeta)}} g_G(w,z_0) \cdot k(h,w) \right] \qquad \forall \zeta \in \tilde{G} \backslash \mathfrak{P}, \; \zeta \neq z_0, \infty.$$

Theorem 2.5. *Let G, μ, h satisfy all assumptions of Theorem 3. Additionally suppose that $\mu_0 < 1$. Then*

$$| h(\zeta) - h(z) | \exp \left[- \sum_{p \in \mathfrak{P}} (g_G(p,\zeta) + g_G(p,z)) \cdot k(h,p) \right] \leq \mu(| \zeta - z |) \times$$

$$\exp \left[- \sum_{\substack{w \in G \\ h(w)=h(\zeta)}} g_G(w,z) \cdot k(h,w) \right] \quad \forall z, \zeta \in (\tilde{G} \cap \mathbb{C}) \backslash \mathfrak{P}, \ \zeta \neq z. \tag{6}$$

Theorem 2.6. *Let G, μ, h satisfy all assumptions of Theorem 3. If $\infty \in \tilde{G}$, additionally suppose that h can be extended to ∞ in such a way that the extended function $\tilde{h}$ satisfies the following hypothesis: the restriction $\tilde{h}|_{\tilde{T}}$ of $\tilde{h}$ to $\tilde{T}$ is finely continuous (and finite) at ∞ for every finely connected component T of G with $\infty \in \tilde{T}$. Under these assumptions the inequality (13) is true.*

3. Proof of Theorems

We need to recall a number of notations from[1] .

Let L be the class of all functions $\lambda : (0, +\infty) \to [-\infty, +\infty)$ for each of which the set $I_\lambda := \{x : \lambda(x) > -\infty\}$ is connected and the restriction of λ to I_λ is concave with respect to $\log x$. Let L^* be the class of all $\lambda \in L$ for which I_λ is non-empty. For $\lambda \in L^*$ let us denote by x_λ^- and x_λ^+, respectively, the left and the right ends of the interval I_λ. Obviously, $0 \leq x_\lambda^- \leq x_\lambda^+ \leq +\infty$. When $\lambda(\cdot)$ runs through the classes L or L^*, the function $\exp \lambda(\cdot)$ runs through the classes $\mathfrak{M}$ or $\mathfrak{M}^*$, respectively. When $x_\lambda^- < x_\lambda^+$, the mentioned concavity condition is equivalent to the combination of the following conditions: the function $\lambda(x)$ is concave with respect to $\log x$ (and therefore continuous) in the interval $(x_\lambda^-, x_\lambda^+)$ and lower semicontinuous on I_λ. For $\lambda \in L$ the limits

$$\lambda^0 := \lim_{x \to 0} \frac{\lambda(x)}{\log x}, \quad \lambda^\infty := \lim_{x \to +\infty} \frac{\lambda(x)}{\log x}$$

exist, and we have

$$\lambda^0 \geq \lambda^\infty, \quad \lambda^0 > -\infty, \quad \lambda^\infty < +\infty.$$

For every fixed $\alpha \in \mathbb{R}$, $\theta \in \mathbb{R}$ the function $\lambda(x) := \alpha \log x + \theta$ belongs to L^*, and for it we have $\lambda_0 = \lambda_\infty = \alpha$.

Let $a \in CG$ be a fixed point.

Let G be a finely open set in $\widehat{\mathbb{C}}$ and $a \in CG$ be a fixed point. For any function $u : G \to [-\infty, +\infty]$ let us denote:

$$u^a_{G,f} := \begin{cases} \text{fine lim sup}_{\zeta \to a,\ \zeta \in G} \frac{u(\zeta)}{|\log|\zeta - z||} & \text{when} \quad a \in \partial_f G, \\ 0 & \text{when} \quad a \notin \partial_f G. \end{cases}$$

$$u^\infty_{G,f} :== \begin{cases} \text{fine lim sup}_{\zeta \to \infty,\ \zeta \in G} \frac{u(\zeta)}{\log|\zeta|} & \text{when} \ \infty \in \widehat{\partial_f} G, \\ 0 & \text{when} \ \infty \notin \widehat{\partial_f} G\,. \end{cases}$$

Let $\lambda \in L$. Consider the following conditions:

(A$'$, ∞) $\quad \infty \in b(CG)$ and for every finely connected component T of G with $\infty \in b(T)$ there holds $\quad u^\infty_{T,f} < +\,\infty$;

(B$'$, ∞) $\quad \infty \notin b(CG)$ and there exist a constant $t \in \mathbb{R}$ and a fine neighbourhood V of ∞ for which

$$u(\zeta) \le \lambda(|\zeta - a|) + t \quad \forall \zeta \in G \cap V;$$

(A$'$, a) $\quad a \in b(CG)$ and for every finely connected component T of G with $a \in b(T)$ there holds $\quad u^a_{T,f} < +\,\infty$;

(B$'$, a) $\quad a \notin b(CG)$ and there exist a constant $t \in \mathbb{R}$ and a fine neighbourhood V of a for which

$$u(\zeta) \le \lambda(|\zeta - a|) + t \quad \forall \zeta \in G \cap V.$$

The validity of Theorem 1 follows from Theorem 2.

In [1, pp. 355-357] the following statement was actually proved, although not formulated.

Lemma 3.1. *Let $a \in \mathbb{C}$ be a fixed point; $G \backslash \{a\}$ be a finely open set; $\mu \in \mathfrak{M}$; $h : \tilde{G} \to \widehat{\mathbb{C}}$ be a finely meromorphic function.*

Denote $z_1 := a$, $z_2 := \infty$ and suppose that for G, h, μ and some $s = 1, 2$ one of the conditios (A, z_s) or (B, z_s) is satisfied. Then for G, the functions $\log|h(\zeta)|$, $\lambda(x) := \log \mu(x)$ and the same s the corresponding of the conditions $(A^{'}, z_s)$ or (B', z_s) is fulfilled, and z_s is finely distincted from the set $\mathfrak{P}$ of all poles of h in G and h is finely holomorphically extendable into z_s.

Notice that the presenting in[1] fine holomorphicity of h in the context of the given statement was used only in a fine neighbourhoods of the points a and ∞, not in the whole G.

Proof. [Proof of Theorem 2] Applying Lemma 7 to the set G and the function h we obtain, that for G, the functions $u(\zeta) := \log|h(\zeta)|$, $\lambda(x) := \log \mu(x)$, and the same s the corresponding of the conditions $(A^{'}, z_s)$ or (B', z_s) is fulfilled.

We omit from consideration those connected components of the set G at each point of which the series $\sum_{p\in\mathfrak{P}\cap G(\zeta)} g_G(p,\zeta)\cdot k(h,p)$diverges. Therefore, without loss of generality, we shal assume that it converges on the set $G\backslash\mathfrak{P}$. Assume now that Z is an arbitrary finite set of zeroes of the function h in G. We consider in G the function

$$v_Z(\zeta) := u(\zeta) + \sum_{z\in Z} g_G(w,\zeta)k(h,w), \qquad \zeta \in G.$$

The set $Q_Z := Q \cup (FG)_i$ is nearly negligible relative to every finely connected component T of G. The function $v_Z(\zeta)$ is finely hypoharmonic in G and

$$\check{u}_{G,f}(z) = (\check{v}_Z)_{G,f}(z) \qquad \forall z \in (\partial_f G)\backslash Q_Z.$$

So, the function $v_Z(\zeta)$ fulfils all assumptations of Theorem 8.2 from[1] , and we obtain

$$v_Z(\zeta) \leq \lambda(|\zeta - a|)$$

and finally

$$| h(\zeta) | \exp\left[-\sum_{p\in\mathfrak{P}} g_G(p,\zeta)\cdot k(h,p)\right] \leqslant$$

$$\mu(| \zeta - a |) \ \exp\left[-\sum_{z\in Z} g_G(w,\zeta)\cdot k(h,w)\right] \quad \forall\zeta \in G\backslash\ \mathfrak{P}.$$

Since Z is an arbitrary finite subset of the set of zeroes of h, from here we obtain (8). ■

Proof. [Proof of Theorem 3] Fix an arbitrary point $w \in (\partial_f G)_r$. Let us check that we may apply Theorem 2 with w as a , and the function

$$\phi(\zeta) := h(\zeta) - h(w) \qquad (\zeta \in G)$$

instead of $h(\zeta)$. Indeed from (4) we get (3) with w as a , and ϕ as h. Since for every finely connected component T of G the function $h|_{(\tilde{T}\cap\mathbb{C})\backslash\ \mathfrak{P}}$ is finely continuous and $w \in (\partial_f G)_r \subset b(CG)$, therefore h satisfies the condition (A, w), and the same is true for ϕ.

One of the conditions (A,∞) or (B_0,∞) is assumed to hold for h, and it implies the same condition for ϕ.

Hence, Theorem 2 is applicable in the situation mentioned above. Since $w \in (\partial_f G)_r$, we have $G \neq \mathbb{C}\setminus\{w\}$ and the exceptional case of Theorem 1 is impossible in the situation under consideration. Thus we get (3) for ϕ instead of h, and the estimate of (5) with w as z.

Because of the choice of w, it gives us (5).

Theorem 3 is proved. ■

The proofs of Theorems 4-6 use our Theorem 3 and are similar to the proofs of Theorems 4-6 in[1] .

References

1. P.M. Tamrazov, Finely holomorphic and finely subharmonic functions in contour-solid problems. Annales Academia Scientiarum Fennica. Mathematica, Vol. 26, 2001, 325-360.
2. P.M. Tamrazov and T. Aliyev, A contour-solid problem for meromorphic functions taking into account zeroes and nonunivalence. Dokl. Akad. Nauk SSSR 288 (1986), No. 2, 304-308 (Russian); English translations in Soviet Math. Dokl. Vol. 33(1986), No. 3, 670-674.
3. T.G. Aliyev and P.M. Tamrazov, *A contour–solid problem for meromorphic functions, taking into account nonunivalence*, Ukrainian Math. Zh. 39 (1987), 683–690 (Russian).
4. T. Aliyev and P. Tamrazov, Contour-solid theorems for meromorphic functions taking multivalence into account. In: Progress in Analysis. Proceedings of 3-d International ISAAC Congress, Vol. I, 469-476. World Scientific, Singapore, 2003.
5. T.G. Aliyev, *Irregular boundary zeroes of analytic functions in contour-solid theorems*, Bull. Soc. Sci. Lettres Lodz 49 Ser. Rech. Deform. 28(1999), 45-53.
6. T.G. Aliyev, *Contour-solid theorems*, Preprint 85.84, Math. Inst. Acad. Sci. Ukrainian SSR, Kiev, 1985 (Russian).
7. M. Brelot, On Topologies and Boundaries in Potential Theory. -Lecture Notes in Math. 175, Springer-verlag,1971.
8. B.Fuglede, Finely harmonic Functions.-Lecture Notes in Math. 175, Springer-verlag,1971.
9. B.Fuglede, Finely holomorphic Functions.A survey.- Rev. Roumaine Math. Pures Appl. 33, 1988, 283-295.

A GENERALIZED SCHWARTZ LEMMA AT THE BOUNDARY

T. ALIYEV AZEROĞLU

University Department, University Name,
Gebze Institute of Technology,Department of Mathematics,
Faculty of Sciences Gebze, Kocaeli, TURKEY 141-41400
aliyev@gyte.edu.tr

BÜLENT N. ÖRNEK

University Department, University Name,
Gebze Institute of Technology,Department of Mathematics,
Faculty of Sciences Gebze, Kocaeli, TURKEY 141-41400
nornek@gyte.edu.tr

In the recent decade in publications of Burns and Krantz [1, pp. 662-664], Dov Chelst[2] and a few other authors papers there was studied the uniqueness portion of the Schwarz Lemma namely:

Let the function $\phi : D \to D$ be a holomorphic in the unit disk D. From property

$$\phi(z) = z + O(|z-1|^4), \qquad z \to 1, \qquad z \in D, \tag{1}$$

at the point 1 follows that $\phi(z) = z, \quad z \in D$.

In the present work this circle of problems are generalized in the following aspects:

a) more general majaront are taken instead of the habitual power majorant in (1);

b) in (1) condition $z \to 1$ was usually treated as tending from inside of the unit disk; we take the behavior of ϕ at the point 1 along the boundary instead;

c) instead of $f(z) = z$ in the right-hand side of (1), the finite Blaschke product is taken(see[2]).

Let $\mathfrak{M}$ be the class of functions $\mu : (0, +\infty) \to (0, +\infty)$ for each of which $\log \mu(x)$ is concave with respect to $\log x$ (see[3]).

For each function $\mu \in \mathfrak{M}$ the limit

$$\mu_0 := \lim_{x \to 0} \frac{\log \mu(x)}{\log x},$$

exists, and

$$-\infty < \mu_0 \leqslant +\infty\,.$$

Let $\mathfrak{N}$ be the class of sets with zero capacities.

Let $d(z,E)$ and $U(z,r)$ be the distance of the point z from the set E and open disk with centre z and radius r respectively.

We prove the following results.

Theorem 1. Let $\mu \in \mathfrak{M}$, $\mu_0 > 3$; $\phi : D \to D$ be a holomorphic function that is continuous on $\overline{D} \cap U(1,\delta_0)$ for some $\delta_0 > 0$ and it satisfies condition:

$$\phi(\zeta) = \zeta + O\left(\mu(|\zeta - 1|\right) \qquad \zeta \in \partial D,\ \zeta \to 1. \tag{2}$$

Then $\phi(\zeta) = \zeta$ on D.

Theorem 2. Let $Q \in \mathfrak{N}$, $\mu \in \mathfrak{M}$; $\mu_0 > 3$; $\phi : D \to D$ be a holomorphic function, which satisfies the following condition

$$\limsup_{\zeta \to z,\ \zeta \in D} |\phi(\zeta) - \zeta| \leqslant \mu(|z-1|) \qquad \forall z \in (\partial D \backslash Q) \cap U(1,\delta_0),$$

for some $\delta_0 > 0$.Then $\phi(\zeta) = \zeta$ on D.

Theorem 3. Let $f : D \to D$ be a finite Blaschke product which equals $\tau \in \partial D$ on a finite set $A_f \subset \partial D$ and $\phi : D \to D$ be a holomorphic function that is continuous on $\overline{D} \cap \{z : d(z, A_f) < \delta_0\}$ for some $\delta_0 > 0$; $\mu^1, \mu^2 \in \mathfrak{M}$, $\mu_0^1 > 3, \mu_0^2 > 2$. Suppose the following conditions are satisfied

(i) for a given $\gamma_0 \in A_f$

$$\phi(z) = f(z) + O(\mu^1(|z - \gamma_0|)), \qquad z \in \partial D,\ z \to \gamma_0,$$

(ii) for all $\forall \gamma \in A_f - \{\gamma_0\}$

$$\phi(z) = f(z) + O(\mu^2(|z - \gamma|)), \qquad z \in \partial D,\ \ z \to \gamma.$$

Then $\phi(\zeta) = f(\zeta)$ on D.

Theorem 4. Let $f : D \to D$ be a finite Blaschke product which equals $\tau \in \partial D$ on a finite set $A_f \subset \partial D$ and $\phi : D \to D$ be a holomorphic function; $Q \in \mathfrak{N}$,$\mu^1, \mu^2 \in \mathfrak{M}$, $\mu_0^1 > 3, \mu_0^2 > 2$. Let the following conditions are satisfied (i) for a given $\gamma_0 \in A_f$

$$\limsup_{\zeta \to z,\ \zeta \in D} |\phi(\zeta) - f(\zeta)| = O(\mu^1(|z - \gamma_0|)), \qquad z \in (\partial D \backslash Q),\ \ z \to \gamma_0, \tag{3}$$

(ii) for all

$$\forall \gamma \in A_f \backslash \{\gamma_0\} \ \limsup_{\zeta \to z,\ \zeta \in D} |\phi(\zeta) - f(\zeta)| = O(\mu^2(|z - \gamma_0|)), \qquad z \in (\partial D \backslash Q),\ z \to \gamma_0. \tag{4}$$

Then $\phi(\zeta) = f(\zeta)$ on D.

For proof of our theorems we use the following propositions:

Lemma 1. (Hopf's lemma on the disc) [4, p.34] Let u be a nonconstant real-valued harmonic function in D. Let $\gamma \in \partial D$ be such that:

(i) u is continuous at γ;

(ii) $u(\gamma) \geq u(z)$ for all z$\in \partial D$.

Then the outer normal derivative $\frac{\partial u}{\partial v}$ of u at γ, if it exists, satisfies the strict inequality

$$\frac{\partial u}{\partial v}(\gamma) > 0.$$

Remark 1.[2] Let $f : D \to D$ be analytic function and have a continous limit at some $\gamma \in \partial D$, and let $f(\gamma) = 1$. Then f is not $o(z-\gamma)$.

Proof. of Theorem 1. Let the assumptions of Theorem 1 is satisfied. By (2) there exists a number $C_1 > 0$ such that

$$|\phi(\zeta) - \zeta| \leqslant C_1 \mu(|\zeta - 1|), \quad \forall \zeta \in \partial D \cap U(1, \delta_0).$$

Let us denote k and C_2 as follows;

$$k := \sup_{\substack{|\zeta - 1| = \delta_0 \\ \zeta \in G}} |\phi(\zeta) - \zeta|,$$

$$C_2 := \max\left\{ \frac{k}{\mu(\delta_0)}, C_1 \right\}.$$

It can be easily seen that for all boundary points of the set $D \cap U(1, \delta_0)$ the inequality

$$|\phi(\zeta) - \zeta| \leqslant C_2 \mu(|\zeta - 1|)$$

is satisfied.

Applying Theorem 3 of [3] (see also[5], [6] and[7]) to the set $D \cap U(1, \delta_0)$ and the function $\phi(\zeta) - \zeta$ one receives

$$|\phi(\zeta) - \zeta| \leqslant C_2 \mu(|\zeta - 1|) \qquad \forall \zeta \in D \cap U(1, \delta_0). \tag{5}$$

From $\mu_0 > 3$ follows that there exist some positive constants ε and $\sigma < \min(\delta_0, 1)$ such that

$$\frac{\log \mu(x)}{\log(x)} \geqslant 3 + \epsilon \quad \forall x \in (0, \sigma)$$

and

$$\log \mu(x) \leqslant (3 + \epsilon) \log x, \ \forall x \in (0, \sigma).$$

In other words

$$\mu(x) \leqslant x^{3+\epsilon}, \ \forall x \in (0, \sigma). \tag{6}$$

Now from (5) and (6) we obtain

$$|\phi(\zeta) - \zeta| \leqslant C_2 \left(|\zeta - 1|^{3+\epsilon} \right). \quad \forall \zeta \in D \cap U(1, \sigma). \tag{7}$$

We introduce the harmonic function

$$v(\zeta) = \Re\left(\frac{1+\phi(\zeta)}{1-\phi(\zeta)}\right) - \Re\left(\frac{1+\zeta}{1-\zeta}\right).$$

The function $\frac{1+\phi(\zeta)}{1-\phi(\zeta)}$ maps the disk D to the right half plane and hence the first term of v is nonnegative, the second term is zero on $\partial D\backslash\{1\}$. Consequently,

$$\liminf_{\zeta \to z, \zeta \in D} v(\zeta) \geq 0 \quad \forall z \in \partial D\backslash\{1\}.$$

By means of elementary transform; one can receive that

$$v(\zeta) = \Re\left(\frac{2(\phi(\zeta)-\zeta)}{(1-\phi(\zeta))(1-\zeta)}\right).$$

From Remark follows that, the deminator of the last fraction can decrease no more quickly than $O(|\zeta-1|^2)$ at the point 1. According (7), the numerator approaches 0 as $O\left(|\zeta-1|^{3+\epsilon}\right)$. Thus, $v(\zeta)$ is $O\left(|\zeta-1|^{1+\epsilon}\right)$ in some neighborhood of 1. We obtain from the maximum principle (see, for example [8, p.48]) either

$$v(\zeta) > 0 \qquad \forall \zeta \in D,$$

or $v \equiv 0$. If v is not a constant, it takes minimum at the point $\zeta = 1$, and it is $O\left(|\zeta-1|^{1+\epsilon}\right)$ there, as well. This contradicts Hopf's lemma statement. Consequently $v \equiv 0$. This implies $\phi(\zeta) \equiv \zeta$. ■

Theorem 2 is generalization of Theorem 1, and its proof is similar to the proof of Theorem 1 given above. Theorem 4 is generalization of Theorem 3.

Proof. of Theorem 4. Due to (3), there exist numbers $C_3 > 0$, $\delta_0 \in (0,1)$ such that

$$\limsup_{\zeta\to z,\ \zeta\in D} |\phi(\zeta) - f(\zeta)| \leq C_3\mu^1(|z-\gamma_0|), \qquad \forall\, z \in (\partial D\backslash Q),\ \ |z-\gamma_0| \leq \delta_0$$

Let us denote k and C_4 as follows;

$$k := \sup_{\substack{|\zeta-\gamma_0| = \delta_0 \\ \zeta \in D}} |\phi(\zeta) - f(\zeta)|,$$

$$C_4 := \max\left\{\frac{k}{\mu^1(\delta_0)}, C_3\right\}.$$

It can be easily seen that for all boundary points of the set $D \cap U(\gamma_0, \delta_0)$ the inequality

$$\limsup_{\zeta\to z,\ \zeta\in D} |\phi(\zeta) - f(\zeta)| \leqslant C_4\mu^1\left(|z-\gamma_0|\right)$$

is satisfied.

Applying Theorem 3 of [3] to the set $D \cap U(\gamma_0, \delta_0)$ and to the function $\phi(\zeta) - f(\zeta)$ yields

$$|\phi(\zeta) - f(\zeta)| \leqslant C_4\mu^1\left(|\zeta-\gamma_0|\right) \qquad \forall \zeta \in D \cap U(\gamma_0, \delta_0). \tag{8}$$

From $\mu_0 > 3$ follows that there are some positive constants ε and $\sigma < \min(\delta_0, 1)$ such that inequality (6) is satisfied . Combining (8) and (6), we obtain

$$|\phi(\zeta) - f(\zeta)| \leqslant C_4 \left(|\zeta - \gamma_0|^{3+\epsilon} \right). \quad \forall \zeta \in D \cap U(\gamma_0, \sigma). \tag{9}$$

Analogously, for any point $\gamma \in A_f \backslash \{\gamma_0\}$, from conditions $\mu_0^2 > 2$ and (4) we have

$$|\phi(\zeta) - f(\zeta)| \leqslant C_5 \left(|\zeta - \gamma|^{2+\epsilon} \right). \quad \forall \zeta \in D \cap U(\gamma, \sigma_5) \tag{10}$$

with some constants C_5 and σ_5.

We introduce the harmonic function

$$v(\zeta) := \mathfrak{R} \left(\frac{1 + \phi(z)}{1 - \phi(z)} \right) - \mathfrak{R} \left(\frac{1 + f(z)}{1 - f(z)} \right).$$

Since a finite Blaschke product f is analytic function on $\overline{D}$ and that $|f| = 1$ on ∂D, we have that the second term of v is zero on $\partial D - A_f$. The first term of v is nonnegative. Consequently, when taking limits to any boundary point in $(\partial D \backslash Q) - A_f$, one always obtains a nonnegative(possible infinite) value.

Now, let's examine behavior of the function v at points of set A_f. Let us represent $v(\zeta)$ in the form

$$v(\zeta) = \mathfrak{R} \left(\frac{2(\phi(\zeta) - f(\zeta))}{(1 - \phi(\zeta))(1 - f(\zeta))} \right).$$

Let's take any point $\gamma \in A_f \backslash \gamma_0$. Due to (10), the numerator of the last fraction approaches zero as $O\left(|\zeta - 1|^{2+\epsilon}\right)$. From Remark the dominator can decrease no more quickly than $O(|\zeta - 1|^2)$. Thus, v must have a liminf at γ. As in the proof of Theorem 1, from (9) we obtain $v(\zeta)$ is $O\left(|\zeta - \gamma_0|^{1+\epsilon}\right)$ in some neighborhood of γ_0. Thus, from the Phragmen-Lindelof principle (see, for example [8, p.232]) we obtain, either

$$v(\zeta) > 0 \qquad \forall \zeta \in D,$$

or $v \equiv 0$. If v is not constant, it takes minimum at the point $\zeta = 1$ and is $O\left(|\zeta - 1|^{1+\epsilon}\right)$ there, as well. This contradicts Hopf's lemma statement. Consequently $v \equiv 0$. But $v \equiv 0$ implies $\phi \equiv f$. ■

References

1. Daniel M. Burns and Steven G. Krantz, *Rigidity of homorphic mapping and a new schwarz lemma at the boundary*, Journal of the AMS **7** (1994), no. 3, p.661-676.
2. Dov Chelst, *A generalized Schwarz lemma at the boundary*, Proceedings of the American Mathematical Society, Volume **129**, Number 11, Pages 3275- 3278
3. P.M. Tamrazov, *Holomorphic functions and mappings in the contour–solid problem*, Dokl. Akad. Nauk SSSR **279** (1984), no. 1, English transl. in Soviet Math. Dokl. Vol. **30** (1984), no. 3.

4. D. Gilbarg and N.S. Trudinger, *Eliptic partial differential equations of second order*, 2nd ed., Springer-Verlag, Berlin, 1983.
5. P.M. Tamrazov and T.G. Aliyev, *A "contour–solid" problem for meromorphic functions, taking into account zeros and non-univalence*, Dokl. Akad. Nauk SSSR **228** (1986), no. 2; English transl. in Soviet Math. Dokl. Vol. **33** (1986), no. 3.
6. T.G. Aliyev and P.M. Tamrazov, *A contour–solid problem for meromorphic functions, taking into account nonunivalence*, Ukrainian Math. Zh. **39** (1987), 683–690 (Russian).
7. T.G. Aliyev, *Irregular boundary zeros of analytic functions in contour-solid theorems*, Bull. Soc. Sci. Lettres Lodz **49** Ser. Rech. Deform. **28**(1999), 45-53.
8. W.K. Hayman and P.B. Kennedy, *Subharmonic Function*, Academic Press, London, 1976

SINGULAR PERTURBATION PROBLEMS IN POTENTIAL THEORY AND APPLICATIONS

MASSIMO LANZA DE CRISTOFORIS
Dipartimento di Matematica Pura ed Applicata,
Università di Padova
Via Trieste 63,
35121 Padova, Italy.
mldc@math.unipd.it

Keywords: Nonlinear boundary value problem, singularly perturbed domain, Laplace operator, real analytic continuation in Banach space.

1. Introduction.

This paper is devoted to present applications of a functional analytic approach to the analysis of nonlinear boundary value problems on a domain with a small hole.

We consider a bounded open connected subset $\mathbb{I}^o$ of $\mathbb{R}^n$ of class $C^{1,\alpha}$ with unit outward normal ν^o and with $0 \in \mathbb{I}^o$, and with $\mathbb{R}^n \setminus \mathrm{cl}\mathbb{I}^o$ connected, and we assume that the boundary value problem

$$\begin{cases} \Delta u = 0 & \text{in } \mathbb{I}^o, \\ B^o\left(t, u(t), \frac{\partial u}{\partial \nu^o}(t)\right) = 0 & \forall t \in \partial\mathbb{I}^o, \end{cases} \tag{1}$$

admits at least a solution $\tilde{u} \in C^{1,\alpha}(\mathrm{cl}\mathbb{I}^o)$. Here α is a given element of $]0,1[$, $\mathrm{cl}\mathbb{I}^o$ denotes the closure of $\mathbb{I}^o$, B^o is a given function of $\partial\mathbb{I}^o \times \mathbb{R}^2$ to $\mathbb{R}$, and $C^{m,\alpha}(\mathrm{cl}\mathbb{I}^o)$ denotes the space of real-valued m times continuously differentiable functions of $\mathrm{cl}\mathbb{I}^o$ to $\mathbb{R}$ with α-Hölder continuous m-th order derivatives, for each natural $m \geq 0$.

Problems such as (1) are known to have a solution for a large class of B^o's. Here we mention the work of Carleman [5], Nakamori and Suyama [21], Klingelhöfer [7], [8], [9], [10], Efendiev, Schmitz and Wendland [6]. We also mention the contribution of Begehr and Hsiao [3], Begehr and Hsiao [4], Begehr and Hile [2] for problems in the plane.

Next we make a hole in the domain $\mathbb{I}^o$. To do so, we consider another bounded open connected subset $\mathbb{I}^i$ of $\mathbb{R}^n$ of class $C^{1,\alpha}$ with unit outward normal ν^i and with $0 \in \mathbb{I}^i$, and with $\mathbb{R}^n \setminus \mathrm{cl}\mathbb{I}^i$ connected, and we consider the annular domain

$$\mathbb{A}(\epsilon) \equiv \mathbb{I}^o \setminus \epsilon\mathrm{cl}\mathbb{I}^i .$$

for all $|\epsilon| \leq \epsilon_0$, with $\epsilon_0 > 0$ sufficiently small so that $\epsilon\mathrm{cl}\mathbb{I}^i \subseteq \mathbb{I}^o$. Obviously,

$$\partial\mathbb{A}(\epsilon) = \epsilon\partial\mathbb{I}^i \cup \partial\mathbb{I}^o .$$

Next we introduce the boundary conditions on the boundary $\epsilon\partial\mathbb{I}^i$ of the hole $\epsilon\mathrm{cl}\mathbb{I}^i$ by assigning a function B^i of $]0,\epsilon_0[\times\partial\mathbb{I}^i\times\mathbb{R}^2$ to $\mathbb{R}$, and we consider the boundary value problem

$$\begin{cases} \Delta u = 0 & \text{in } \mathbb{A}(\epsilon), \\ B^i\left(\epsilon, t\epsilon^{-1}, u(t), \frac{\partial u}{\partial \nu^i}(t)\right) = 0 & \forall t \in \epsilon\partial\mathbb{I}^i, \\ B^o\left(t, u(t), \frac{\partial u}{\partial \nu^o}(t)\right) = 0 & \forall t \in \partial\mathbb{I}^o, \end{cases} \tag{2}$$

for each $\epsilon \in]0,\epsilon_0[$. Under suitable conditions on the data, our first goal is to identify a family of solutions $\{u(\epsilon,\cdot)\}_{\epsilon\in]0,\epsilon_0[}$ of (2) which in a sense approaches the given solution $\tilde{u}$ when ϵ approaches 0^+. One could try to solve (1) or (2) by representing the solutions as a sum of a simple and of a double layer potential, and by solving the corresponding integral equations. Here we shall assume that one can do so by using a simple layer potential, and we illustrate in the next section an example where one can actually do so at least for a certain class of nonlinear boundary conditions. We denote by Υ_n the function of $]0,+\infty[$ to $\mathbb{R}$ defined by

$$\Upsilon_n(r) \equiv \begin{cases} \frac{1}{s_n}\log r & \forall r \in]0,+\infty[, \text{ if } n = 2, \\ \frac{1}{(2-n)s_n} r^{2-n} & \forall r \in]0,+\infty[, \text{ if } n > 2, \end{cases} \tag{3}$$

where s_n denotes the $(n-1)$ dimensional measure of $\partial\mathbb{B}_n$. We denote by S_n the function of $\mathbb{R}^n\setminus\{0\}$ to $\mathbb{R}$ defined by $S_n(\xi)\equiv\Upsilon_n(|\xi|)$ for all $\xi\in\mathbb{R}^n\setminus\{0\}$. S_n is well-known to be the fundamental solution of the Laplace operator. Then we consider solutions of (1) in the form

$$\tilde{u}(t) = \int_{\partial\mathbb{I}^o} S_n(t-s)\tilde{\tau}(s)\,d\sigma_s \qquad \forall t \in \mathrm{cl}\mathbb{I}^o,$$

and of (2) in the form

$$u(t) = \int_{\partial\mathbb{I}^o} S_n(t-s)\tau(s)\,d\sigma_s + \int_{\epsilon\partial\mathbb{I}^i} S_n(t-s)\tau(s)\,d\sigma_s \qquad \forall t \in \mathrm{cl}\mathbb{A}(\epsilon),$$

where $\tilde{\tau}$, τ are unknown densities (or moments). Since ϵ is a parameter in our problem, we find convenient to get rid of its presence in the domain of integration of the simple layer. It turns out that a good way to do so is to introduce the function $\tilde{\rho}\in C^{0,\alpha}(\partial\mathbb{I}^o)$ and the pair of functions $(\eta,\rho)\in C^{0,\alpha}(\partial\mathbb{I}^i)\times C^{0,\alpha}(\partial\mathbb{I}^o)$ by setting

$$\begin{aligned} &\tilde{\tau}(s) \equiv \tilde{\rho}(s) \quad \text{if } s \in \partial\mathbb{I}^o, \\ &\tau(s) \equiv \rho(s) \quad \text{if } s \in \partial\mathbb{I}^o, \quad \tau(s) \equiv \epsilon^{1-n}\eta(s/\epsilon) \quad \text{if } s \in \epsilon\partial\mathbb{I}^i. \end{aligned}$$

Then we have

$$\tilde{u}(t) = \int_{\partial\mathbb{I}^o} S_n(t-s)\tilde{\rho}(s)\,d\sigma_s \qquad \forall t \in \mathrm{cl}\mathbb{I}^o,$$

and

$$u(t) = \int_{\partial\mathbb{I}^o} S_n(t-s)\rho(s)\,d\sigma_s + \int_{\partial\mathbb{I}^i} S_n(t-s\epsilon)\eta(s)\,d\sigma_s \qquad \forall t \in \partial\mathbb{A}(\epsilon)\,, \tag{4}$$

$$\begin{aligned}\frac{\partial u}{\partial \nu^i}(\epsilon t)\epsilon^{n-1} = \epsilon^{n-1} &\int_{\partial\mathbb{I}^o} \nu^i(t)\cdot D_t S_n(t\epsilon - s)\rho(s)\,d\sigma_s \\ &+\frac{1}{2}\eta(t) + \int_{\partial\mathbb{I}^i} \nu^i(t)\cdot D_t S_n(t-s)\eta(s)\,d\sigma_s \qquad \forall t\in\partial\mathbb{I}^i\,,\end{aligned} \tag{5}$$

$$\begin{aligned}\frac{\partial u}{\partial \nu^o}(t) = -\frac{1}{2}\rho(t) + &\int_{\partial\mathbb{I}^o} \nu^o(t)\cdot D_t S_n(t-s)\rho(s)\,d\sigma_s \\ &+\int_{\partial\mathbb{I}^i} \nu^o(t)\cdot D_t S_n(t-s\epsilon)\eta(s)\,d\sigma_s \qquad \forall t\in\partial\mathbb{I}^o\,,\end{aligned} \tag{6}$$

where D_t denotes the gradient with respect to the t-variable. By plugging (4)–(6) into the boundary conditions of (2), one obtains a system of two integral equations in the unknowns η, ρ, which depends on the parameter $\epsilon \in]0,\epsilon_0[$. We rewrite such a system into the abstract form

$$M[\epsilon,\eta,\rho] = 0\,, \tag{7}$$

where M is a suitable nonlinear map of $]0,\epsilon_0[\times C^{0,\alpha}(\partial\mathbb{I}^i)\times C^{0,\alpha}(\partial\mathbb{I}^o)$ to $C^{0,\alpha}(\partial\mathbb{I}^i)\times C^{0,\alpha}(\partial\mathbb{I}^o)$. In many instances M may be seen to be the restriction of some regular map (still denoted M) of $]-\epsilon_0,\epsilon_0[\times C^{0,\alpha}(\partial\mathbb{I}^i)\times C^{0,\alpha}(\partial\mathbb{I}^o)$ to $C^{0,\alpha}(\partial\mathbb{I}^i)\times C^{0,\alpha}(\partial\mathbb{I}^o)$, and equation $M[0,0,\rho]=0$ corresponds to the integral equation relative to problem (1) for $\tilde{\rho}$. Under suitable conditions on B^o, B^i, we may be able to solve locally around $(0,0,\tilde{\rho})$ equation (7), and to prove that the set of solutions of (7) is the graph of a nonlinear operator $\epsilon\mapsto(E[\epsilon],R[\epsilon])$ of $]-\epsilon_0,\epsilon_0[$ to $C^{0,\alpha}(\partial\mathbb{I}^i)\times C^{0,\alpha}(\partial\mathbb{I}^o)$ for a possibly smaller ϵ_0. If B^o, B^i are real analytic, one would expect that $(E[\cdot],R[\cdot])$ is also real analytic. Then by setting

$$u(\epsilon,t) \equiv \int_{\partial\mathbb{I}^o} S_n(t-s)R[\epsilon](s)\,d\sigma_s + \int_{\partial\mathbb{I}^i} S_n(t-s\epsilon)E[\epsilon](s)\,d\sigma_s \qquad \forall t\in \mathrm{cl}\mathbb{A}(\epsilon)\,, \tag{8}$$

one obtains a solution of (2) which converges in the $C^{1,\alpha}$-norm on the compact subsets of $\mathrm{cl}\mathbb{I}^o\setminus\{0\}$ to $\tilde{u}$ as ϵ tends to 0.

Once the family of solutions $\{u(\epsilon,\cdot)\}_{\epsilon\in]0,\epsilon_0[}$ is established, two questions appear as natural.

(j) Let $\tilde{\Omega}$ be a bounded open subset of $\mathbb{I}^o\setminus\{0\}$ such that $0\notin \mathrm{cl}\tilde{\Omega}$. What can be said on the map $]0,\epsilon_0[\ni\epsilon\mapsto u(\epsilon,t)_{|\mathrm{cl}\tilde{\Omega}}\in C^{1,\alpha}(\mathrm{cl}\tilde{\Omega})$ around $\epsilon=0$?

(jj) What can be said on the map $]0,\epsilon_0[\ni\epsilon\mapsto \mathcal{E}[\epsilon]\equiv\int_{\mathbb{A}(\epsilon)}|Du(\epsilon,t)|^2\,dt\in\mathbb{R}$ around $\epsilon=0$?

Problems of this type with linear boundary conditions have long been investigated with the methods of Asymptotic Analysis, which aims at giving complete asymptotic expansions of the solutions in terms of the parameter ϵ. It is perhaps difficult to provide a complete list of the contributions. Here, we mention the work of Kozlov, Maz'ya and Movchan [11], Maz'ya, Nazarov and Plamenewskii [19],

Kühnau [12], Movchan [20], Ozawa [22], Ward and Keller [27]. For nonlinear problems on domains with small holes far less seems to be known, we mention the results which concern the existence of a limiting value of the solutions or of their energy integral as the holes degenerate to points, as those of J. Ball [1], Sivaloganathan, Spector and Tilakraj [23], and the literature of homogenization theory. We also mention the computation of the expansions in the case of quasilinear equations of Ward, Henshaw and Keller [25], Ward and Keller [26], Titcombe and Ward [24].

The goal of asymptotic analysis for problem (2) would be to write asymptotic expansions for the maps in (j), (jj). Thus for example an expansion of the form

$$\mathcal{E}[\epsilon] = \sum_{j=0}^{r} a_j \epsilon^j + o(\epsilon^r) \qquad \text{as } \epsilon \to 0^+ \,, \tag{9}$$

for suitable coefficients a_j. Our goal is instead to represent $u(\epsilon,\cdot)_{|\mathrm{cl}\tilde{\Omega}}$ and $\mathcal{E}[\epsilon]$ for $\epsilon \in]0,\epsilon_0[$ by means of

(a) real analytic maps defined on a whole neighborhood of $\epsilon = 0$;
(b) possibly singular at $\epsilon = 0$, but known functions of ϵ (such as ϵ^{-1}, $\log \epsilon$, etc. ...).

We observe that our approach does have its advantages. Indeed, if for example we can prove that there exists a real analytic real valued function $\mathcal{F}(\cdot)$ defined in a whole neighborhood of 0 such that $\mathcal{F}[\epsilon] = \mathcal{E}[\epsilon]$ for $\epsilon \in]0,\epsilon_0[$, then we know that an asymptotic expansion such as (9) for all r would necessarily generate a convergent series $\sum_{j=0}^{\infty} a_j \epsilon^j$, and that the sum of such a series would be $\mathcal{E}[\epsilon]$ for $\epsilon > 0$.

Now let $\tilde{\Omega}$ be as in (j). Under conditions in which $(E[\cdot], R[\cdot])$ is real analytic, the map $U_{\tilde{\Omega}}$ of $]-\epsilon_0,\epsilon_0[$ to $C^{1,\alpha}(\mathrm{cl}\tilde{\Omega})$ defined by

$$U_{\tilde{\Omega}}[\epsilon] \equiv \int_{\partial\mathbb{I}^o} S_n(t-s) R[\epsilon](s)\, d\sigma_s + \int_{\partial\mathbb{I}^i} S_n(t-s\epsilon) E[\epsilon](s)\, d\sigma_s \qquad \forall t \in \mathrm{cl}\tilde{\Omega} \,,$$

for all $\epsilon \in]-\epsilon_0,\epsilon_0[$ offers a real analytic continuation for the map $\epsilon \mapsto u(\epsilon,\cdot)_{|\mathrm{cl}\tilde{\Omega}}$, which is defined only for $\epsilon \in]0,\epsilon_0[$. Thus in this case the answer for question (j) is that

$$u(\epsilon,\cdot)_{|\mathrm{cl}\tilde{\Omega}} = U_{\tilde{\Omega}}[\epsilon](\cdot) \qquad \forall \epsilon \in]0,\epsilon_0[\,. \tag{10}$$

In order to analyze question (jj), we write

$$\mathcal{E}[\epsilon] = -\int_{\epsilon\partial\mathbb{I}^i} u(\epsilon,t) \frac{\partial}{\partial \nu^i(t)} u(\epsilon,t)\, d\sigma_t + \int_{\partial\mathbb{I}^o} u(\epsilon,t) \frac{\partial}{\partial \nu^o(t)} u(\epsilon,t)\, d\sigma_t \,,$$

and we note that

$$\int_{\epsilon\partial\mathbb{I}^i} u(\epsilon,t)\frac{\partial}{\partial\nu^i(t)}u(\epsilon,t)\,d\sigma_t \tag{11}$$
$$= \int_{\partial\mathbb{I}^i}\left\{\int_{\partial\mathbb{I}^o} S_n(t\epsilon - s)R[\epsilon](s)\,d\sigma_s + \int_{\partial\mathbb{I}^i} S_n(t\epsilon - s\epsilon)E[\epsilon](s)\,d\sigma_s\right\}$$
$$\cdot\left\{\epsilon^{n-1}\int_{\partial\mathbb{I}^o}\nu^i(t)\cdot D_t S_n(t\epsilon - s)R[\epsilon](s)\,d\sigma_s\right.$$
$$\left.+\frac{1}{2}E[\epsilon](t) + \int_{\partial\mathbb{I}^i}\nu^i(t)\cdot D_t S_n(t-s)E[\epsilon](s)\,d\sigma_s\right\}d\sigma_t\,.$$

Now we note that for $n > 2$, we have $S_n(\epsilon(t-s)) = (2-n)s_n\Upsilon_n(\epsilon)S_n(t-s)$, and that $S_2(\epsilon(t-s)) = \Upsilon_2(\epsilon) + S_2(t-s)$. Hence, we can prove that there exist two real analytic operators $\mathcal{F}_1$, $\mathcal{F}_2$ of $]-\epsilon_0,\epsilon_0[$ to $\mathbb{R}$ such that

$$-\int_{\epsilon\partial\mathbb{I}^i} u(\epsilon,t)\frac{\partial}{\partial\nu^i(t)}u(\epsilon,t)\,d\sigma_t = \mathcal{F}_1[\epsilon] + \Upsilon_n(\epsilon)\mathcal{F}_2[\epsilon] \qquad \forall\epsilon\in]0,\epsilon_0[\,, \tag{12}$$

$\mathcal{F}_2[0] = 0$. Next we note that

$$\int_{\partial\mathbb{I}^o} u(\epsilon,t)\frac{\partial}{\partial\nu^o(t)}u(\epsilon,t)\,d\sigma_t \tag{13}$$
$$= \int_{\partial\mathbb{I}^o}\left\{\int_{\partial\mathbb{I}^o} S_n(t-s)R[\epsilon](s)\,d\sigma_s + \int_{\partial\mathbb{I}^i} S_n(t - s\epsilon)E[\epsilon](s)\,d\sigma_s\right\}$$
$$\cdot\left\{-\frac{1}{2}R[\epsilon](t) + \int_{\partial\mathbb{I}^o}\nu^o(t)\cdot D_t S_n(t-s)R[\epsilon](s)\,d\sigma_s\right.$$
$$\left.+\int_{\partial\mathbb{I}^i}\nu^o(t)\cdot D_t S_n(t-s\epsilon)E[\epsilon](s)\,d\sigma_s\right\}d\sigma_t\,.$$

Hence, one can prove that there exists a real analytic operator $\mathcal{F}_3$ of $]-\epsilon_0,\epsilon_0[$ to $\mathbb{R}$ such that

$$\int_{\partial\mathbb{I}^o} u(\epsilon,t)\frac{\partial}{\partial\nu^o(t)}u(\epsilon,t)\,d\sigma_t = \mathcal{F}_3[\epsilon] \qquad \forall\epsilon\in]0,\epsilon_0[\,. \tag{14}$$

By (12) and (14) we conclude that

$$\mathcal{E}[\epsilon] = (\mathcal{F}_1[\epsilon] + \mathcal{F}_3[\epsilon]) + \Upsilon_n[\epsilon]\mathcal{F}_2[\epsilon] \qquad \forall\epsilon\in]0,\epsilon_0[\,, \tag{15}$$

an equality which answers question (jj). We note that our conclusions (10), (15) are relative to the various simplifications we have made so far to carry on the elementary presentation of this introduction, such as that the solutions of (1) and of (2) can be represented as simple layers, and that those boundary value problems are solavable exactly when the corresponding integral equations are solvable, and that one can solve locally equation (7) and obtain an implicitly defined operator $(E[\cdot], R[\cdot])$. All such circumstances are not always present. Hence, under different circumstances, formulas (10), (15) may have a different form. This is the case for example when both B^o, B^i correspond to the linear Dirichlet boundary conditions, as shown in [13], [16], [17], [18].

One could extend the ideas exposed above in order to consider also pertubations of $\partial\mathbb{I}^i$, $\partial\mathbb{I}^o$ by representing $\partial\mathbb{I}^i$, $\partial\mathbb{I}^o$ by global parametrizations say ϕ^i, ϕ^o defined for example on the unit sphere $\partial\mathbb{B}_n$ of $\mathbb{R}^n$, and analyze the corresponding singular perturbation problem in the complex of variables $(\epsilon, \phi^i, \phi^o)$ considered as a point in the Banach space $\mathbb{R} \times \left(C^{1,\alpha}(\partial\mathbb{I}^i)\right)^n \times \left(C^{1,\alpha}(\partial\mathbb{I}^o)\right)^n$, as done for linear problems in [16] – [18] and for problems related to the Riemann mapping in [13], [14]. In this case the difficulties would increase. Thus for example the analysis of an equation such as (7) is more complicated and the presentation is necessarily longer, and the corresponding treatment is not illustrated here.

2. A concrete case: the case of the nonlinear Robin boundary conditions

The material of this section is entirely based on the paper [15], where the special case $B^i\left(\epsilon, t\epsilon^{-1}, u(t), \frac{\partial u}{\partial \nu^i}(t)\right) = -\frac{\partial u}{\partial \nu^i}(t)$, $B^o\left(t, u(t), \frac{\partial u}{\partial \nu^o}(t)\right) = \frac{\partial u}{\partial \nu^o}(t) - G^o(t, u(t))$ has been considered. Here G^o is a continuous map of $\partial\mathbb{I}^o \times \mathbb{R}$ to $\mathbb{R}$. We first have the following Theorem, which asserts the existence and *local* uniqueness of the family of solutions $u(\epsilon, \cdot)$.

Theorem 1. Let $\alpha \in]0,1[$. Let $\mathbb{I}^i$, $\mathbb{I}^o$ be bounded open connected subsets of $\mathbb{R}^n$ of class $C^{1,\alpha}$. Let $\mathbb{R}^n \setminus \mathrm{cl}\mathbb{I}^i$, $\mathbb{R}^n \setminus \mathrm{cl}\mathbb{I}^o$ be connected. Let $0 \in \mathbb{I}^i$, $0 \in \mathbb{I}^o$. Let $G^o \in C^0(\partial\mathbb{I}^o \times \mathbb{R})$ be such that the operator T_{G^o} of $C^{0,\alpha}(\partial\mathbb{I}^o)$ to $C^0(\partial\mathbb{I}^o)$ defined by

$$T_{G^o}[v](t) \equiv G^o(t, v(t)) \qquad \forall t \in \partial\mathbb{I}^o,\ \forall v \in C^{0,\alpha}(\partial\mathbb{I}^o)$$

map $C^{0,\alpha}(\partial\mathbb{I}^o)$ real analytically to itself and map bounded sets of $C^{0,\alpha}(\partial\mathbb{I}^o)$ to bounded sets of $C^{0,\alpha}(\partial\mathbb{I}^o)$. Assume that there exists a solution $\tilde{u} \in C^{1,\alpha}(\mathrm{cl}\mathbb{I}^o)$ of problem

$$\begin{cases} \Delta u = 0 & \text{in } \mathbb{I}^o, \\ \frac{\partial u}{\partial \nu^o}(t) = G^o(t, u(t)) & \forall t \in \partial\mathbb{I}^o, \end{cases} \tag{1}$$

such that

$$G^o_u(t, \tilde{u}(t)) \leq 0 \qquad \forall t \in \partial\mathbb{I}^o, \qquad \int_{\partial\mathbb{I}^o} G^o_u(s, \tilde{u}(s))\, d\sigma_s \neq 0\,. \tag{2}$$

Then the following statements hold.

(i) There exists $\epsilon' \in]0, \epsilon_0[$ and a family $\{u(\epsilon, \cdot)\}_{\epsilon \in]0,\epsilon'[}$ of solutions $u(\epsilon, \cdot) \in C^{1,\alpha}(\mathrm{cl}\mathbb{A}(\epsilon))$ of problem

$$\begin{cases} \Delta u = 0 & \text{in } \mathbb{A}(\epsilon), \\ -\frac{\partial u}{\partial \nu^i} = 0 & \text{on } \epsilon\partial\mathbb{I}^i, \\ \frac{\partial u}{\partial \nu^o}(t) = G^o(t, u(t)) & \forall t \in \partial\mathbb{I}^o, \end{cases} \tag{3}$$

such that $\lim_{\epsilon \to 0} u(\epsilon, \cdot)_{|\mathrm{cl}\tilde{\Omega}} = \tilde{u}_{|\mathrm{cl}\tilde{\Omega}}(\cdot)$ in $C^{1,\alpha}(\mathrm{cl}\tilde{\Omega})$ for all bounded open subsets $\tilde{\Omega}$ of $\mathbb{I}^o \setminus \{0\}$ such that $0 \notin \mathrm{cl}\tilde{\Omega}$.

(ii) If $\{\epsilon_j\}_{j\in\mathbb{N}}$ is a sequence of $]0,+\infty[$ converging to 0 and if $\{u_j\}_{j\in\mathbb{N}}$ is a sequence of functions such that

$$u_j \in C^{1,\alpha}(\mathrm{cl}\mathbb{A}(\epsilon_j)),$$
$$u_j \text{ solves (3) for } \epsilon = \epsilon_j,$$
$$\lim_{j\to\infty} u_j = \tilde{u} \text{ in } C^{1,\alpha}(\mathrm{cl}\tilde{\Omega}) \text{ for all bounded open subsets } \tilde{\Omega} \text{ of } \mathbb{I}^o \setminus \{0\} \text{ such that } 0 \notin \mathrm{cl}\tilde{\Omega},$$

then there exists $j_0 \in \mathbb{N}$ such that $u_j(\cdot) = u(\epsilon_j, \cdot)$ for all $j \geq j_0$.

Then we answer questions (j), (jj) of section 1 by means of the following.

Theorem 2. Let the assumptions of Theorem 2 hold. Then the following statements hold.

(i) Let $\tilde{\Omega}$ be a bounded open subset of $\mathbb{I}^o \setminus \{0\}$ such that $0 \notin \mathrm{cl}\tilde{\Omega}$. Then there exists $\epsilon'' \in]0, \epsilon'[$ and a real analytic operator $U_{\tilde{\Omega}}$ of $]-\epsilon'', \epsilon''[$ to $C^{1,\alpha}(\mathrm{cl}\tilde{\Omega})$ such that $\mathrm{cl}\tilde{\Omega} \subseteq \mathbb{A}(\epsilon)$ for all $\epsilon \in]-\epsilon'', \epsilon''[$ and such that

$$U_{\tilde{\Omega}}[\epsilon](\cdot) = u(\epsilon, \cdot)_{|\mathrm{cl}\tilde{\Omega}} \qquad \text{if } \epsilon \in]0, \epsilon''[\,.$$

Moreover, $U_{\tilde{\Omega}}[0] = \tilde{u}_{|\mathrm{cl}\tilde{\Omega}}$.

(ii) There exists a real analytic operator $\mathcal{F}$ of $]-\epsilon'', \epsilon''[$ to $\mathbb{R}$ such that

$$\mathcal{F}[\epsilon] = \int_{\mathbb{A}(\epsilon)} |D_t u(\epsilon, t)|^2 \, dt \qquad \forall \epsilon \in]0, \epsilon''[\,.$$

Moreover, $\mathcal{F}[0] = \int_{\mathbb{I}^o} |D_t \tilde{u}(t)|^2 \, dt$.

Acknowledgments

The author is indebted to Prof. B. Dittmar and to Prof. E. Wegert for pointing out a number of references concerning the existence of solutions for problems (1), (1), and to Prof. A.B. Movchan, and to Prof. J. Sivaloganathan, and to Prof. M.J. Ward, for pointing out a number of references on nonlinear singular perturbation problems on domains with small holes.

References

1. J.M. Ball, *Discontinuous equilibrium solutions and cavitation in nonlinear elasticity*, Philos. Trans. Roy. Soc. London Ser. A, **306**, (1982), 557–611.
2. H. Begehr and G.N. Hile, *Nonlinear Riemann boundary value problems for a nonlinear elliptic system in the plane*, Math. Z., **179**, (1982), 241–261.
3. H. Begehr and G.C. Hsiao, *Nonlinear boundary value problems for a class of elliptic systems*, Komplexe Analysis und ihre Anwendung auf partielle Differentialgleichungen, Martin-Luther-Univeristät, Halle-Wittenberg, (1980), 90–102.
4. H. Begehr and G.C. Hsiao, *Nonlinear boundary value problems of Riemann-Hilbert type*, Contemporary Mathematics, **11**, (1982), 139–153.
5. T. Carleman, *Über eine nichtlineare Randwertaufgabe bei der Gleichung* $\Delta u = 0$, Math. Z., **9**, (1921), 35–43.

6. M.A. Efendiev, H. Schmitz and W. Wendland, *On some nonlinear potential problems*, Electron. J. Differential Equations, **1999**, (1999), 1–17 .
7. K. Klingelhöfer, *Modified Hammerstein integral equations and nonlinear harmonic boundary value problems*, J. Math. Anal. Appl. **28**, (1969), 77–87.
8. K. Klingelhöfer, *Nonlinear harmonic boundary value problems. I*, Arch. Rational Mech. Anal. 31, (1968)/(1969), 364–371.
9. K. Klingelhöfer, *Nonlinear harmonic boundary value problems. II. Modified Hammerstein integral equations*, J. Math. Anal. Appl. **25**, (1969), 592–606.
10. K. Klingelhöfer, *Über nichtlineare Randwertaufgaben der Potentialtheorie*, Mitt. Math. Sem. Giessen Heft **76**, (1967), 1–70.
11. V.A. Kozlov, V.G. Maz'ya and A.B. Movchan, *Asymptotic analysis of fields in multi-structures*, Oxford Mathematical Monographs, The Clarendon Press, Oxford university Press, New York, 1999.
12. R. Kühnau, *Die Kapazität dünner Kondensatoren*, Math. Nachr., **203**, (1999), 125–130.
13. M. Lanza de Cristoforis, *Asymptotic behaviour of the conformal representation of a Jordan domain with a small hole, and relative capacity*, in 'Complex Analysis and Dynamical Systems', edited by M. Agranovsky, L. Karp, D. Shoikhet, and L. Zalcman, Contemp. Math., **364**, (2004) pp. 155-167.
14. M. Lanza de Cristoforis, *Asymptotic behaviour of the conformal representation of a Jordan domain with a small hole in Schauder spaces*, Computat. Methods Funct. Theory, **2**, 2002, pp. 1–27.
15. M. Lanza de Cristoforis, *Asymptotic behaviour of the solutions of a nonlinear Robin problem for the Laplace operator in a domain with a small hole. A functional analytic approach*, submitted, 2006.
16. M. Lanza de Cristoforis, *Asymptotic behaviour of the solutions of the Dirichlet problem for the Laplace operator in a domain with a small hole. A functional analytic approach*, submitted, 2004.
17. M. Lanza de Cristoforis, *A singular domain perturbation problem for the Poisson equation*, submitted, 2005.
18. M. Lanza de Cristoforis, *A singular perturbation Dirichlet boundary value problem for harmonic functions on a domain with a small hole*, Proceedings of the 12th International Conference on Finite or Infinite Dimensional Complex Analysis and Applications, Tokyo July 27–31 2004, edited by H. Kazama, M. Morimoto, C. Yang, Kyushu University Press, (2005), 205–212.
19. V.G. Mazya, S.A. Nazarov and B.A. Plamenewskii, *Asymptotic theory of elliptic boundary value problems in singularly perturbed domains*, **I, II**, (translation of the original in German published by Akademie Verlag 1991), Oper. Theory Adv. Appl., **111**, **112**, Birkhäuser Verlag, Basel, 2000.
20. A.B. Movchan, *Contributions of V.G. Maz'ya to analysis of singularly perturbed boundary value problems*, The Maz'ya anniversary collection, 1 (Rostock, 1998), Oper. Theory Adv. Appl., **109**, Birkhäuser, Basel, 1999, pp. 201–212.
21. K. Nakamori and Y. Suyama, *On a nonlinear boundary problem for the equations $\Delta u = 0$ and $\Delta u = f(x, y)$* (Esperanto) Mem. Fac. Sci. Kyûsyû Univ. A., **5**, (1950), 99–106.
22. S. Ozawa, *Electrostatic capacity and eigenvalues of the Laplacian*, J. Fac. Sci. Univ. Tokyo Sect. IA Math., **30**, (1983), 53–62 .
23. J. Sivaloganathan, S.J. Spector and V. Tilakraj, *The convergence of regularized minimizers for cavitation problems in nonlinear elasticity*, SIAM J. Appl. Math., **66**, (2006), 736–757.

24. M.S. Titcombe, M.J. Ward, *Summing logarithmic expansions for elliptic equations in multiply-connected domains with small holes*, Canad. Appl. Math. Quart., **7**, (1999), 313-343.
25. M.J. Ward, W. Henshaw, and J. Keller, *Summing logarithmic expansions for singularly perturbed eigenvalue problems*, SIAM J. Appl. Math., **53**, (1993), pp. 799-828.
26. M.J. Ward, J. Keller, *Nonlinear eigenvalue problems under strong localized perturbations with applications to chemical reactors*, Stud. Appl. Math., **85**, (1991), 1-28.
27. M.J. Ward and J.B. Keller, *Strong localized perturbations of eigenvalue problems*, SIAM J. Appl. Math., **53** (1993), 770–798.

RESIDUES ON A KLEIN SURFACE

ARTURO FERNÁNDEZ ÁRIAS
Dpto. de Matemáticas
Fundamentales Facultad de
Ciencias UNED C/Senda del Rey s/n
Madrid 28040 Spain
afernan@mat.uned.es

JAVIER PÉREZ ALVAREZ
Dpto. de Matemáticas
Fundamentales Facultad de
Ciencias UNED C/Senda del Rey s/n
Madrid 28040 Spain
j1235a@hotmail.com

The reconstruction of a Riemann surface starting from the meromorphic function field K, comes from Dedekind and Weber who developed an algebraic function theory in one variable over an algebraically closed field k. Alling and Greenleaf present a counterpart to this approach starting from a real algebraic curve. From this point of view, the residues theorem is a classical result which depends strongly on the algebraically closed character of the base field. In this paper, via the complex double, we translate this fact to the case where we start from a function field in one variable over $\mathbf{R}$.

Keywords: Klein surface, valuation ring, residue.

1. Klein surface of a real function field

Let K be a field and A be a subring of K; we shall say that A is a valuation ring of K if for each $x \in K$ then $x \in A$ or $x^{-1} \in A$; it holds that A is a local ring: if it has two different maximal ideals p_1 and p_2, then, taking $a \in p_1 \setminus p_2$ and $b \in p_2 \setminus p_1$, we have $a/b \notin A$ and $b/a \notin A$

Definition 1. Let $\mathbf{R}$ be the field of real numbers and K be a finite extension of transcendence degree 1 over $\mathbf{R}$ in which -1 is not a square. We shall call Klein surface of the field K, the set S of valuations rings of K which contain $\mathbf{R}$. In the same way, the set S_c of valuation rings of $K[i]$ which contain $\mathbf{C}$ shall be called the double cover of S.

We shall denote a point P of S by A_p when we intend to indicate the ring that it represents, denoting then by p or m_p its only maximal ideal.

Let us fix an element $z \in K \setminus \mathbf{R}$, let n be the degree of the extension $\mathbf{R}(z) \hookrightarrow K$ and let us denote with O the integral clausure of $\mathbf{R}[z]$ in K, that is, the subring of K formed by the elements which verify a monic polynomial with coefficients in $\mathbf{R}[z]$.

For each element $\lambda = \lambda_1 + i\lambda_2$ in $K[i]$, let us write $\overline{\lambda} = \lambda_1 - i\lambda_2$.

Proposition 1. *O is a noetherian ring.*

Proof. We shall see that O is a free $\mathbf{R}[z]-$module of rank n. If t is a primitive element of the extension $\mathbf{R}(z) \hookrightarrow K$, we can suppose, multiplying the minimal polynomial of t over $\mathbf{R}[z]$ by an adecuate element of $\mathbf{R}[z]$, that $t \in O$. In this way, if $y \in O$, we can write

$$y = a_0 + a_1 t + a_2 t^2 + ... + a_{n-1} t^{n-1}, \ a_i \in \mathbf{R}(z)$$

If $\langle \sigma_i \rangle_{i=1}^n$ are the $\mathbf{R}(z)-$isomorphisms of K in the minimal normal algebraic extension L of K over $\mathbf{R}(z)$, we obtain the n following equations:

$$\sigma_i(y) = a_0 + a_1\sigma_i(t) + a_2\sigma_i(t)^2 + ... + a_{n-1}\sigma_i(t)^{n-1} \quad (1 \leq i < n).$$

In this way $a_j = \frac{A_j}{D}$, where D is the Vandermonde determinant of the system and A_j is a polynomial with integral coefficients in $\sigma_i(t)$, $\sigma_i(y)$. Since D^2 is invariant by $\langle \sigma_i \rangle_{i=1}^n$ and integral over $\mathbf{R}[z]$, we have $D^2 \in \mathbf{R}[z]$. As on the other hand $D^2 a_j = DA_j$ is an integral element of $\mathbf{R}(z)$ over $\mathbf{R}[z]$, we get $D^2 a_j \in \mathbf{R}[z]$, hence $D^2 O \subset \langle 1, ..., t^{n-1} \rangle_{\mathbf{R}[z]} \subset O$, and we are done.

Proposition 2. *Every prime ideal $p \neq 0$ of O is maximal.*

Proof. We shall see that for each maximal ideal $m \subset O$, the local ring O_m has dimension 1; this is a consequence of the dimension theory for local noetherian rings, since the dimension of O_m is the maximun number of parametres in m which are algebraically independent over $\mathbf{R}$. See [2].

In this way, for every point P of S, the quotient ring A_p/p is a finite algebraic extension of $\mathbf{R}$, whereby $A_p/p = \mathbf{R}$ or $A_p/p = \mathbf{C}$. Therefore we shall call degree of the point P of S, the integer $gr(P) = [A_p/p : \mathbf{R}]$.

The free group generated by the points of S shall be called group of divisors on S. In this way, given a divisor $D = \sum n_i p_i$ on S, we shall call degree of D the integer $gr(D) = \sum n_i gr(p_i)$.

Let us define the maps $\sigma : S_c \longrightarrow S_c$, $\pi_2 : S_c \longrightarrow S$, such that to each valuation ring V_q of $K[i]$

$$\sigma(V_q) = \overline{V_q} = \{\overline{\lambda} : \lambda \in V_q\}$$

$$\pi_2(V_q) = V_q \cap K.$$

Definition 2. *Given $f \in K$ and $P \in S$, we shall say that f has a zero of order $n \in \mathbf{N}$ at p if $f \in p^n \setminus p^{n+1}$ in A_p and that f has a pole of order $s \in \mathbf{N}$ at p if $1/f$ has a zero of order s at p. If f is invertible in the local ring A_p then f has neither a zero nor a pole at P.*

From now on we shall denote with $Spec(O)$ the set of prime ideals of O.

It is easy to see that the valuation rings which contain z, contain also O [2] and that those are the local rings O_p for $p \in Spec(O)$. On the other hand if we denote with O' the integral clausure of $\mathbf{R}[1/z]$ in K, it is clear that the valuations rings of K that are not in $Spec(O)$ are those points in $Spec(O')$ which contain $1/z$ in its maximal ideal; since these are the prime ideals of O' which appear in the factorization of $(1/z)O'$, they are a finite set; particularly every element of K has finitely many zeros and poles in S, and in a language that we shall not use again, we can exhibite S as a scheme structure:

$$S = Spec(O) \cup Spec(O').$$

Proposition 3. *The fiber by π_2 of a point p of S has one or two points depending on whether $A_p/p = \mathbf{R}$ or $A_p/p = \mathbf{C}$.*

Proof. See [5].

The following lemma is easy to prove.

Lemma. *Let P be a point of S with $A_p/p = \mathbf{C}$; let t be a parameter in p and A_q, $A_{\sigma(q)}$ the points of the fiber by π_2 of P, then $(t)A_q = m_q$, and $(t)A_{\sigma(q)} = m_{\sigma(q)}$.*

2. Abelian Differentials

We shall call derivation of K over $\mathbf{R}$ any application $D : K \longrightarrow K$ verifying

$$D(a+b) = Da + Db, D(ab) = aDb + dDa$$
$$D\lambda = 0 \text{ if } \lambda \in \mathbf{R}.$$

Let us denote with $Der_R(K)$ the set of derivations thus defined. If D_z is the derivation such that $D_z(z) = 1$ and on any element $t \in K$, $D_z(t)$ is such that

$$g_z(z,t) + g_t(z,t)D_z(t) = 0,$$

where $g(z,t)$ is the irreducible polynomial of t over $\mathbf{R}(z)$, then any other derivation D' verifies $D' = D'(z)D_z$.

Thereby we have

Proposition 4. *$Der_R(K)$ is a R-vector space of dimension 1.*

Let us also define $Der_{\mathbf{C}}(K[i])$ and conclude similarly that this set is a $\mathbf{C}$–vector space of dimension 1.

Proposition 5. *Every derivation D de K over $\mathbf{R}$ extends to a derivation D^* of $K[i]$ over $\mathbf{C}$.*

Proof. It suffices to write $D^*(i) = 0$.

Definition 3. *We shall adpot the following notation*

$$\Omega_{\mathbf{R}}(K) = Hom_{\mathbf{R}}(Der_{\mathbf{R}}(K), \mathbf{R}),$$
$$\Omega_{\mathbf{C}}(K) = Hom_{\mathbf{C}}(Der_{\mathbf{C}}(K[i]), \mathbf{C})$$

and these shall be called the spaces of Abelian differentials on S and S_c respectively.

We define the map $K \longrightarrow \Omega_{\mathbf{R}}(K)$ by $z \mapsto dz$ where dz is such that $dz(D) = Dz$ $(D \in Der_{\mathbf{R}}(K))$. In this way, by Proposition 4, $\Omega_{\mathbf{R}}(K) = Kdz$; thereafter any element of $\Omega_{\mathbf{R}}(K)$ shall be considered as a differential on $K[i]$ via the extension of scalars

$$\Omega_{\mathbf{R}}(K) \to \Omega_{\mathbf{R}}(K)_{\mathbf{R}}\mathbf{C}.$$

Let w be an Abelian differential on S. For every point P of S such that $A_p/p = \mathbf{R}$, let us consider a generator t in p and let us write $w = fdt$ $(f \in K)$. If f has in p a pole of order s then $g = t^s f \in A_p$. There exists $a_{-s} \in \mathbf{R}$ such that $g - a_{-s} \in p$, whereby we can write $g - a_{-s} = t\lambda_0$, $\lambda_0 \in A_p$. Repeating the process with λ_0, and so on, we have the following expansion in a power series in A_p

$$f = \frac{a_{-s}}{t^s} + \ldots + \frac{a_{-1}}{t} + a_0 + a_1 t + a_2 t + \ldots \tag{1}$$

We shall define the residue of w in P as the coeficient a_{-1}; an easy computation [3] proves that this definition does not depend on the parameter chosen in the ring A_p.

If $P' \in S$ is such that $A_{p'}/p' = \mathbf{C}$, let us choose a parameter $z \in p'$ and let us write $w = hdz$, there exists a power of z such that $z^l h \in A_{p'}$. If $Q' \in \pi_2^{-1}(P')$ the previous lemma allows us to write the following series expansion

$$f = \frac{a_{-l}}{z^l} + \ldots + \frac{a_{-1}}{z} + a_0 + a_1 z + a_2 z^2 + \ldots \text{ in } A_{q'}, \tag{2}$$

In that way, as $a_{-1} \in K[i]$, we shall define the residue of w in P', 'the part in K' of a_{-1}.

Definition 4. *Let $w \in \Omega_{\mathbf{R}}(K)$, we shall call residue divisor of w, the divisor*

$$D_r(w) = \sum_{P_i \in S} a_i P_i,$$

where P_i is a pole of w with residue a_i.

Theorem. *The degree of the residues divisor of an Abelian differential on S is* 0.

Proof. First of all we shall check that the definition of residue of an abelian differential w on S at a point P' does not depend on the chosen point of the fiber of P' for the projection $\pi_2 : S_c \to S$. Making use of the previous notations, if $\sigma(Q) \in \pi_2^{-1}(P)\backslash\{Q\}$, and $w = hdz$ on S, we have the following series expansion

$$f = \frac{b_{-l}}{z^l} + \ldots + \frac{b_{-1}}{z} + b_0 + b_1 z + b_2 z^2 + \ldots \text{ in } A_{\sigma(q')}. \tag{3}$$

Let us see that the coefficients of the series expansions (2) and (3) are conjugate (in $K[i]$). In fact, in $A_{q'}$ we have $z^l f = a_{-l} + z\lambda_0$, whereby in $A_{\sigma(q')}$ we shall have

$$z^l f = \sigma(z^l f) = \sigma(a_{-l}) + z\sigma(\lambda_0),$$

whence $b_{-l} = \sigma(a_{-l})$, and following with the same argument, we have $b_i = \sigma(a_i)$, $\forall i \geq -l$.

Now, let us determine the degree of the divisor $D_r(w) = \sum_{P_i \in S} a_i P_i$. If P is a point of S of degree 1 and $Q = \pi_2^{-1}(P)$ it is clear that the reside of w considered as a differential on S_c is real, whereby it coincides with the residue of w in P.

If $P' \in S$ is such that $\pi_2^{-1}(P') = \{Q', \sigma(Q')\}$ in which w has a residue $r \in K$, then

$$2r = res(w, Q') + res(w, \sigma(Q')).$$

In this way, by the classical fact

$$\sum_{Q \in S_c} res(w, Q) = 0,$$

we conlude that

$$gr\left(\sum_{P \in S} P \cdot res(w, P)\right) = 0.$$

References

1. Alling, N. L. and Greenleaf, N. Foundations of the theory of Klein surfaces. Lecture Notes in Mathematics 212 (1971).
2. Atiyah, M. F. and Macdonald, I. G. Introduction to Commutative Algebra. Addison-Wesley (1969).
3. Bujalance, E. Gamboa, J. Gromadzki, G. Automorphism groups of compact bordered Klein surfaces. Lecture Notes in Mathematics, 1439.
4. Lang, S. Introduction to Algebraic and Abelian Functions. Graduate Texts in Mathematics 89 (1972).
5. Iwasawa, K. Algebraic Functions. Translations of Mathematical Monographs, 118 AMS (1993).
6. Gamboa, J. M. Compact Klein surfaces with boundary viewed as real compact smooth algebraic curves. Memorias de la Real Academia de Ciencias de Madrid. Vol XXVII (1991).

COMBINATORIAL THEOREMS OF COMPLEX ANALYSIS

YU.B. ZELINSKII

Institute of Mathematics of the National Academy of Sciences of Ukraine,
3 Tereshchenkivska Street,Kiev, Ukraine
zel@imath.kiev.ua

Combinatorial theorems of real convex and complex analysis are considered. We give some generalizations of classical Caratheodory and Helly theorems.

Keywords: Euclidean space, simplex, convex envelop, Mayer-Victoris exact cohomological sequence, complex line, hyperplane.

We consider classical combinatorial theorems of convex analysis under point view of generalization and application to complex analysis.

The main goal of the first result is to show that under some additional restrictions on the set, the number of points in Caratheodory's theorem which defines convex envelop can be decreased.

Theorem 1. *Let E be subset of n-dimensional Euclidean space $\mathbb{R}^n$ which consists of not more than n connected components. Then any point of convex envelop of E can be presented as convex combination of not more than n points of E.*

Proof. Let we suppose that the theorem is not true. Than there exists a point x in convex envelop of E, which by Caratheodory's theorem can be represented as convex combination of $n+1$ points $x_1, x_2, \ldots, x_{n+1}$ of E.

Those points defined n-dimensional simplex Δ. According to our supposition the point x can not be convex combination of any n points of E. We consider all $(n-1)$-dimensional planes across point x and every subset of $(n-1)$ vertex of simplex Δ. Every hyperplane of this family contains convex cone with apex x generated by $(n-2)$-dimensional side of Δ. Let we consider symmetrical relatively to point x cones to cones defined above. It is easy to see that no one of the last cones can not contain points of the set E. If the last is not true and let x_1 be point in one of cited cone, let C_1, then point x can be represented as convex combination of point x_1 and $(n-1)$ vertexes of $(n-2)$-dimensional simplex opposite to cone C_1. But this is impossible by reasons of theorem. The union of cones C_i, $i = 1, \ldots, \frac{n(n+1)}{2}$ divides Euclidean space $\mathbb{R}^n$ into $(n+1)$ parts, every of which contains one vertex of simplex Δ (part of the set E). The last contradicts to supposition that E has no more than

n connected components.

The classical Helly theorem does not admit to obtain information concerning a family of convex compact sets in the Euclidean n-dimensional space if it is known that only subfamilies consisting of k elements, $0 < k \leq n$, possess nonempty intersections. Below we consider the variant of Helly theorem for this case and also investigate behavior of generalized convex families.

Theorem 2. *Let $A = \{A_i\}$ be a family of convex compact sets in $\mathbb{R}^n$ for which every subset from k elements has a common point. Then either every subset from $k+1$ elements has a common point or there exist $k+1$ compacts A_i^1, $i = 1, 2, \ldots, k+1$ for which $H^{k-1}(\bigcup_{i=1}^{k+1} A_i^1) \neq 0$ (where $H^{k-1}(*)$ is $(k-1)$-dimensional cohomology group).*

The proof follows from two technical lemmas.

Lemma 1. *Let $\{A_i\}$ $i = 1, 2, \ldots, m, \ldots$ be a family of convex compact sets for which every subset from k elements has a common point. Then*

$$H^j(A_1 \cup A_2 \cup \ldots \cup A_m \cap A_{m+1} \cap \ldots \cap A_k) = 0$$

for every j, m, $0 \leq m \leq k$.

Lemma 2. *Let$\{A_i\}$ $i = 1, 2, \ldots, k+1, \ldots$ be a family of convex compact sets for which every subset from k elements has a common point. Then*

$$H^{j-2}[(A_1 \cup A_2 \cup \ldots \cup A_j) \cap A_{j+1} \cap \ldots \cap A_{k+1}]$$

$$\approx H^{j-1}[(A_1 \cup A_2 \cup \ldots \cup A_j \cup A_{j+1}) \cap A_{j+2} \cap \ldots \cap A_{k+1}],\ 2 \leq j \leq k.$$

Proof of two lemmas follows from Mayer-Victoris exact cohomological sequence.

Definition 1. A set $E \subset \mathbb{C}^n$ is called linearly convex if for every $z \in \mathbb{C}^n \setminus E$ there exists a complex hyperplane L such that $z \in L \subset \mathbb{C}^n \setminus E$.

Definition 2. A set $E \subset \mathbb{C}^n$ is called $\mathbb{C}$-convex if for every complex line γ sets $\gamma \cap E$ and $\gamma \setminus \gamma \cap E$ are connected.

Definition 3. Linearly convex set $E \subset \mathbb{C}^n$ is called $\mathbb{C}$-quasiconvex if for every complex line γ intersection $\gamma \cap E$ is simpleconnected.

Theorem 3. *The class of $\mathbb{C}$-quasiconvex sets is closed relatively to the intersection of subsets.*

Proof. Let K_1, K_2 be $\mathbb{C}$-quasiconvex compacts. For an arbitrary complex line γ, $\gamma \cap K_1 \cap K_2 \neq \emptyset$ we consider Mayer-Victoris exact cohomological sequence

$$H^1(\gamma \cap K_1) \oplus H^1(\gamma \cap K_2) \to H^1(\gamma \cap K_1 \cap K_2) \to H^2(\gamma \cap (K_1 \cap K_2)).$$

The first element of sequence is trivial because K_1, K_2 are $\mathbb{C}$-quasiconvex. The last one is trivial because $\gamma \cap (K_1 \cap K_2)$ is a proper compact subset of really two dimensional complex line γ.

As conclusion follows triviality of middle element and simpleconnectedness of intersection.

References

1. Yu. Zelinskii *Multivalued mappings in analysis*, Kyiv: Naukova dumka, 1993, 264 p. [in Russian]

GEOMETRIC APPROACH IN THE THEORY OF GENERALIZED QUASICONFORMAL MAPPINGS

ANATOLY GOLBERG

Department of Mathematics, Bar-Ilan University,
Ramat-Gan, 52900, Israel
golbera@math.biu.ac.il
and
Holon Institute of Technology,
52 Golomb St., P.O.Box 305, Holon 58102, Israel
golberga@hit.ac.il

The paper presents a geometric approach for studying properties of mappings. We establish new conditions provided the analyticity of continuous functions of complex variable. We also extend this method for investigation of mappings with finite mean dilatations in $\mathbb{R}^n$.

Keywords: Bohr's and Menshoff's theorems, local univalent functions, quasiconformal mappings, normal neighborhood systems.

1. The Cauchy theorems for univalent functions. In this paper, we present a somewhat new geometric approach which provides the analyticity of local univalent functions as well as of arbitrary continuous functions. In the particular cases these results generalize the classical theorems of Bohr, Menshoff and Trokhimchuk. Of course, all such theorem can be regarded as the Cauchy theorem:

Theorem A. *If a function $f(z)$ of the complex variable z is continuous and monogenic in a domain $G \subset \mathbb{C}$, then it is analytic in G.*

We recall that a function $f(z)$ is **monogenic** if there exist the limit

$$\lim_{h \to 0} \frac{f(z+h) - f(z)}{h} =: f'(z),$$

which is called the derivative of f at the point z.

In fact, Cauchy had proved this theorem under assumption that the derivative $f'(z)$ is continuous in the domain. Later Goursat [5] showed that the assumption of continuity of $f'(z)$ can be omitted.

The further attempts to generalize the Cauchy theorem have been naturally related to replace the rather rigid assumption of monogeniety of $f(z)$ by a condition as weak as possible.

Note that the monogeneity of $f(z)$ is equivalent to existence of the both limits:

$$\lim_{h\to 0}\left|\frac{f(z+h)-f(z)}{h}\right|, \tag{1}$$

which geometrically means the independence of stretching in the given direction and

$$\lim_{h\to 0} Arg\frac{f(z+h)-f(z)}{h}, \tag{2}$$

which means preserving the angles at the points, where $f'(z)\neq 0$. The next natural step is to find the characterizations of analytic functions either only in terms of stretching (1) or only in terms of preserving the angles (2).

The first step in this direction was the following theorem of Bohr [1]:

Theorem B. *If $w = f(z)$ is a continuous univalent mapping of a domain G, for which a finite limit (1) exists and differs from 0 at almost every point of G, then either the function $f(z)$ or the conjugate function $\overline{f(z)}$ is analytic in G.*

The Bohr example

$$f(z)=\begin{cases} z, & Imz\geq 0,\\ \overline{z}, & Imz<0,\end{cases}$$

shows that for functions which are not univalent this theorem is, in general, false.

The next important result is the following theorem of Menshoff [8] based on the second fundamental property of monogenic function (preserving the angles).

Theorem C. *If a mapping $w = f(z)$ is continuous and univalent in a domain C and if at almost every point of C, finite limit (2) exists, then the function $f(z)$ is analytic in G.*

Of course, it suffices to require in both theorems the local univalence of the functions.

Using the quasiconformal mappings, Menshoff has obtained in [9] another generalization of the Bohr theorem. Namely, let us consider a continuous and locally univalent mapping $w = f(z)$ of a domain G of the z-plane onto a domain G^* of the w-plane. For an arbitrary point $z_0 \in G$, we take the circle $C(z_0,r)=\{z:|z-z_0|=r\}\subset D$ and put

$$H(z_0,r)=\frac{\max\limits_{|z'-z_0|=r}|f(z')-f(z_0)|}{\min\limits_{|z''-z_0|=r}|f(z'')-f(z_0)|}.$$

We say that *the continuous univalent function $f(z)$ maps the infinitesimal circle $C(z_0,r)$ into an infinitesimal circle, if*

$$\lim_{r\to 0} H(z_0,r)=1.$$

Obviously, the last condition is more general than the constancy of (1). The important Menshoff generalization of Bohr's theorem is the following result.

Theorem D. *Let a function $f(z)$ be continuous and locally univalent in a domain G, and let map the infinitesimal circles $C(z, r)$ into infinitesimal circles for almost all points $z \in G$. Then either $f(z)$ or $\overline{f(z)}$ is analytic in G.*

2. Main result for locally univalent functions. We shall use the following notations.

Let z be an arbitrary point in $\mathbb{C}$. Assume that some closed neighborhood $\mathcal{G}_t(z)$ of z is defined for any $t \in (0, 1]$. We say that a set of the neighborhoods $\mathcal{G}_t(z)$ of the point z constitutes a **normal system**, if there exists a continuous function $v : \mathbb{C} \to \mathbb{R}$ such that $v(z) = 0$, $v(\zeta) > 0$ for any $\zeta \neq z$. Here $\mathcal{G}_t(z) = \{\zeta \in \mathbb{C} : v(\zeta) \leq t\}$ for any $t \in (0, 1]$. Let $\Gamma_t(z) = \{\zeta \in \mathbb{C} : v(\zeta) = t\}$ denote the boundary of $\mathcal{G}_t(z)$. The function v is called the **generating function** for a given normal system $\{\mathcal{G}_t(z)\}$ (see, e.g., [11]).

Denote

$$r(z,t) = \inf_{\zeta \in \Gamma_t(z)} |\zeta - z|, \quad \mathcal{R}(z,t) = \sup_{\zeta \in \Gamma_t(z)} |\zeta - z|.$$

These values $r(z, t)$ and $\mathcal{R}(z, t)$ are equal, respectively, to the minimal and the maximal radii of the neighborhood $\mathcal{G}_t(z)$. The limit

$$\Delta(z) = \limsup_{t \to 0} \frac{\mathcal{R}(z,t)}{r(z,t)}$$

is called the **regularity parameter** of the family $\{\mathcal{G}_t(z), 0 < t \leq 1\}$. Any such system $\{\mathcal{G}_t(z)\}$ is called the **regular normal system**, provided $\Delta(z) < \infty$.

Let now $f : G \to G^*$ be a homeomorphism of two bounded domains in $\mathbb{C}$, and let $\{\mathcal{G}_t(z)\}$ be a normal system of neighborhoods of $z \in G$. One can introduce similarly the minimal and the maximal radii for the image of $\mathcal{G}_t(z)$ by

$$r^*(z,t) = \inf_{\zeta \in \Gamma_t(z)} |f(\zeta) - f(z)|, \quad \mathcal{R}^*(z,t) = \sup_{\zeta \in \Gamma_t(z)} |f(\zeta) - f(z)|$$

and

$$\Delta^*(z) = \limsup_{t \to 0} \frac{\mathcal{R}^*(z,t)}{r^*(z,t)}.$$

We now present an essential strengthening of the above theorems.

Let p be a fixed real number such that $1 \leq p < \infty$. Denote by $I_p(z, t)$ and $O_p(z, t)$ the following quantities

$$I_p(z,t) = \frac{\mathcal{R}^*(z,t)}{r(z,t)} \left(\frac{\mathcal{R}(z,t)}{r^*(z,t)} \right)^{p-1}, \quad O_p(z,t) = \left(\frac{\mathcal{R}^*(z,t)}{r(z,t)} \right)^{p-1} \frac{\mathcal{R}(z,t)}{r^*(z,t)}, \tag{3}$$

and put

$$\Theta(z) = \lim_{t \to 0} \frac{mf(B(z, \mathcal{R}(z,t)))}{\pi \mathcal{R}^2(z,t)},$$

where $B(z, h)$ is the disc $\{\zeta \in \mathbb{C} : |\zeta - z| < h\}$, and mA is the Lebesgue two-dimensional measure of the set A.

Theorem 1. *Let a function $f(z)$ be continuous and locally univalent in a domain G, and let for almost every point $z \in G$ there exists a normal regular system of neighborhoods $\{\mathcal{G}_t(z)\} \subset G$ such that either the inequality*

$$\limsup_{t\to 0} I_p(z,t) \le \left[\Theta(z)\right]^{\frac{2-p}{2}}, \tag{4}$$

holds for $1 \le p \le 2$ or the inequality

$$\limsup_{t\to 0} O_p(z,t) \le \left[\Theta(z)\right]^{\frac{p-2}{2}}, \tag{5}$$

holds for $2 \le p < \infty$. Then either $f(z)$ or the conjugate function $\overline{f(z)}$ is analytic in G.

We shall establish the assertion of the theorem for the function $f(z)$ itself, assuming that f is orientation preserving. The proof for $\overline{f(z)}$ is accomplished in a similar way.

Proof of Theorem 1. The proof is based on the following lemmas given in [3].

Lemma 1. *Under the assumptions of Theorem 1, the function $f(z)$ is differentiable almost everywhere in the domain G, and for any Borel set $E \subset G$, we have*

$$\int_E |f'(z)|^2 dxdy < \infty.$$

Lemma 2. *If f is orientation preserving and satisfies the condition of Theorem 1, then for almost all points $z \in G$, we have the equality*

$$\frac{|f_z| + |f_{\bar z}|}{|f_z| - |f_{\bar z}|} = 1.$$

Here f_z and $f_{\bar z}$ denote the formal derivatives of f at z.

Lemma 3. *The function f is ACL, i.e., absolutely continuous on lines.*

Lemma 4. *For any rectangle $\Pi \Subset G$ with the sides parallel to the coordinate axes, we have the equality*

$$\oint_{\partial\Pi} f(z)\, dz = 0. \tag{6}$$

Now the proof of Theorem 1 is completed in the following way. Lemma 1 and Lemma 2 yield together that the function $f(z)$ is $\mathbb{C}$-differentiable in the domain G and has there the derivative $f'(z)$; moreover, this derivative is square integrable over G. In addition, f satisfies by Lemma 3 the N-property on almost all lines parallel to the coordinate axes. By Lemma 4 we have the equality (6) for arbitrary rectangle Π. It follows now from the classical Morera theorem that $f(z)$ is complex analytic on the domain G.

3. Generalizations of the Bohr and Menshoff theorems for continuous functions. The assumption of univalence of continuous functions was essential in the investigations by Menshoff and by other mathematicians. The further investigation of analyticity of continuous functions has been made by Trokhimchuk. He used the notion of the interior mapping and the fundamental theorem of interior mappings obtained by Stoïlow. Another ingredient of the method of Trokhimchuk is the theory of the sets of monogeniety.

Following [13], we call a continuous mapping $w = f(z)$ of the domain G **interior** if it satisfies the conditions:
(i) no continuum of the domain is mapped onto a point,
(ii) the image of an open set of the z-plane is open in the w-plane.

We now consider the mappings preserving orientation for the case of continuous functions.

Let z_0 be a point of a domain G, where the function $w = f(z)$ is continuous, and let $w_0 = f(z_0)$.

The point z_0 will be called a ***U*-point** of the mapping $w = f(z)$ if there exist two sequences $\{z_i'\}$, $\{z_i''\}$ $(i = 1, 2, \ldots)$ of points converging to z_0, such that the semitangents t', t'' at z_0 lie on the different straight lines and all the points $w_i' = f(z_i')$, $w'' = f(z_i'')$ are different from $w_0 = f(z_0)$.

We say that the mapping w is **orientation preserving at the *U*-point** z_0 if the sequences $\{w_i'\}$, $\{w_i''\}$ have semitangents T', T'' at w_0 with the following property: if $0 < \{t' \widehat{,} t''\} < \pi$ then $0 \leq \{T' \widehat{,} T''\} < \pi$.

Yu. Yu. Trokhimchuk [14] gave the generalizations of the above Bohr and Menshoff theorem for arbitrary continuous functions.

Theorem E. *Let $w = f(z)$ be a continuous mapping of the domain G with constant stretching at almost every point z in G, and let this mapping be orientation preserving at almost every U-point. Then the function $f(z)$ is analytic in G. In addition, if there are no U-points of f in G, then $f(z) \equiv const$.*

In this section we give an extended version of Theorem 1 for arbitrary continuous functions $w = f(z)$. We use the local geometric characteristics introduced in (3).

Since in the case of the continuous functions the characteristic neighborhood $\mathcal{G}_t(z)$ can be shrunk to a point or to an arc, the quantities (3) are not determined in the general case unlike the case of univalent functions. But for analytic functions the values $I_p(z,t)$ and $O_p(z,t)$ are finite al least for sufficiently small t.

Theorem 2. *Let a function $f(z)$ be continuous in a domain G. Assume that for almost every point $z \in G$ there exists a normal regular system of neighborhoods $\{\mathcal{G}_t(z)\} \subset G$ such that the quantities $I_p(z,t)$ and $O_p(z,t)$ are finite for sufficiently small $t > 0$, and let f be orientation preserving at every U-point. If either the inequality (4) holds for $1 \leq p \leq 2$ or the inequality (5) holds for $2 \leq p < \infty$. Then the function $f(z)$ is analytic in the domain G and this function is not constant.*

The proof of Theorem 2 is based on certain estimates of (3) obtained in Lemma 3 (see, [3]). We also apply Theorem 9 of Trokhimchuk [14, p. 60] for interior mappings.

4. Generalized quasiconformal mappings in $\mathbb{R}^n$. The above results can be extended to quasiconformal mappings (even to the mappings quasiconformal in the integral sense) of multiply-dimensional domains.

Let $A: \mathbb{R}^n \to \mathbb{R}^n$ be a linear bijection. Consider the quantities (see, e.g., [4], [2])

$$H_{I,\alpha}(A) = \frac{|J(A)|}{l^\alpha(A)}, \quad H_{O,\alpha}(A) = \frac{L^\alpha(A)}{|J(A)|}, \quad \alpha \geq 1.$$

Here $l(A) = \min_{|h|=1} |Ah|$, $L(A) = \max_{|h|=1} |Ah|$ and $detA$ is the determinant of A.

For $\alpha = n$, the values of $H_{I,\alpha}(A)$ and $H_{O,\alpha}(A)$ coincide with the inner and outer dilatations of A, respectively (see, e.g., [15]).

Let G and G^* be two bounded domains in $\mathbb{R}^n$, $n \geq 2$, and let a mapping $f: G \to G^*$ be differentiable at a point $x \in G$. This means there exists a linear mapping $f'(x): \mathbb{R}^n \to \mathbb{R}^n$, called the (strong) derivative of the mapping f at x, such that $f(x+h) = f(x) + f'(x)h + \omega(x,h)|h|$, where $\omega(x,h) \to 0$ as $h \to 0$.

Now we consider the homeomorphisms f differentiable almost everywhere in G and fix the real numbers α, β satisfying $1 \leq \alpha < \beta < \infty$. Define

$$HI_{\alpha,\beta}(f) = \int_G H_{I,\alpha}^{\frac{\beta}{\beta-\alpha}}(x,f)\,dx, \quad HO_{\alpha,\beta}(f) = \int_G H_{O,\beta}^{\frac{\alpha}{\beta-\alpha}}(x,f)\,dx,$$

where $H_{I,\alpha}(x,f) = H_{I,\alpha}(f'(x))$, $H_{O,\beta}(x,f) = H_{O,\beta}(f'(x))$. We call these values the **inner** and the **outer** mean dilatations of a mapping f of a given domain G.

Define for the fixed real numbers $\alpha, \beta, \gamma, \delta$ such that $1 \leq \alpha < \beta < \infty$, $1 \leq \gamma < \delta < \infty$, the **class** $B(G, G^*, \alpha, \beta, \gamma, \delta)$ of such homeomorphisms $f: G \to G^*$ which satisfy:

(iii) f and f^{-1} are ACL-homeomorphisms,

(iv) f and f^{-1} are differentiable with the Jacobians $J(x,f) \neq 0$ and $J(y, f^{-1}) \neq 0$ a.e. in G and G^*, respectively,

(v) the inner and the outer mean dilatations $HI_{\alpha,\beta}(f)$ and $HO_{\gamma,\delta}(f)$ are finite.

In other words $B(G, G^*, \alpha, \beta, \gamma, \delta)$ consists of **mappings with finite mean dilatations**. The case $\beta = \delta = n$ was explicitly studied in [6].

It follows that for any $f \in B(G, G^*, \alpha, \beta, \gamma, \delta)$, we have the equalities

$$HI_{\alpha,\beta}(f^{-1}) = HO_{\alpha,\beta}(f), \quad HO_{\alpha,\beta}(f^{-1}) = HI_{\alpha,\beta}(f).$$

Yu. G. Reshetnyak [11] has investigated quasiconformal mappings of the space domains using the radii of the normal regular system of neighborhoods. He called a mapping f **quasiconformal** at a point $x \in G$ if there exists a normal regular system $\{\mathcal{G}_t(x)\}$ of neighborhoods of x such that $\Delta(x)\Delta^*(x) < \infty$.

To give the corresponding geometric description for generalized quasiconformal mappings we consider also certain set functions.

Let Φ be a finite nonnegative function in domain G defined for open subsets E of G so that $\sum_{k=1}^m \Phi(E_k) \leqslant \Phi(E)$ for any finite collection $\{E_k\}_{k=1}^m$ of nonintersecting open sets $E_k \subset E$. We denote the class of such set functions Φ by $\mathcal{F}$.

The **upper** and **lower** derivatives of a set function $\Phi \in \mathcal{F}$ at a point $x \in G$ are defined by

$$\overline{\Phi'}(x) = \lim_{h\to 0} \sup_{d(Q)<h} \frac{\Phi(Q)}{mQ}, \qquad \underline{\Phi'}(x) = \lim_{h\to 0} \inf_{d(Q)<h} \frac{\Phi(Q)}{mQ},$$

where Q ranges over all open cubes and open balls such that $x \in Q \subset G$ and $d(Q) = Q$. Here mA denotes the n-dimensional Lebesgue measure of the set A. Due to [10], these derivatives have the following properties:
(vi) $\overline{\Phi'}(x)$ and $\underline{\Phi'}(x)$ are Borel's functions;
(vii) $\overline{\Phi'}(x) = \underline{\Phi'}(x) < \infty$ a.e. in G;
(viii) for each open set $V \subset G$, $\int\limits_V \overline{\Phi'}(x)\,dx \le \Phi(V)$.

Using these set functions, we define for the fixed real numbers $\alpha, \beta, \gamma, \delta$ such that $1 \le \alpha < \beta < \infty$ and $1 \le \gamma < \delta < \infty$, the **class** $\mathcal{H}(G, G^*, \alpha, \beta, \gamma, \delta)$ of homeomorphisms $f : G \to G^*$ which satisfy:
(ix) there exist $\Phi, \Psi \in \mathcal{F}$ in G,
(x) for any point $x \in G$ there exists $\{\mathcal{G}_t(x)\} \subset G$,
(xi) the inequalities

$$\limsup_{t\to 0} \frac{mf(B(x, \mathcal{R}(x,t)))\mathcal{R}^{\alpha-n}(x,t)}{\Omega_n r^{*\alpha}(x,t)} \le \left[\Phi'(x)\right]^{\frac{\beta-\alpha}{\beta}},$$

$$\limsup_{t\to 0} \frac{\Omega_n \mathcal{R}^{*\delta}(x,t)}{mf(B(x, r(x,t)))r^{\delta-n}(x,t)} \le \left[\Psi'(x)\right]^{\frac{\delta-\gamma}{\gamma}}.$$

hold for all points $x \in G$ at which the derivatives $\Phi'(x)$ and $\Psi'(x)$ exist; $B(x,h) = \{y \in \mathbb{R}^n : |y - x| < h\}$.

5. Equivalence of analytic and geometric descriptions. We now are enable to establish that the classes $\mathcal{B}(G, G^*, \alpha, \beta, \gamma, \delta)$ and $\mathcal{H}(G, G^*, \alpha, \beta, \gamma, \delta)$ are coincide.

Lemma 5. *Let $f \in \mathcal{H}(G, G^*, \alpha, \beta, \gamma, \delta)$, then f is ACL-mapping, which is differentiable a.e. in G.*

Sketch of the proof. First, we show that f is ACL-mapping in G. Fix for each $x \in G$ a normal regular system $\{\mathcal{G}_t(x)\}$ of neighborhoods such that $\mathcal{G}_t(x) \subset G$ for any $t \in (0,1]$. Consider an arbitrary point $a = (a_1, \ldots, a_n) \in G$ and $h > 0$ so that the cube $\bar{Q} = \bar{Q}(a,h)$ belongs to G, where $Q(a,h) = \{x : |x_i - a_i| < h, i = 1, \ldots, n\}$. Denote by C_k, $k = 1, \ldots, n$, the intersection of $\bar{Q}$ with the plane $P_k(a) = \{x \in \mathbb{R}^n : x_k = a_k\}$, and let $p(x)$ be the segment $|x_k - a_k| \le h/2$ of the line passing through x ($x \in C_k$) parallel to the kth coordinate axis x_k. Similarly, let $p(A)$ denote the union of all segments $p(x)$, when $A \subset C_k$ and $x \in A$.

We show that for almost all $x \in C_k$ (with respect to $(n-1)$-dimensional Lebesgue measure) the restriction of f to $p(x)$ admits the Lusin N-property.

Further, for $x, \tilde{x} \in G, \tilde{x} \ne x$, we define

$$k(x) = \limsup_{\tilde{x}\to x} \frac{|f(\tilde{x}) - f(x)|}{|\tilde{x} - x|}.$$

Then $k(x) < \infty$ for almost all $x \in G$, and for any open set $E \subset G$ $k(x)$ is integrable with the powers $\beta/(\beta - n + 1)$ and γ. This conclusion allows us to the classical Stepanov theorem [12] on differentiability and we obtain that f is differentiable almost everywhere in G.

Lemma 6. *Let $f \in \mathcal{H}(G, G^*, \alpha, \beta, \gamma, \delta)$, then $HI_{\alpha,\beta}(f)$ and $HO_{\gamma,\delta}(f)$ are finite.*

Lemma 7. *If $f \in \mathcal{B}(G, G^*, \alpha, \beta, \gamma, \delta)$, then $f \in \mathcal{H}(G, G^*, \alpha, \beta, \gamma, \delta)$.*

The proofs of Lemmas 6 and 7 are similar to the corresponding lemmas from [2].

Now we can formulate the following theorem:

Theorem 3. *The classes $\mathcal{B}(G, G^*, \alpha, \beta, \gamma, \delta)$ and $\mathcal{H}(G, G^*, \alpha, \beta, \gamma, \delta)$ coincide.*

Thus we can characterize geometrically in terms of radii of normal neighborhood systems many well-known classes of mappings (quasiconformal, quasiconformal in the mean, enc). For the planar mappings which are quasiconformal in the mean, we refer to [7].

References

1. H. Bohr, *Ueber streckentreue und konforme Abbildung*, Math. Ztschr. **1** (1918), 3–19.
2. A. Golberg, *Geometric characteristics of mappings with first generalized derivatives*, Revue Roumaine de Math. Pure et Appl. **47** (2002), 671–680.
3. A. Golberg, *On generalization of Menshoff's theorem*, Israel J. Math. **156** (2006), 243–254.
4. V. Gol'dshteyn, L. Gurov, and A. Romanov, *Homeomorphisms that induce monomorphisms of Sobolev spaces*, Israel J. Math. **91** (1995), no. 1-3, 31–60.
5. E. Goursat, *Sur la définition générale des fonctions analytiques d'aprés Cauchy*, Trans. Amer. Math. Soc., **1** (1900), 14–16.
6. V. I. Kruglikov, *Capacities of condensors and quasiconformal in the mean mappings in space*, Mat. Sb. **130** (1986), no. 2, 185–206.
7. V. S. Kud'yavin, *Local structure of plane mappings that are quasiconformal in the mean* (Russian), Dokl. Akad. Nauk Ukrain. SSR 1991, no. 3, 10–12, 164.
8. D. Menshoff, *Sur la representation conforme des domaines plans*, Math. Ann. **95** (1926), 640–670.
9. D. Menshoff, *Sur une generalisation d'un theoreme de M. H. Bohr*, Mat. sb. **2** (1937), 339–356.
10. T. Rado and P. Reichelderfer, *Continuous Transformations in Analysis*, Springer-Verlag, 1955.
11. Yu. G. Reshetnyak, *Space Mappings with Bounded Distortion*, Amer. Math. Soc., Providence, RI, 1989.
12. V. V. Stepanov, *Sur les conditions de l'existence de la differentielle totale*, Mat. sb. **30** (1924), 487-489.
13. S. Stoïlow, *Lecons sur les principes topologiques de la théorie des fonctions analytiques*, Paris, 1938.
14. Yu. Yu. Trokhimchuk, *Continuous mappings and conditions of monogeneity*, Translated from Russian, Israel Program for Scientific Translations, Jerusalem; Daniel Davey & Co., Inc., New York, 1964.
15. J. Väisälä, *Lectures on n-dimensional Quasiconformal Mappings*, Springer-Verlag, 1971.

SEPARATELY QUASI-NEARLY SUBHARMONIC FUNCTIONS

JUHANI RIIHENTAUS

Department of Mathematics, University of Joensuu
P.O. Box 111, FI-80101 Joensuu, Finland
riihentaus@member.ams.org

After recalling some properties of separately subharmonic functions, we present the corresponding counterparts for separately quasi-nearly subharmonic functions. In addition, considering the subharmonicity of functions subharmonic in the first variable and harmonic in the second, we generalize some results of Cegrell and Sadullaev and Kolodziej and Thornbiörnson.

Keywords: Separately subharmonic, harmonic, quasi-nearly subharmonic, Harnack, integrability condition.

1. Introduction

1.1. Separately subharmonic functions. Solving a long standing problem, Wiegerinck [Wi88], see also [WZ91, Theorem 1, p. 246], showed that a separately subharmonic function need not be subharmonic. On the other hand, Armitage and Gardiner [AG93, Theorem 1, p. 256] showed that a separately subharmonic function u on a domain Ω of $\mathbb{R}^{m+n}$, $m \geq n \geq 2$, is subharmonic provided $\phi(\log^+ u^+)$ is locally integrable, where $\phi : [0, +\infty) \to [0, +\infty)$ is an increasing function such that

$$\int_1^{+\infty} s^{(n-1)/(m-1)} (\phi(s))^{-1/(m-1)} \, ds < +\infty. \qquad (1)$$

For related previous results of Lelong, Avanissian, Arsove and Riihentaus, see e.g. [Le45], [Le61], [Le69], [Av61], [Ar66], [He71], [Ri89] and the references therein. One of these previous results was ours:

Theorem A. ([Ri89, Theorem 1, p. 69]) *Let Ω be a domain in $\mathbb{R}^{m+n}$, $m, n \geq 2$. Let $u : \Omega \to [-\infty, +\infty)$ be such that*

(a) *for each $y \in \mathbb{R}^n$ the function*

$$\Omega(y) \ni x \mapsto u(x, y) \in [-\infty, +\infty)$$

is subharmonic,

(b) *for each $x \in \mathbb{R}^m$ the function*

$$\Omega(x) \ni y \mapsto u(x, y) \in [-\infty, +\infty)$$

is subharmonic,

(c) *for some* $p > 0$ *there is a function* $v \in \mathcal{L}^p_{\mathrm{loc}}(\Omega)$ *such that* $u \leq v$.

Then u *is subharmonic.*

Though the cited result of Armitage and Gardiner includes our Theorem A, and in fact their result is even "almost" sharp, we present below in Theorem 1 a generalization to Theorem A. This is justified because of two reasons. First, our $\mathcal{L}^p_{\mathrm{loc}}$ integrability condition, $p > 0$, is, unlike the condition of Armitage and Gardiner (1), very simple, and second, our generalization to Theorem A is stated for quasi-nearly subharmonic functions, and as such, it is very general, see **2.1.** below.

1.2. Functions subharmonic in one variable and harmonic in the other. An open problem is, whether a function which is subharmonic in one variable and harmonic in the other, is subharmonic. For results on this area, see e.g. [WZ91], [CS93] and [KT96] and the references therein. We consider here a result of Cegrell and Sadullaev, Theorem B below, and a result of Kolodziej and Thornbiörson, Theorem C below.

Theorem B. ([CS93, Theorem 3.1, p. 82]) *Let* Ω *be a domain in* $\mathbb{R}^{m+n}$, $m, n \geq 2$. *Let* $u : \Omega \to \mathbb{R}$ *be such that*

(a) *for each* $y \in \mathbb{R}^n$ *the function*

$$\Omega(y) \ni x \mapsto u(x, y) \in \mathbb{R}$$

is subharmonic,

(b) *for each* $x \in \mathbb{R}^m$ *the function*

$$\Omega(x) \ni y \mapsto u(x, y) \in \mathbb{R}$$

is harmonic,

(c) *there is a nonnegative function* $\varphi \in \mathcal{L}^1_{\mathrm{loc}}(\Omega)$ *such that* $-\varphi \leq u$.

Then u *is subharmonic.*

Cegrell and Sadullaev use Poisson modification in their proof. In [Ri06$_2$] we give a new proof, which avoids the use of Poisson modification, and is based simply on mean value operators and on Theorem 1 below.

Cegrell and Sadullaev state also a corollary [CS93, Corollary, p. 82] to their result, just choosing $\varphi = 0$. In Theorem 2 below we give a similar counterpart to Cegrell's and Sadullaev's Corollary for quasi-nearly subharmonic functions. Again our result is very general.

Kolodziej and Thorbiörnson gave the following result. Their proof uses, among other things, the above result of Cegrell and Sadullaev, see [CS93, proof of Theorem 3.2, p. 83].

Theorem C. ([KT96, Theorem 1, p. 463]) *Let* Ω *be a domain in* $\mathbb{R}^{m+n}$, $m, n \geq 2$. *Let* $u : \Omega \to \mathbb{R}$ *be such that*

(a) *for each* $y \in \mathbb{R}^n$ *the function*

$$\Omega(y) \ni x \mapsto u(x,y) \in \mathbb{R}$$

is subharmonic and $\mathcal{C}^2$,

(b) *for each* $x \in \mathbb{R}^m$ *the function*

$$\Omega(x) \ni y \mapsto u(x,y) \in \mathbb{R}$$

is harmonic.

Then u *is subharmonic and continuous.*

Below in Theorem 3, Theorem 4 and Corollary we give generalizations to the above result of Kolodziej and Thornbiörnson. Instead of the standard Laplacians of $\mathcal{C}^2$ functions we use generalized Laplacians, that is the Blascke-Privalov operators.

2. Definitions and notation

2.1. Our notation is rather standard, see e.g. [Ri89], [Ri00],[Ri06$_1$], [Ri06$_2$] and [He71].

Let D be a domain in the Euclidean space $\mathbb{R}^N$, $N \geq 2$. A Lebesgue measurable function $u : D \to [0, +\infty]$ is *quasi-nearly subharmonic*, if $u \in \mathcal{L}^1_{\text{loc}}(D)$ and if there is a constant $K = K(N, u, D) > 0$ such that

$$u(x) \leq \frac{K}{r^N} \int\limits_{B(x,r)} u(y)\, dm(y) \tag{2}$$

for any ball $B^N(x,r) \subset D$. For the Lebesgue measure in $\mathbb{R}^N$, $N \geq 2$, we use both m and m_N. (Below m will be used also for the dimension of the Euclidean space $\mathbb{R}^m$, but this will surely cause no confusion.) This function class of quasi-nearly subharmonic functions is natural, it has important and interesting properties and, at the same time, it is large, see e.g. [Pa94], [Ri00] (where they were called pseudosubharmonic functions), [PR05], [Ri06$_1$] and [Ri06$_2$]. We recall here only that it includes, among others, nonnegative subharmonic functions, nonnegative nearly subharmonic functions (see e.g. [He71]), functions satisfying certain natural growth conditions, especially certain eigenfunctions, and polyharmonic functions. Also, any Lebesgue measurable function $u : D \to [m, M]$, where $0 < m \leq M < +\infty$, is quasi-nearly subharmonic.

Constants will be denoted by C and K. They will be nonnegative and may vary from line to line.

2.2. As a counterpart to nonnegative harmonic functions, we recall the definition of Harnack functions, see [Vu82, p. 259]. A continuous function $u : D \to [0, +\infty)$ is a *Harnack function*, if there are constants $\lambda \in (0,1)$ and $C = C(\lambda) \geq 1$ such that

$$\max_{z \in B(x,\lambda r)} u(z) \leq C \min_{z \in B(x,\lambda r)} u(z)$$

whenever $B(x,r) \subset D$. It is well-known that for each compact set F in D there exists a smallest constant $C(F) \geq C$ depending only on N, λ, C and F such that for all u satisfying the above condition,

$$\max_{z \in F} u(z) \leq C(F) \min_{z \in F} u(z).$$

One sees easily that Harnack functions are quasi-nearly subharmonic. Also the class of Harnack functions is very wide. It includes, among others, nonnegative harmonic functions as well as nonnegative solutions of some elliptic equations. Also, any continuous function $u : D \to [m, M]$, where $0 < m \leq M < +\infty$, is a Harnack function. See [Vu82, pp. 259, 263].

2.3. A function $\psi : [0, +\infty) \to [0, +\infty)$ is *permissible*, if there exists an increasing (strictly or not), convex function $\psi_1 : [0, +\infty) \to [0, +\infty)$ and a strictly increasing surjection $\psi_2 : [0, +\infty) \to [0, +\infty)$ such that $\psi = \psi_2 \circ \psi_1$ and such that the following conditions are satisfied:

(a) ψ_1 satisfies the Δ_2-condition.
(b) ψ_2^{-1} satisfies the Δ_2-condition.
(c) The function $t \mapsto \frac{\psi_2(t)}{t}$ is *quasi-decreasing*, i.e. there is a constant $C = C(\psi_2) > 0$ such that $\frac{\psi_2(s)}{s} \geq C \frac{\psi_2(t)}{t}$ for all $0 \leq s \leq t$.

See also [PR05, Lemma 1 and Remark 1]. Recall that a function $\varphi : [0, +\infty) \to [0, +\infty)$ satisfies the Δ_2-condition, if there is a constant $C = C(\varphi) \geq 1$ such that $\varphi(2t) \leq C \varphi(t)$ for all $t \in [0, +\infty)$.

3. Separately subharmonic functions

3.1. The counterpart to Theorem A is:

Theorem 1. *Let Ω be a domain in $\mathbb{R}^{m+n}$, $m, n \geq 2$. Let $u : \Omega \to [0, +\infty)$ be a Lebesgue measurable function such that*

(a) *for each $y \in \mathbb{R}^n$ the function*

$$\Omega(y) \ni x \mapsto u(x, y) \in [0, +\infty)$$

is quasi-nearly subharmonic,

(b) *for each $x \in \mathbb{R}^m$ the function*

$$\Omega(x) \ni y \mapsto u(x, y) \in [0, +\infty)$$

is quasi-nearly subharmonic,

(c) *there exists a non-constant permissible function $\psi : [0, +\infty) \to [0, +\infty)$ such that $\psi \circ u \in \mathcal{L}^1_{\text{loc}}(\Omega)$.*

Then u is quasi-nearly subharmonic.

Proof. Since permissible functions are continuous, $\psi \circ u$ is measurable. To see that $\psi \circ u$ is locally bounded, take $(a, b) \in \Omega$ and $R > 0$ such that $B^{m+n}((a, b), R) \subset \Omega$.

We show that there is a constant $K = K(m,n,u,\psi,\Omega) > 0$ such that

$$\psi(u(x_0,y_0)) \leq \frac{K}{R^{m+n}} \int\limits_{B^{m+n}((a,b),R)} \psi(u(x,y))\, dm_{m+n}(x,y)$$

for all $(x_0,y_0) \in B^m(a,\frac{R}{4}) \times B^n(b,\frac{R}{4})$. For this purpose choose $(x_0,y_0) \in B^m(a,\frac{R}{4}) \times B^n(b,\frac{R}{4})$ arbitrarily. Then

$$\psi(u(x_0,y_0)) \leq \frac{K}{(\frac{R}{4})^m} \int\limits_{B^m(x_0,\frac{R}{4})} \psi(u(x,y_0))\, dm_m(x)$$

$$\leq \frac{K}{(\frac{R}{4})^m} \int\limits_{B^m(x_0,\frac{R}{4})} \Big[\frac{K}{(\frac{R}{4})^n} \int\limits_{B^n(y_0,\frac{R}{4})} \psi(u(x,y))\, dm_n(y)\Big] dm_m(x)$$

$$\leq \frac{4^{m+n}K^2}{R^{m+n}} \int\limits_{B^{m+n}((a,b),R)} \psi(u(x,y))\, dm_{m+n}(x,y).$$

Using then properties of the permissible function ψ one sees easily that also u is locally bounded above, thus locally integrable. Proceeding then as above, but now ψ replaced with the identity mapping and choosing $(x_0,y_0) = (a,b)$, one sees that u satisfies the condition (2), and is thus quasi-nearly subharmonic on Ω.

3.2. Remark. Unlike in Theorem A, the measurability assumption is now necessary. Indeed, with the aid of Sierpinski's nonmeasurable function, given e.g. in [Ru79, 7.9 (c), pp. 152-153], one easily constructs a nonmeasurable, separately quasi-nearly subharmonic function $u : \mathbb{C}^2 \to [1,2]$. Indeed, let $\tilde{Q} = \{0\} \times Q \times \{0\} \subset \mathbb{R} \times \mathbb{R}^2 \times \mathbb{R}$, where $Q \subset \mathbb{R}^2$ is the set of Sierpinski, see [Ru79, pp. 152-153]. Then the function $v(z_1,z_2) = v(x_1,y_1,x_2,y_2) := \chi_{\tilde{Q}}(z_1,z_2) + 1$ is clearly nonmeasurable, but still separately quasi-nearly subharmonic.

4. The result of Cegrell and Sadullaev

Then a counterpart to Cegrell's and Sadullaev's Corollary of their result, Theorem B above, see [CS, Corollary, p. 82]:

Theorem 2. *Let Ω be a domain in $\mathbb{R}^{m+n}$, $m,n \geq 2$. Let $u : \Omega \to [0,+\infty)$ be such that*

(a) *for each* $y \in \mathbb{R}^n$ *the function*

$$\Omega(y) \ni x \mapsto u(x, y) \in [0, +\infty)$$

is quasi-nearly subharmonic,

(b) *for each* $x \in \mathbb{R}^m$ *the function*

$$\Omega(x) \ni y \mapsto u(x, y) \in [0, +\infty)$$

is a Harnack function.

Then u *is quasi-nearly subharmonic.*

Proof. It is well-known that u is Lebesgue measurable. Let $(a, b) \in \Omega$ and $R > 0$ be such that $B^{m+n}((a, b), R) \subset \Omega$. Choose $(x_0, y_0) \in B^m(a, \frac{R}{4}) \times B^n(b, \frac{R}{4})$ arbitrarily. Since $u(\cdot, y_0)$ is quasi-nearly subharmonic, one has

$$u(x_0, y_0) \leq \frac{K}{(\frac{R}{4})^m} \int\limits_{B^m(x_0, \frac{R}{4})} u(x, y_0)\, dm_m(x).$$

On the other hand, since the functions $u(x, \cdot)$, $x \in B^m(a, \frac{R}{2})$, are Harnack functions in $B^n(b, \frac{R}{2})$, there is a constant $C = C(n, \lambda, C_\lambda, R)$ (here λ and C_λ are the constants in **2.2**) such that

$$\frac{1}{C} \leq \frac{u(x, y_0)}{u(x, b)} \leq C$$

for all $x \in B^m(a, \frac{R}{4})$. See e.g. [ABR01, proof of 3.6, pp. 48–49]. Therefore

$$u(x_0, y_0) \leq \frac{K}{(\frac{R}{4})^m} \int\limits_{B^m(x_0, \frac{R}{4})} C\, u(x, b)\, dm_m(x) = \frac{C \cdot K}{(\frac{R}{4})^m} \int\limits_{B^m(a, \frac{R}{2})} u(x, b)\, dm_m(x)$$

$$= \frac{K}{R^m} \int\limits_{B^m(a, \frac{R}{2})} u(x, b)\, dm_m(x) < \infty.$$

Thus u is locally bounded above in $B^m(a, \frac{R}{4}) \times B^n(b, \frac{R}{4})$, and therefore the result follows from Theorem 1 above.

5. The result of Kolodziej and Thornbiörnson

5.1. In our generalization to the cited result of Kolodziej and Thorbiörnson, we use the generalized Laplacian, defined with the aid of the Blaschke-Privalov operators, see e.g. [Sa41], [Ru50], [Sh71] and [Sh78]. Let D be a domain in $\mathbb{R}^N$, $N \geq 2$, and $f : D \to \mathbb{R}$, $f \in \mathcal{L}^1_{\rm loc}(D)$. We write

$$\Delta^* f(x) := \limsup_{r \to 0} \frac{2(N+2)}{r^2} \cdot \Big[\frac{1}{\nu_N r^N} \int\limits_{B^N(x,r)} f(x') dm_N(x') - f(x)\Big],$$

$$\Delta_* f(x) := \liminf_{r \to 0} \frac{2(N+2)}{r^2} \cdot \Big[\frac{1}{\nu_N r^N} \int\limits_{B^N(x,r)} f(x') dm_N(x') - f(x)\Big].$$

If $\Delta^* f(x) = \Delta_* f(x)$, then write $\Delta f(x) := \Delta^* f(x) = \Delta_* f(x)$. If $f \in \mathcal{C}^2(D)$, then

$$\Delta f(x) = (\sum_{j=1}^{N} \frac{\partial^2 f}{\partial x_j^2})(x),$$

the standard Laplacian with respect to the variable $x = (x_1, \dots, x_N)$. More generally, if $x \in D$ and $f \in t_2^1(x)$, i.e. f has an $\mathcal{L}^1$ total differential at x of order 2, then $\Delta f(x)$ equals with the pointwise Laplacian of Δf at x, i.e.

$$\Delta f(x) = \sum_{j=1}^{N} D_{jj} f(x).$$

Here $D_{jj} f$ represent a generalization of the usual $\frac{\partial^2 f}{\partial x_j^2}$, $j = 1, \dots, N$. See e.g. [CZ61, p. 172], [Sh71, p. 369] and [Sh78, p. 29].

Recall that there are functions which are not $\mathcal{C}^2$ but for which the generalized Laplacian is nevertheless continuous. The following function gives a simple example:

$$f(x) = \begin{cases} -1, & \text{when } x_N < 0, \\ 0, & \text{when } x_N = 0, \\ 1, & \text{when } x_N > 0. \end{cases}$$

If f is subharmonic on D, it follows from [Sa41, p. 451] (see also [Ru50, Lemma 2.2, p. 280]) that $\Delta^* f(x) = \Delta_* f(x)$ for almost all $x \in D$.

Below the following notation is used. Let Ω is a domain in $\mathbb{R}^{m+n}$, $m, n \geq 2$, and $u : \Omega \to \mathbb{R}$. If $y \in \mathbb{R}^n$ is such that the function

$$\Omega(y) \ni x \mapsto f(x) := u(x, y) \in \mathbb{R}$$

is in $\mathcal{L}^1_{\text{loc}}(\Omega(y))$, then we write $\Delta_1^* u(x, y) := \Delta^* f(x)$, $\Delta_{1*} u(x, y) := \Delta_* f(x)$, and $\Delta_1 u(x, y) := \Delta f(x)$.

5.2. Then a generalization to [KT96, Theorem 1, p. 463]:

Theorem 3. *Let Ω be a domain in $\mathbb{R}^{m+n}$, $m, n \geq 2$. Let $u : \Omega \to \mathbb{R}$ be such that for each $(x_0, y_0) \in \Omega$ there is $r_0 > 0$ such that $\overline{B^m(x_0, r_0) \times B^n(y_0, r_0)} \subset \Omega$ and such that the following conditions are satisfied:*

(a) *for each $y \in B^n(y_0, r_0)$ the function*

$$\overline{B^m(x_0, r_0)} \ni x \mapsto u(x, y) \in \mathbb{R}$$

is continuous and subharmonic in $B^m(x_0, r_0)$,

(b) *for each $x \in B^m(x_0, r_0)$ the function*

$$\overline{B^n(y_0, r_0)} \ni y \mapsto u(x, y) \in \mathbb{R}$$

is continuous and harmonic in $B^n(y_0, r_0)$,

(c) *for each $y \in B^n(y_0, r_0)$ one has $\Delta_{1*} u(x, y) < +\infty$ for all $x \in B^m(x_0, r_0)$, possibly with the exception of a polar set in $B^m(x_0, r_0)$,*

(d) *there is a set* $H \subset B^n(y_0, r_0)$, *dense in* $B^n(y_0, r_0)$, *such that for each* $y \in H$ *the function*

$$B^m(x_0, r_0) \ni x \mapsto \Delta_1 u(x, y) \in \mathbb{R}$$

is defined and continuous,

(e) *for each* $\tilde{y}_0 \in B^n(y_0, r_0)$, *for almost all* $\tilde{x}_0 \in B^m(x_0, r_0)$, *and for all sequences* $x_j \in B^m(x_0, r_0)$, $j = 1, 2, \ldots$, *such that* $x_j \to \tilde{x}_0$, *the sequence* $\Delta_{1*}u(x_j, \tilde{y}_0)$ *has a convergent subsequence, converging to* $\Delta_{1*}u(\tilde{x}_0, \tilde{y}_0)$.

Then u *is subharmonic.*

Since the proof is too long to be presented here, we just refer to [Ri06$_2$].

5.3. Another variant of the above result is the following, where the assumption (e) is replaced with a certain "continuity" condition of $v(\cdot, \cdot)$ in the second variable.

Theorem 4. *Let* Ω *be a domain in* $\mathbb{R}^{m+n}$, $m, n \geq 2$. *Let* $u : \Omega \to \mathbb{R}$ *be such that for each* $(x_0, y_0) \in \Omega$ *there is* $r_0 > 0$ *such that* $\overline{B^m(x_0, r_0) \times B^n(y_0, r_0)} \subset \Omega$ *and such that the following conditions are satisfied:*

(a) *for each* $y \in B^n(y_0, r_0)$ *the function*

$$\overline{B^m(x_0, r_0)} \ni x \mapsto u(x, y) \in \mathbb{R}$$

is continuous and subharmonic in $B^m(x_0, r_0)$,

(b) *for each* $x \in B^m(x_0, r_0)$ *the function*

$$\overline{B^n(y_0, r_0)} \ni y \mapsto u(x, y) \in \mathbb{R}$$

is continuous and harmonic in $B^n(y_0, r_0)$,

(c) *for each* $y \in B^n(y_0, r_0)$ *one has* $\Delta_{1*}u(x, y) < +\infty$ *for all* $x \in B^m(x_0, r_0)$, *possibly with the exception of a polar set in* $B^m(x_0, r_0)$,

(d) *there is a set* $H \subset B^n(y_0, r_0)$, *dense in* $B^n(y_0, r_0)$, *such that for each* $y \in H$ *the function*

$$B^m(x_0, r_0) \ni x \mapsto \Delta_1 u(x, y) \in \mathbb{R}$$

is defined and continuous,

(e) *for each* $x \in B^m(x_0, r_0)$ *the function*

$$B^n(y_0, r_0) \ni y \mapsto v(x, y) := \int G_{B^m(x_0, r_0)}(x, z)\, \Delta_1 u(z, y)\, dm(z) \in \mathbb{R}$$

is continuous.

Then u *is subharmonic.*

For the proof we refer again to [Ri06$_2$].

5.4. The assumptions of Theorems 3 and 4 above, especially the (e)-assumptions, are undoubtedly somewhat technical. However, just replacing Kolodziej's and

Thornbiörnson's C^2 assumption of the functions $u(\cdot, y)$ by the continuity requirement of the generalized Laplacians $\Delta_1 u(\cdot, y)$, we obtain the following concise corollary to Theorem 3:

Corollary. *Let Ω be a domain in $\mathbb{R}^{m+n}$, $m, n \geq 2$. Let $u : \Omega \to \mathbb{R}$ be such that*

(a) *for each $y \in \mathbb{R}^n$ the function*

$$\Omega(y) \ni x \mapsto u(x, y) \in \mathbb{R}$$

is continuous and subharmonic,

(b) *for each $x \in \mathbb{R}^m$ the function*

$$\Omega(x) \ni y \mapsto u(x, y) \in \mathbb{R}$$

is harmonic,

(c) *for each $y \in \mathbb{R}^n$ the function*

$$\Omega(y) \ni x \mapsto \Delta_1 u(x, y) \in \mathbb{R}$$

is defined and continuous.

Then u is subharmonic.

References

[ABR01] Axel, S., Bourdon, P., Ramey, W. "'Harmonic Function Theory", Springer-Verlag, New York, 2001 (Second Edition).

[AG93] Armitage, D.H., Gardiner, S.J. "Conditions for separately subharmonic functions to be subharmonic", Pot. Anal., **2** (1993), 255–261.

[Ar66] Arsove, M.G. "On subharmonicity of doubly subharmonic functions", Proc. Amer. Math. Soc., **17** (1966), 622–626.

[Av61] Avanissian, V. "Fonctions plurisousharmoniques et fonctions doublement sousharmoniques", Ann. Sci. cole Norm. Sup., **78** (1961), 101–161.

[CZ61] Calderon, A.P., Zygmund, A. "Local properties of solutions of elliptic partial differential equations", Studia Math., **20** (1961), 171–225.

[CS93] Cegrell, U., Sadullaev, A. "Separately subharmonic functions", Uzbek. Math. J., **1** (1993), 78–83.

[He71] Herv, M. "Analytic and Plurisubharmonic Functions in Finite and Infinite Dimensional Spaces", Lecture Notes in Mathematics 198, Springer-Verlag, Berlin, 1971.

[KT96] Kolodziej, S., Thorbiörnson, J. "Separately harmonic and subharmonic functions", Pot. Anal., **5** (1996), 463–466.

[Le45] Lelong, P. "Les fonctions plurisousharmoniques", Ann. Sci. cole Norm. Sup., **62** (1945), 301–328.

[Le61] Lelong, P. "Fonctions plurisousharmoniques et fonctions analytiques de variables relles", Ann. Inst. Fourier, Grenoble, **11** (1961), 515–562.

[Le69] Lelong, P. "Plurisubharmonic Functions and Positive Differential Forms", Gordon and Breach, London, 1969.

[Pa94] Pavlović, M. "On subharmonic behavior and oscillation of functions on balls in $\mathbb{R}^n$", Publ. Inst. Math. (Beograd), **55 (69)** (1994), 18–22.

[PR05] Pavlović, M., Riihentaus, J. "Classes of quasi-nearly subharmonic functions", preprint, 2005.

[Ri89] Riihentaus, J. "On a theorem of Avanissian–Arsove", Expo. Math., **7** (1989), 69–72.

[Ri00] Riihentaus, J. "Subharmonic functions: non-tangential and tangential boundary behavior" in: Function Spaces, Differential Operators and Nonlinear Analysis (FSDONA'99), Proceedings of the Syöte Conference 1999, Mustonen, V., Rákosnik, J. (eds.), Math. Inst., Czech Acad. Science, Praha, 2000, pp. 229–238 (ISBN 80-85823-42-X).

[Ri06$_1$] Riihentaus, J. "A weighted boundary limit result for subharmonic functions", Adv. Algebra and Analysis, **1** (2006), 27–38.

[Ri06$_2$] Riihentaus, J. "Separately subharmonic functions", manuscript, 2006.

[Ru50] Rudin, W. "Integral representation of continuous functions", Trans. Amer. Math. Soc., **68** (1950), 278–286.

[Ru79] Rudin, W. *Real and Complex Analysis*, Tata McGraw-Hill, New Delhi, 1979.

[Sa41] Saks, S. "On the operators of Blaschke and Privaloff for subharmonic functions", Rec. Math. (Mat. Sbornik), **9 (51)** (1941), 451–455.

[Sh71] Shapiro, V.L. "Removable sets for pointwise subharmonic functions", Trans. Amer. Math. Soc., **159** (1971), 369–380.

[Sh78] Shapiro, V.L. "Subharmonic functions and Hausdorff measure", J. Diff. Eq., **27** (1978), 28–45.

[Vu82] Vuorinen, M. "On the Harnack constant and the boundary behavior of Harnack functions", Ann. Acad. Fenn., Ser. A I, Math., **7** (1982), 259–277.

[Wi88] Wiegerinck, J. "Separately subharmonic functions need not be subharmonic", Proc. Amer. Math. Soc., **104** (1988), 770-771.

[WZ91] Wiegerinck, J., Zeinstra, R. "Separately subharmonic functions: when they are subharmonic" in: Proceedings of Symposia in Pure Mathematics, vol. **52**, part **1**, Eric Bedford, John P. D'Angelo, Robert E. Greene, Steven G. Krantz (eds.), Amer. Math. Soc., Providence, Rhode Island, 1991, pp. 245–249.

HARMONIC COMMUTATIVE BANACH ALGEBRAS AND SPATIAL POTENTIAL FIELDS

S. A. PLAKSA

Institute of Mathematics of the National Academy of Sciences of Ukraine,
3 Tereshchenkivska Street, Kiev, Ukraine
plaksa@imath.kiev.ua

For investigation of equations with partial derivatives we develop a method analogous to the analytic function method in the complex plane. We have obtained expressions of solutions of elliptic equations degenerating on an axis via components of analytic functions taking values in a commutative associative Banach algebra.

Keywords: Laplace equation; harmonic commutative Banach algebras; monogenic function; axial-symmetric potential; Stokes flow function.

1. Introduction

Analytic function methods in the complex plane for plane potential fields inspire searching of analogous methods for spatial potential solenoid fields. The problem to construct such methods for spatial potential solenoid fields was posed by M.A. Lavrentyev [1, p. 205]). Besides, even independently of relation with applications in mathematical physics, for a long time a variety and an effectiveness of the analytic function methods stimulate developing of analogous methods for equations with partial derivatives.

Apparently, Hamilton made the first attempts to construct an algebra associated with the three-dimensional Laplace equation

$$\Delta u(x,y,z) \equiv \left(\frac{\partial^2}{\partial x^2} + \frac{\partial^2}{\partial y^2} + \frac{\partial^2}{\partial z^2}\right) u(x,y,z) = 0 \tag{1}$$

in that sense that components of differentiable functions taking values in this algebra satisfy Eq. (1). However, after constructing quaternion algebra he had not studied a problem about constructing any other algebra (see [2]).

In the paper [3] I. Mel'nichenko considered the problem to construct commutative associative Banach algebra such that monogenic (i.e. differentiable in accordance with Gateaux) functions taking values in this algebra have components satisfying Eq. (1). It is obvious that the specified problem is appeared as an attempt to generalize the fundamental relation between the algebra of complex numbers and the two-dimensional Laplace equation. As it is well known, this relation means that, on the one hand, analytic functions of complex variable satisfy the two-dimensional

Laplace equation, and, on the other hand, plane harmonic functions conjugate with Cauchy-Riemann conditions are components of certain analytic function of complex variable.

Because monogenic functions taking values in a commutative Banach algebra form a functional algebra, note that a relation between these functions and solutions of Eq. (1) is important for a development of effective methods for constructing mentioned solutions. It is quite natural that on such a way a quantity of fulfilled operations will be minimal in an algebra of third rank.

But in the paper [3] it is established that there does not exist commutative associative algebra of third rank with the main unit over the field of real numbers in which monogenic functions would satisfy Eq. (1). At the same time, for commutative associative algebras of third rank over the field of complex numbers in the papers [3, 4] I. Mel'nichenko developed a method for extracting bases such that hypercomplex monogenic functions constructed in these bases have components satisfying Eq. (1).

However, it is impossible to obtain all solutions of Eq. (1) in the form of components of monogenic functions taking values in commutative algebras of third rank that were constructed in the papers [3,4]. In particular, for each mentioned algebra there exist spherical functions which are not components of specified hypercomplex monogenic functions.

Below, we consider an infinite-dimensional commutative Banach algebra $\mathbb{F}$ over the field of real numbers and establish that any spherical function is a component of some monogenic function taking values in this algebra. Thus, monogenic functions taking values in $\mathbb{F}$ form the widest of known functional algebras associated with Eq. (1).

2. A problem about extracting harmonic triad of vectors

Let $\mathbb{A}$ be a commutative associative Banach algebra over the field of real numbers $\mathbb{R}$ with the basis $\{e_k\}_{k=1}^n$, $3 \leq n \leq \infty$. Consider in $\mathbb{A}$ the linear subspace E_m generated by vectors $e_1, e_2, \ldots, e_m$, where $m \leq n$.

Let G be a domain in E_m. We say that a function $\Phi : G \to \mathbb{A}$ is *monogenic* in the domain G if Φ is differentiable in accordance with Gateaux in every point of G, i.e. for every $\xi \in G$ there exists an element $\Phi'(\xi) \in \mathbb{A}$ such that

$$\lim_{\varepsilon \to 0+0} \left[\Phi(\xi + \varepsilon h) - \Phi(\xi)\right] \varepsilon^{-1} = h\Phi'(\xi) \qquad \forall\, h \in E_m\,.$$

For a domain Q of the three-dimensional space $\mathbb{R}^3$ consider the domain $Q_\xi := \{\xi = xe_1 + ye_2 + ze_3 : (x, y, z) \in Q\} \subset E_3$ which are congruent to Q.

Note that if there exists a twice differentiable in accordance with Gateaux function $\Phi : Q_\xi \to \mathbb{A}$ which satisfies Eq. (1) and the inequality $\Phi''(\xi) \neq 0$ at least in one point $\xi := xe_1 + ye_2 + ze_3 \in Q_\xi$, then in this case the basic elements e_1, e_2, e_3 satisfy the condition

$$e_1^2 + e_2^2 + e_3^2 = 0\,. \tag{2}$$

We say that an algebra $\mathbb{A}$ is *harmonic* if in $\mathbb{A}$ there exists a triad of linearly independent vectors $\{e_1, e_2, e_3\}$ satisfying the equality (2) provided that $e_j^2 \neq 0$ for $j = 1, 2, 3$. We say also that such a triad $\{e_1, e_2, e_3\}$ is *harmonic*.

In the paper [3] it is proved that there does not exist the three-dimensional harmonic algebra over the field of real numbers.

3. Monogenic functions in an infinite-dimensional harmonic algebra

Consider as harmonic algebra $\mathbb{A}$ an infinite-dimensional commutative associative Banach algebra

$$\mathbb{F} := \{g = \sum_{k=1}^{\infty} c_k e_k : c_k \in \mathbb{R}, \sum_{k=1}^{\infty} |c_k| < \infty\}$$

over the field $\mathbb{R}$ with the norm $\|g\|_{\mathbb{F}} := \sum\limits_{k=1}^{\infty} |c_k|$ and the basis $\{e_k\}_{k=1}^{\infty}$, where the multiplication table for elements of basis is of the following form:

$$e_n e_1 = e_n, \qquad e_{2n+1} e_{2n} = \frac{1}{2}\, e_{4n} \quad \forall n \geq 1\,,$$

$$e_{2n+1} e_{2m} = \frac{1}{2} \Big(e_{2n+2m} - (-1)^m e_{2n-2m} \Big) \quad \forall n > m \geq 1\,,$$

$$e_{2n+1} e_{2m} = \frac{1}{2} \Big(e_{2n+2m} + (-1)^n e_{2m-2n} \Big) \quad \forall m > n \geq 1\,,$$

$$e_{2n+1} e_{2m+1} = \frac{1}{2} \Big(e_{2n+2m+1} + (-1)^m e_{2n-2m+1} \Big) \quad \forall n \geq m \geq 1\,,$$

$$e_{2n} e_{2m} = \frac{1}{2} \Big(-e_{2n+2m+1} + (-1)^m e_{2n-2m+1} \Big) \quad \forall n \geq m \geq 1\,.$$

It is evident that here e_1, e_2, e_3 form a harmonic triad of vectors.

Theorem 3.1. *In order that a function $\Phi : Q_\xi \longrightarrow \mathbb{F}$ of the form*

$$\Phi(xe_1 + ye_2 + ze_3) = \sum_{k=1}^{\infty} U_k(x, y, z)\, e_k, \tag{3}$$

where $U_k : Q \longrightarrow \mathbb{R}$, be monogenic in the domain $Q_\xi \subset E_3$, it is necessary and sufficient that the functions U_k, $k = 1, 2, \ldots$, be differentiable in the domain Q, and that the conditions

$$\frac{\partial U_1(x,y,z)}{\partial y} = -\frac{1}{2}\, \frac{\partial U_2(x,y,z)}{\partial x}\,,$$

$$\frac{\partial U_2(x,y,z)}{\partial y} = \frac{\partial U_1(x,y,z)}{\partial x} + \frac{1}{2}\, \frac{\partial U_5(x,y,z)}{\partial x}\,,$$

$$\frac{\partial U_3(x,y,z)}{\partial y} = -\frac{1}{2}\,\frac{\partial U_4(x,y,z)}{\partial x},$$

$$\frac{\partial U_{2k}(x,y,z)}{\partial y} = \frac{1}{2}\,\frac{\partial U_{2k-1}(x,y,z)}{\partial x} + \frac{1}{2}\,\frac{\partial U_{2k+3}(x,y,z)}{\partial x}, \qquad k = 2,3,\dots,$$

$$\frac{\partial U_{2k+1}(x,y,z)}{\partial y} = -\frac{1}{2}\,\frac{\partial U_{2k-2}(x,y,z)}{\partial x} - \frac{1}{2}\,\frac{\partial U_{2k+2}(x,y,z)}{\partial x}, \quad k = 2,3,\dots, \tag{4}$$

$$\frac{\partial U_1(x,y,z)}{\partial z} = -\frac{1}{2}\,\frac{\partial U_3(x,y,z)}{\partial x},$$

$$\frac{\partial U_2(x,y,z)}{\partial z} = -\frac{1}{2}\,\frac{\partial U_4(x,y,z)}{\partial x},$$

$$\frac{\partial U_3(x,y,z)}{\partial z} = \frac{\partial U_1(x,y,z)}{\partial x} - \frac{1}{2}\,\frac{\partial U_5(x,y,z)}{\partial x},$$

$$\frac{\partial U_k(x,y,z)}{\partial z} = \frac{1}{2}\,\frac{\partial U_{k-2}(x,y,z)}{\partial x} - \frac{1}{2}\,\frac{\partial U_{k+2}(x,y,z)}{\partial x}, \qquad k = 4,5,\dots,$$

be satisfied in Q, and that the following relations be fulfilled:

$$\sum_{k=1}^{\infty}\left|\frac{\partial U_k(x,y)}{\partial x}\right| < \infty, \tag{5}$$

$$\lim_{\varepsilon\to 0+0}\sum_{k=1}^{\infty}\Bigg| U_k(x+\varepsilon h_1, y+\varepsilon h_2, z+\varepsilon h_3) - U_k(x,y,z) - \frac{\partial U_k(x,y,z)}{\partial x}\,\varepsilon h_1 - $$

$$ - \frac{\partial U_k(x,y,z)}{\partial y}\,\varepsilon h_2 - \frac{\partial U_k(x,y,z)}{\partial z}\,\varepsilon h_3\Bigg|\,\varepsilon^{-1} = 0 \quad \forall\, h_1,h_2,h_3 \in \mathbb{R}. \tag{6}$$

The proof of Theorem 2.1 is similar to the proof of the corresponding classical theorem in the theory of analytic functions of complex variables. Note that the conditions (4) are similar by nature to the Cauchy–Riemann conditions for monogenic functions of complex variables.

It is clear that if the Gateaux derivative Φ' of monogenic function $\Phi : Q_\xi \longrightarrow \mathbb{F}$, in turn, is monogenic function in the domain Q_ξ, then all components U_k of expansion (3) satisfy Eq. (1) in Q in consequence of condition (2). At the same time, the following statement is true even independently of relation between solutions of the system of equations (4) and monogenic functions.

Theorem 3.2. *If the functions $U_k : Q \longrightarrow \mathbb{R}$ have continuous second-order partial derivatives in the domain Q and satisfies the conditions (4), then they satisfies Eq. (1) in Q.*

Note that the algebra $\mathbb{F}$ is isomorphic to the algebra $\mathbf{F}$ of absolutely convergent trigonometric Fourier series

$$g(\tau) = a_0 + \sum_{k=1}^{\infty}\Big(a_k\, i^k \cos k\tau + b_k\, i^k \sin k\tau\Big)$$

with real coefficients a_0, a_k, b_k and the norm $\|g\|_{\mathbf{F}} := |a_0| + \sum\limits_{k=1}^{\infty}\Big(|a_k| + |b_k|\Big)$. In this case, we have the isomorphism $e_{2k-1} \longleftrightarrow i^{k-1}\cos(k-1)\tau$, $e_{2k} \longleftrightarrow i^k \sin k\tau$ between basic elements.

Let us write the expansion of a power function of the variable $\xi = xe_1 + ye_2 + ze_3$ in the basis $\{e_k\}_{k=1}^{\infty}$, using spherical coordinates ρ, θ, ϕ which have the following relations with x, y, z:

$$x = \rho\cos\theta, \quad y = \rho\sin\theta\sin\phi, \quad z = \rho\sin\theta\cos\phi\,. \tag{7}$$

In view of the isomorphism of the algebras $\mathbb{F}$ and $\mathbf{F}$, the construction of expansions of this sort is reduced to the determination of relevant Fourier coefficients. So we have

$$\xi^n = \rho^n\Bigg(P_n(\cos\theta)\; e_1 +$$

$$+2\sum_{m=1}^{n}\frac{n!}{(n+m)!}\; P_n^m(\cos\theta)\Big(\sin m\phi\; e_{2m} + \cos m\phi\; e_{2m+1}\Big)\Bigg), \tag{8}$$

where n is a positive integer, P_n and P_n^m are Legendre polynomials and associated Legendre polynomials, respectively, namely:

$$P_n(t) := \frac{1}{2^n\; n!}\,\frac{d^n}{dt^n}\,(t^2 - 1)^n, \qquad P_n^m(t) := (1-t^2)^{m/2}\,\frac{d^m}{dt^m}\,P_n(t)\,.$$

Thus, $2n+1$ linearly independent spherical functions of the n-th power are components of the expansion (8) of the function ξ^n. Using the expansion (8) and rules of multiplication for basic elements of the algebra $\mathbb{F}$, it is easy to prove the following statement.

Theorem 3.3. *Every spherical function*

$$\rho^n\Bigg(a_{n,0}P_n(\cos\theta) + \sum_{m=1}^{n}\Big(a_{n,m}\cos m\phi + b_{n,m}\sin m\phi\Big)P_n^m(\cos\theta)\Bigg)$$

where $a_{n,0}, a_{n,m}, b_{n,m} \in \mathbb{R}$, is the first component of expansion of the monogenic function

$$\Bigg(a_{n,0}\, e_1 + \sum_{m=1}^{n}(-1)^m\,\frac{(n+m)!}{n!}\,\Big(b_{n,m}\, e_{2m} + a_{n,m}\, e_{2m+1}\Big)\Bigg)\,\xi^n$$

in the basis $\{e_k\}_{k=1}^{\infty}$, where $\xi := xe_1 + ye_2 + ze_3$, and x, y, z have the relations (7) with the spherical coordinates ρ, θ, ϕ.

4. Monogenic functions associated with axial-symmetric potential fields

Now, let us consider a subalgebra $\mathbb{H} := \{a = \sum_{k=1}^{\infty} a_k e_{2k-1} \ : \ a_k \in \mathbb{R}, \ \sum_{k=1}^{\infty} |a_k| < \infty\}$ of the algebra $\mathbb{F}$. In the paper [5] I. Mel'nichenko offered the algebra $\mathbb{H}$ for describing spatial axial-symmetric potential fields.

A spatial potential solenoid field symmetric with respect to the axis Ox is described in meridian plane xOr in terms of the axial-symmetric potential φ and the Stokes flow function ψ satisfying the following system of equations:

$$r\,\frac{\partial\varphi(x,r)}{\partial x} = \frac{\partial\psi(x,r)}{\partial r}\,, \qquad r\,\frac{\partial\varphi(x,r)}{\partial r} = -\frac{\partial\psi(x,r)}{\partial x}\,, \tag{9}$$

where $r^2 = y^2 + z^2$.

As in the papers [6, 7], consider a comlexification $\mathbb{H}_{\mathbb{C}} := \mathbb{H} \oplus i\mathbb{H} \equiv \{c = a + ib \ : \ a, b \in \mathbb{H}\}$ of the algebra $\mathbb{H}$ such that the norm of element $g := \sum_{k=1}^{\infty} c_k e_{2k-1} \in \mathbb{H}_{\mathbb{C}}$ is given by means the equality $\|g\|_{\mathbb{H}_{\mathbb{C}}} := \sum_{k=1}^{\infty} |c_k|$. Consider the set $\mathcal{I}_0 := \{g \in \mathbb{H}_{\mathbb{C}} : \sum_{k=1}^{\infty} (-1)^k (\operatorname{Re} c_{2k-1} - \operatorname{Im} c_{2k}) = 0, \sum_{k=1}^{\infty} (-1)^k (\operatorname{Re} c_{2k} + \operatorname{Im} c_{2k-1}) = 0\}$ which is a maximum ideal of the algebra $\mathbb{H}_{\mathbb{C}}$.

Let $f_{\mathcal{I}_0} : \mathbb{H}_{\mathbb{C}} \longrightarrow \mathbb{C}$ be the linear functional such that $\mathcal{I}_0$ is its kernel.

Consider the Cartesian plane $\mu := \{\zeta = xe_1 + re_3 \ : \ x, r \in \mathbb{R}\}$. For a domain $D \subset \mathbb{R}^2$ we consider the domains $D_z := \{z = x + ir \ : \ (x,r) \in D\} \subset \mathbb{C}$ and $D_\zeta := \{\zeta = xe_1 + re_3 \ : \ (x,r) \in D\} \subset \mu$ which are congruent to the domain D.

Let A be the linear operator which assigns to every function $\Phi : D_\zeta \longrightarrow \mathbb{H}_{\mathbb{C}}$ the function $F : D_z \longrightarrow \mathbb{C}$ by the formula $F(z) := f_{\mathcal{I}_0}(\Phi(\zeta))$, where $z = x + ir$ and $\zeta = xe_1 + re_3$. It is easy to prove that if Φ is a monogenic function in D_ζ, then F is an analytic function in D_z.

In the paper [6] we established necessary and sufficient conditions for monogenety of function $\Phi : D_\zeta \longrightarrow \mathbb{H}_{\mathbb{C}}$ in the form similar to (4) – (6). We established also relations between monogenic functions taking values in the algebra $\mathbb{H}_{\mathbb{C}}$ and solutions of the system (9) in so-called proper domains.

We call D_z a *proper* domain in $\mathbb{C}$, provided that for every $z \in D_z$ with $\operatorname{Im} z \neq 0$ the domain D_z contains the segment connecting points z and $\bar{z}$. In this case D_ζ is also called a *proper* domain in μ.

Theorem 4.1. [6]. *If D_ζ is a proper domain in μ, then every monogenic function $\Phi : D_\zeta \longrightarrow \mathbb{H}_{\mathbb{C}}$ is expressed in the form*

$$\Phi(\zeta) = \frac{1}{2\pi i} \int_\gamma (te_1 - \zeta)^{-1} (A\Phi)(t)\,dt + \Phi_0(\zeta) \qquad \forall\, \zeta \in D_\zeta\,, \tag{10}$$

where γ is an arbitrary closed Jordan rectifiable curve in D_z that embraces the segment connecting the points $z = f_{\mathcal{I}_0}(\zeta)$ and $\bar{z}$, and the function Φ_0 is a monogenic function taking values in the ideal $\mathcal{I}_0$.

Note that under the conditions of Theorem 4.1 the integral in the equality (10) is the principal extension of analytic function $F = A\Phi$ into the domain D_ζ. Thus, the algebra of monogenic functions in D_ζ can be decomposed into the direct sum of the algebra of principal extensions of complex-valued analytic functions and the algebra of monogenic functions taking values in the ideal I_0.

In the paper [6], for every function $F : D_z \to \mathbb{C}$ analytic in a proper domain D_z we constructed explicitly the expansion in the basis $\{e_{2k-1}\}_{k=1}^{\infty}$ of the principal extension of F into the domain D_ζ:

$$\frac{1}{2\pi i}\int\limits_{\gamma}(te_1-\zeta)^{-1}F(t)\,dt = U_1(x,r)\,e_1 + 2\sum_{k=2}^{\infty}U_k(x,r)\,e_{2k-1}\,, \tag{11}$$

where

$$U_k(x,r) := \frac{1}{2\pi i}\int\limits_{\gamma}\frac{F(t)}{\sqrt{(t-z)(t-\bar{z})}}\left(\frac{\sqrt{(t-z)(t-\bar{z})}-(t-x)}{r}\right)^{k-1}dt\,, \tag{12}$$

$\zeta = xe_1 + re_3$ and $z = x + ir$ for $(x,r) \in D$, and the curve γ has the same properties as in Theorem 4.1, and $\sqrt{(t-z)(t-\bar{z})}$ is a continuous branch of this function analytic with respect to t outside of the segment mentioned in Theorem 4.1. Note that for every $z \in D_z$ with $\operatorname{Im} z = 0$, we define $\sqrt{(t-z)(t-\bar{z})} := t - z$.

Theorem 4.2. [6]. *If D_z is a proper domain in $\mathbb{C}$ and F is an analytic function in D_z, then the first and the second components of principal extension (11) of function F into the domain D_ζ generate the solutions φ and ψ of system (9) in D by the formulas*

$$\varphi(x,r) = U_1(x,r), \quad \psi(x,r) = r\,U_2(x,r)\,. \tag{13}$$

From the relations (11) – (13) it follows that the functions

$$\varphi(x,y) = \frac{1}{2\pi i}\int\limits_{\gamma}\frac{F(t)}{\sqrt{(t-z)(t-\bar{z})}}\,dt\,, \tag{14}$$

$$\psi(x,y) = -\frac{1}{2\pi i}\int\limits_{\gamma}\frac{F(t)\,(t-x)}{\sqrt{(t-z)(t-\bar{z})}}\,dt\,, \qquad z = x + ir\,, \tag{15}$$

are solutions of system (9) in the domain D.

5. Integral expressions for axial-symmetric potential and the the Stokes flow function

We generalized integral expressions (14) and (15) for the axial-symmetric potential and the Stockes flow function, respectively, to the case of arbitrary simply connected domain symmetric with respect to the axis Ox.

In what follows, D is a bounded simply connected domain symmetric with respect to the axis Ox of the plane xOr. For every $z \in D_z$ with $\operatorname{Im} z \neq 0$, we fix an arbitrary Jordan rectifiable curve $\Gamma_{z\bar{z}}$ in D_z which connects the points z and $\bar{z}$. In this case, let $\sqrt{(t-z)(t-\bar{z})}$ be a continuous branch of this function analytic with respect to t outside of the cut along $\Gamma_{z\bar{z}}$.

Theorem 5.1. [8]. *Suppose that the axial-symmetric potential $\varphi(x,r)$ is even with respect to the variable r in the domain D. Then there exists the unique function F analytic in the domain D_z and satisfying the condition*

$$F(\bar{z}) = \overline{F(z)} \qquad \forall\, z \in D_z \tag{16}$$

such that the equality (14) is fulfilled for all $(x,r) \in D$.

Theorem 5.2. [8]. *Suppose that the Stokes flow function $\psi(x,r)$ is even with respect to the variable r in the domain D and satisfies the additional assumption*

$$\psi(x,0) \equiv 0 \quad \forall\,(x,0) \in D\,. \tag{17}$$

Then there exists a function F_0 analytic in the domain D_z such that the equality (15) is fulfilled with $F = F_0$ for all $(x,r) \in D$. Moreover, any analytic function F which satisfies the condition (16) and the equality (15) for all $(x,r) \in D$ is expressed in the form $F(z) = F_0(z) + C$, where C is a real constant.

Note that the requirement (17) is natural. For example, for the model of steady flow of an ideal incompressible fluid without sources and vortexes it means that the axis Ox is a line of flow.

Using integral expressions (14) and (15), we developed a method for effective solving boundary problems for axial-symmetric potential fields (see [8]).

References

1. M. A. Lavrentyev and B. V. Shabat, *Problems of hydrodynamics and theirs mathematical models*, Moskow: Nauka, 1977, 408 p. [in Russian]
2. F. Klein, *Vorlesungen über die entwicklung der mathematik im 19 jahrhundert*, T.1, Berlin: Verlag von Julius Springer, 1926.
3. I. P. Mel'nichenko, On expression of harmonic mappings by monogenic functions, *Ukr. Math. J.*, **27** (1975), no. 5, 606–613.
4. I. P. Mel'nichenko, Algebras of functionally-invariant solutions of the three-dimensional Laplace equation, *Ukr. Math. J.*, **55** (2003), no. 9, 1284–1290.
5. I. P. Mel'nichenko, On a method of description of potential fields with axial symmetry, in: *Contemporary Questions of Real and Complex Analysis*, Kiev: Institute of Mathematics of Ukrainian Academy of Sciences, 1984, 98–102. [in Russian]
6. I. P. Mel'nichenko and S. A. Plaksa, Potential fields with axial symmetry and algebras of monogenic functions of vector variable, III, *Ukr. Math. J.*, **49** (1997), no. 2, 253–268.
7. S. Plaksa, Algebras of hypercomplex monogenic functions and axial-symmetrical potential fields, in: *Proceedings of the Second ISAAC Congress, Fukuoka, August 16–21, 1999*, Netherlands–USA: Kluwer Academic Publishers, **1** (2000), 613–622.
8. S. Plaksa, Singular and Fredholm integral equations for Dirichlet boundary problems for axial-symmetric potential fields, in: *Factotization, Singular Operators and Related Problems: Proceedings of the Conference in Honour of Professor Georgii Litvinchuk, Funchal, January 28–February 1, 2002*, Netherlands–USA: Kluwer Academic Publishers, 2003, 219–235.

THE PARAMETER SPACE OF ERROR FUNCTIONS OF THE FORM $a\int_0^z e^{-w^2}dw$

SHUNSUKE MOROSAWA

Department of Mathematics and Information Science,
Faculty of Science, Kochi University
780-8520, Japan
morosawa@math.kochi-u.ac.jp

In this note, we consider the parameter space of error functions of the form $a\int_0^z e^{-w^2}dw$. Some properties of hyperbolic components of the parameter space are shown.

Keywords: Complex dynamics; transcendental entire function; error function; hyperbolic component.

1. Introduction

Let f be a transcendental entire function. The maximal open set in $\mathbb{C}$ where the family $\{f^n\}$ is normal is called the Fatou set of f and its complement in $\widehat{\mathbb{C}}$ is called the Julia set of f. When we consider complex dynamics of transcendental entire functions, the simplest one is an exponential map $E_\lambda(z) = \lambda e^z$ $(\lambda \neq 0)$. It has only one asymptotic value 0 as a singular value. Behavior of its orbit under iterate decides the dynamics. The parameter space for E_λ was first studied by Baker and Rippon[1] and Devaney[2] independently. In this note, we consider error functions, which are defined in Section 2. The error function has two asymptotic values. We investigate a parameter space of a certain subclass of error functions. Elementary facts on dynamics of transcendental entire functions we use in this note can be found in Ref. 4.

2. Results

An error function is a transcendental entire function given by the form

$$aEr(z) + b$$

with $a \in \mathbb{C} \setminus \{0\}$ and $b \in \mathbb{C}$, where

$$Er(z) = \int_0^z e^{-w^2}dw.$$

It has two asymptotic values $\pm a\sqrt{\pi}/2 + b$ and has no other singular value. A certain property of Julia sets of error functions given by the form

$$aEr(z) + i\sqrt{B}$$

with $a \in \mathbb{R} \setminus \{0\}$ and $B \in \mathbb{R}$ are shown by Morosawa[3] . In this note, we investigate a subclass of error functions given by the form

$$f_a(z) = aEr(z)$$

with $a \in \mathbb{C} \setminus \{0\}$. We note that f_a is a one complex parameter family.

We say that f_a is hyperbolic if the orbit of each asymptotic value accumulates to attracting cyclic points. A connected component of the set of parameters for which f_a is hyperbolic is called a hyperbolic component. It is known that hyperbolic components are open.

We define subsets in the parameter space of f_a as follows:

$$\begin{aligned} A &= \{a \mid f_a \text{ has a completely invariant component.}\}, \\ B_n &= \{a \mid f_a \text{ has only one attracting cycle with the period } 2n.\}, \\ D_n &= \{a \mid f_a \text{ has two attracting cycles with the period } n.\}, \end{aligned}$$

for $n \in \mathbb{N}$. We have the following theorems.

Theorem 2.1. *Every hyperbolic component is contained in one of A, B_n and D_n. Furthermore, A is also described by $\{a \mid |a| < 1\}$. Each of B_1 and D_1 consists of only one component.*

Theorem 2.2. *Every hyperbolic component except A is simply-connected and unbounded.*

Theorem 2.3. *Each of B_n and D_n contains a component which is tangent to A.*

3. Proofs

In this section, we give proofs for the theorems given in Section 2.

Proof. [Proof of Theorem 2.1] First, we note that

$$f_a(-z) = -f_a(z).$$

Hence, if there exists a cycle $\{z_1, z_2, \cdots z_n\}$, then $\{-z_1, -z_2, \cdots - z_n\}$ is also a cycle. Furthermore, we have

$$(f_a^n)'(z_i) = (f_a^n)'(-z_i).$$

Thus, if it is attracting, repelling or indifferent, then so is the corresponding one, respectively. Therefore, if f_a has only one attracting cycle with the period $n \geq 2$, then n is even.

If f_a has completely invariant component, then it has only one attracting fixed point or only one parabolic fixed point. In either case, the fixed point is the origin. If f_a has only one parabolic fixed point, then $a = 1$. Since we have

$$f_1(z) = Er(z) = z - \frac{z^3}{3} + \cdots,$$

Fig. 1. The parameter space of $f_a(z)$. The range shown is $|\operatorname{Re} a| \leq 2$, $|\operatorname{Im} a| \leq 2$. The disk in the center is A. Hyperbolic components of B_n are colored white and those of D_n are colored black.

f_a has two petals at the origin. Because Fatou sets of transcendental entire functions have at most one completely invariant component, f_a has no completely invariant component. If f_a has only one attracting fixed point at the origin, then the immediate basin of the attracting fixed point contains at least one asymptotic values. From the symmetry, both asymptotic values are contained in it. It follows that f_a has a completely invariant component. Hence f_a has a completely invariant component if and only if f_a has an attracting fixed point at the origin. It is clear that f_a has an attracting fixed point at the origin if and only if a satisfies $|a| < 1$.

If f_a has a fixed point $\zeta \neq 0$, then a is given by $\zeta/Er(\zeta)$. Furthermore, if ζ is attracting, then it satisfies $|\zeta e^{-\zeta^2}/Er(\zeta)| < 1$. Hence, for f_a with a sufficiently small multiplier of ζ, ζ is in the sectors

$$\left\{z \,\middle|\, |\arg z| < \frac{\pi}{4}\right\} \cup \left\{z \,\middle|\, |\arg z - \pi| < \frac{\pi}{4}\right\}$$

and the same holds for a. It follows that D_1 consists of only one component. Similarly, we see that B_1 consists of only one component. ■

Proof. [Proof of Theorem 2.2] The arguments here have already been given in Refs. 1,2.

It is clear that $g_k(a) = f_a^k(a\sqrt{\pi}/2)$ is an entire function of a. Let D be a component of D_n. We see that $\{g_{nk}(a)\}$ converges uniformly on compact sets in D to one of the attracting cyclic point. Hence D is simply-connected. Next, we consider

$$\left\{\lambda_{nk}(a) = (f_a^n)'\left(f_a^{nk}\left(\frac{a\sqrt{\pi}}{2}\right)\right)\right\}$$

which converges to the multiplier $\lambda(a)$ of the cyclic point in $\overline{D}$. It satisfies that $|\lambda(a)| < 1$ in D and $|\lambda(a)| = 1$ on ∂D. Since $\lambda(a) \neq 0$ in D, we see that D is unbounded.

Similarly, we obtain the claim for B_n. ■

Proof. [Proof of Theorem 2.3] From Theorem 2.1, A is described by $\{a \mid |a| < 1\}$. For $a = e^{i\theta}$ ($\theta \in \mathbb{R}$) on ∂A, the origin is indifferent fixed point. As we saw, f_1 has two petals at the origin. The hyperbolic component D_1 is tangent to A at 1.

For $\theta = 2\pi k/n$ with $k,\ n \in \mathbb{N}$, $k < n$ and $(k, n) = 1$, we have

$$f_a(z) = z - \mu_a z^{n+1} + \cdots$$

for $\mu_a \neq 0$, if n is even and

$$f_a(z) = z - \mu_a z^{2n+1} + \cdots$$

for $\mu_a \neq 0$, if n is odd. Hence, if n is even, then f_a has n petals at the origin and has only one cycle with the period n. If n is odd, then f_a has $2n$ petals at the origin and has two cycles with the period n. By the argument similar to those in the proof of Theorem 3 in Ref. 1,

$$(1+\epsilon)e^{i2\pi\frac{k}{n}}Er(z)$$

has an attracting cycle for sufficiently small $\epsilon > 0$. ■

Acknowledgments

The author would like to thank the foundation Grant-in-Aid for Scientific Research, Japan Society for the Promotion of Science, (No. 17540163).

References

1. I. N. Baker and P. J. Rippon, Iteration of exponential functions, *Ann. Acad. Sci. Fenn. Ser. AI Math.*, **9**(1984), 47–77.
2. R. L. Devaney, Complex dynamics and entire functions, in *Complex Dynamical Systems, Proceeding of Symposia in Applied mathematics 49* (American Mathematical Society, Providence, 1994), 181–206.
3. S. Morosawa, Fatou components whose boundaries have a common curve, *Fund. Math.* **183**(2004), 47–57.
4. S. Morosawa, Y. Nishimura, M. Taniguchi, and T. Ueda, Holomorphic Dynamics, Cambridge Studies in Advanced Mathematics 66, (Cambridge University Press 2000).

ON POTENTIAL THEORY ASSOCIATED TO A COUPLED PDE

ALLAMI BENYAICHE

Université Ibn Tofail,
Faculté de Sciences,
B.P: 133, Kénitra-Morocco

In this paper, we give some results concerning the Martin boundary and the restricted mean value property for harmonic functions associated with a harmonic structure given by a coupled partial differential equations. In particular, we obtain such results for biharmonic functions (i.e: $\triangle^2\varphi = 0$) and for $\triangle^2\varphi = \varphi$ equations.

Keywords: Martin boundary, mean value property, biharmonic functions, coupled partial differential equations.

1. Introduction

Let D be a domain in $\mathbb{R}^d$, $d \geq 1$ and let L_i; i=1,2, be two second order elliptic differential operators on D leading to harmonic spaces $(D, \mathcal{H}_{L_i})$ with Green functions G_i (see[10]). Moreover, we assume that every ball $B \subset \bar{B} \subset D$ is a L_i-regular set. Throughout this paper we consider two positive Radon measures μ_1 and μ_2 such that $K_D^{\mu_i} = \int_D G_i(.,y)\mu_i(dy)$ is a bounded continuous real function on D; i=1,2, and $\| K_D^{\mu_1} \|_\infty . \| K_D^{\mu_2} \|_\infty < 1$. We consider the system:

$$(S)\begin{cases} L_1 u = -v.\mu_1 \\ L_2 v = -u.\mu_2. \end{cases}$$

Note that if U is a relatively compact open subset of D, $\mu_1 = \lambda^d$, where λ^d is the Lebesgue measure, $\mu_2 = 0$, and $L_1 = L_2 = \triangle$, then we obtain the classical biharmonic case on U. In the case when $\mu_1 = \mu_2 = \lambda^d$, and $\lambda^d(D) < \infty$, we obtain equations of $\triangle^2\varphi = \varphi$ type. In this work, we give some results concerning the Martin boundary and the restricted mean value property for harmonic functions with respect to the balayage space given by (S). The interested reader can see[2,3] for more results. Let us note that the notion of a balayage space defined by J. Bliedtner and W. Hansen in[4] , is more general than that of a P-harmonic space. It covers harmonic structures given by elliptic or parabolic partial differential equations, Riesz potentials and biharmonic equations (which are a particular case of this work). In the biharmonic case, a similar study can be done using couples of functions as presented in[1,5,12] .

2. Notations and preliminaries

For j=1, 2, let $X_j = D \times \{j\}$, and let $X = X_1 \bigcup X_2$. Moreover, let i_j and π_j the mappings defined by

$$i_j : \begin{cases} D \longrightarrow X_j \\ x \longmapsto (x,j) \end{cases} \text{ and } \pi_j : \begin{cases} X_j \longrightarrow D \\ (x,j) \longmapsto x. \end{cases}$$

We denote also by π the mapping from X to D such that $\pi \mid_{X_j} = \pi_j$
Let $\mathcal{U}_0$ be the set of all balls B such that $B \subset \bar{B} \subset D$, $\mathcal{U}_j$ be the image of $\mathcal{U}_0$ by i_j, j=1,2 and $\mathcal{U} = \mathcal{U}_1 \bigcup \mathcal{U}_2$.

Definition 2.1. Let v be a measurable function on X. For $U \in \mathcal{U}_1$, we define a kernel S_U by:

$$S_U v = (H^1_{\pi_1(U)}(v \; o \; i_1)) \; o \; \pi_1 + (K^{\mu_1}_{\pi_1(U)}(v \; o \; i_2)) \; o \; \pi_1.$$

For $U \in \mathcal{U}_2$, we define a kernel S_U by:

$$S_U v = (H^2_{\pi_2(U)}(v \; o \; i_2)) \; o \; \pi_2 + (K^{\mu_2}_{\pi_2(U)}(v \; o \; i_1)) \; o \; \pi_2.$$

Where $H^j_{\pi_j(U)}$; j=1,2, denote harmonic kernels associated with $(D, \mathcal{H}_{L_j})$ and

$$K^{\mu_i}_{\pi_i(U)}(w) = \int G_i^{\pi_i(U)}(.,y)w(y)\mu_i(dy) \;, \quad i = 1, 2.$$

Where w is a measurable function on D and $G_i^{\pi_i(U)}$ is the Green function associated with the operator L_i on $\pi_i(U)$. Let G_j, $j = 1, 2$, be the Green kernel associated with L_j on D. The family of kernels $(S_U)_{U \in \mathcal{U}}$ yields a balayage space on X as defined in[4,9] .

Let ${}^*\mathcal{H}(X)$ denote the set of all hyperharmonic functions on X, i.e.

$${}^*\mathcal{H}(X) := \{v \in \mathcal{B}(X); \; v \; is \; l.s.c \; and \; S_U v \leq v \;\; \forall U \in \mathcal{U}\}.$$

Where $\mathcal{B}(X)$ denotes the set of all Borel functions on X. Let $\mathcal{S}(X)$ be the set of all superharmonic functions on X, i.e.

$$\mathcal{S}(X) := \{v \in {}^*\mathcal{H}(X); \; (S_U v) \mid_U \in C(U) \;\; \forall U \in \mathcal{U}\},$$

and let $\mathcal{H}(X)$ be the set of all harmonic functions on X :

$$\mathcal{H}(X) := \{h \in \mathcal{S}(X); S_U h = h \;\; \forall U \in \mathcal{U}\}.$$

Denoting $\mathcal{W} := {}^*\mathcal{H}^+(X)$, the space $(X, \mathcal{W})$ is a balayage space (See[4,9]).

Theorem 2.1 (3). *Let v be a function on X such that $K_D^{\mu_j}(v \; o \; i_k)$; $j \neq k$; $j, k \in \{1, 2\}$, is a finite function. Then, the following properties are equivalent*
1. v is harmonic on X,
2. $v_1 := v \; o \; i_1 - K_D^{\mu_1}(v \; o \; i_2)$ and $v_2 := v \; o \; i_2 - K_D^{\mu_2}(v \; o \; i_1)$ are L_1-harmonic and L_2-harmonic function on D, respectively.

Remark 2.1. (1) Note that if v is a positive harmonic function on X then $K_D^{\mu_j}(v\ o\ i_k)$, $j \neq k$; $j,k \in \{1,2\}$, is a finite function ([2]).
(2) If $v \in \mathcal{H}(X)$, then the couple $(v \circ i_1, v \circ i_2)$ is a solution of (S).
(3) The above theorem still hold if we replace the "harmonicity" by the "hyperharmonicity" (See[3])

In the following, we denote $Q := \sum_{n=0}^{+\infty}(K_D^{\mu_1}K_D^{\mu_2})^n$ (resp. $T := \sum_{n=0}^{+\infty}(K_D^{\mu_2}K_D^{\mu_1})^n$) which coincides with $(I - K_D^{\mu_1}K_D^{\mu_2})^{-1}$ (resp. $(I - K_D^{\mu_2}K_D^{\mu_1})^{-1}$) on $\mathcal{B}_b(D)$, where $(I-K_D^{\mu_1}K_D^{\mu_2})^{-1}$ (resp. $(I-K_D^{\mu_2}K_D^{\mu_1})^{-1}$) is the inverse of the operator $(I-K_D^{\mu_1}K_D^{\mu_2})$ (resp.$(I-K_D^{\mu_2}K_D^{\mu_1})$) on $\mathcal{B}_b(D)$, and $\mathcal{B}_b(D)$ denotes the set of all bounded Borel measurable functions on D . We have the following equalities:

$$(K_D^{\mu_1}K_D^{\mu_2})Q = Q(K_D^{\mu_1}K_D^{\mu_2}),\ (K_D^{\mu_1}K_D^{\mu_2})Q + I = Q,$$

$$(K_D^{\mu_2}K_D^{\mu_1})T = T(K_D^{\mu_2}K_D^{\mu_1}),\ (K_D^{\mu_2}K_D^{\mu_1})T + I = T,$$

$$K_D^{\mu_2}Q = TK_D^{\mu_2} \text{ and } K_D^{\mu_1}T = QK_D^{\mu_1}.$$

Remark 2.2. We note that if φ is a finite positive Borel measurable function on D such that $K_D^{\mu_1}K_D^{\mu_2}\varphi$ is bounded, then $Q\varphi < +\infty$.

3. Martin boundary associated with (S)

Let us fix $x_0 \in D$ and set for all $x,\ y \in D$

$$g^1(x,y) := \begin{cases} \frac{G_1(x,y)}{G_1(x_0,y)} & if\ \ x \neq x_0\ \ or\ \ y \neq x_0 \\ 1 & if\ \ x = y = x_0, \end{cases}$$

$$g^2(x,y) := \begin{cases} \frac{G_2(x,y)}{G_2(x_0,y)} & if\ \ x \neq x_0\ \ or\ \ y \neq x_0 \\ 1 & if\ \ x = y = x_0. \end{cases}$$

Let $\mathcal{A}_1 = \{g^1(x,.),\ x \in D\}$, $\mathcal{A}_2 = \{g^2(x,.),\ x \in D\}$ and $\mathcal{A} = \mathcal{A}_1 \bigcup \mathcal{A}_2$. As in,[6] we consider the Martin compactification $\widehat{D}$ of D associated to $\mathcal{A}$. The boundary $\triangle = \widehat{D} \setminus D$ of D is called the Martin boundary of D associated to (S).
The function $g^k(x,.)$, $k = 1,2$, $x \in D$ can be extended, on $\widehat{D}$, to a continuous function denoted $g^k(x,.)$, $k = 1,2$, $x \in D$ as well.

Definition 3.1. (1) A positive L_j-harmonic function h on D is called L_j-minimal if for any positive L_j-harmonic function u on D, $u \leq h$ implies $u = \alpha.h$ with a factor $\alpha > 0$; $j = 1$, 2
(2) A positive harmonic function h on X is called minimal if for any positive harmonic function u on X, $u \leq h$ implies $u = \alpha h$ with a factor $\alpha > 0$.

Denote for $j = 1,\ 2$, $\triangle_j = \{y \in \triangle :\ \ g^j(.,y)\ \ is\ \ L_j - minimal\}$. Note that, for all $y \in \triangle$, the function $g^j(.,y)$ is L_j-harmonic on D.

Theorem 3.1 (3). *Let v be a positive minimal harmonic function defined on X such that the function $K_D^{\mu_k} K_D^{\mu_j}(v \circ i_k)$, $j \neq k$, $j, k \in \{1,2\}$, is bounded. Then, there exist two real numbers α and β such that*

$$v = \alpha w \quad or \quad v = \beta s,$$

Where w and s are two positive harmonic functions defined on X by

$$w := \begin{cases} (Qg^1(.,y)) \; o \; \pi_1 \;\; on \;\; X_1, \;\; y \in \triangle_1, \\ (K_D^{\mu_2} Qg^1(.,y)) \; o \; \pi_2 \;\; on \;\; X_2, \;\; y \in \triangle_1, \end{cases}$$

and

$$s := \begin{cases} (QK_D^{\mu_1} g^2(.,y)) \; o \; \pi_1 \;\; on \;\; X_1, \;\; y \in \triangle_2, \\ (Tg^2(.,y)) \; o \; \pi_2 \;\; on \;\; X_2, \;\; y \in \triangle_2. \end{cases}$$

We prove in[3] that the set

$$B := \{h \in \mathcal{H}^+(X) : (h \; o \; i_1)(x_o) + (h \; o \; i_2)(x_o) = 1\}, \quad x_o \in D.$$

is a compact base of the cone $\mathcal{H}^+(X)$. Let $\mathcal{E}(B)$ denote the set of all extreme points of $\mathcal{H}^+(X)$ belonging to B (see[6]). Using theorem 3.1, we have

$$\mathcal{E}(B) = \mathcal{E}_1(B) \cup \mathcal{E}_2(B)$$

Where

$$\mathcal{E}_1(B) = \left\{ h \in \mathcal{E}(B) : \exists \alpha \in R^+, \; \exists y \in \triangle_1; \; h = \begin{cases} (\alpha.Qg^1(.,y)) \; o \; \pi_1 \;\; on \;\; X_1 \\ (\alpha.K_D^{\mu_2} Qg^1(.,y)) \; o \; \pi_2 \;\; on \;\; X_2 \end{cases} \right\}$$

and

$$\mathcal{E}_2(B) = \left\{ h \in \mathcal{E}(B) : \exists \beta \in R^+, \; \exists y \in \triangle_2; \; h = \begin{cases} (\beta.QK_D^{\mu_1} g^2(.,y)) \; o \; \pi_1 \;\; on \;\; X_1 \\ (\beta.Tg^2(.,y)) \; o \; \pi_2 \;\; on \;\; X_2 \end{cases} \right\}.$$

Theorem 3.2. *If $g^j(x,.)$; $x \in D$, separates $\triangle_j$, then for any positive harmonic function v on X such that the function $K_D^{\mu_k} K_D^{\mu_j}(v \circ i_k)$; $j \neq k$ $j, k \in \{1,2\}$ is bounded, there exist two unique measures ν_1 and ν_2 supported respectively by $\triangle_1$ and $\triangle_2$ such that v can be represented on X_1 by*

$$v = \int_{\triangle_1} (Qg^1(.,y)) \; o \; \pi_1 d\nu_1(y) + \int_{\triangle_2} (QK_D^{\mu_1} g^2(.,y)) \; o \; \pi_1 d\nu_2(y),$$

and on X_2 by:

$$v = \int_{\triangle_1} (K_D^{\mu_2} Qg^1(.,y)) \; o \; \pi_2 \; d\nu_1(y) + \int_{\triangle_2} (Tg^2(.,y)) \; o \; \pi_2 d\nu_2(y)$$

Proof. If $v = 0$, we have $\nu_1 = \nu_2 = 0$. If $v \neq 0$, we may assume without loss of generality that $v \in B$. Consider the mapping:

$$\Psi : \begin{cases} \triangle_1 \cup \triangle_2 \longrightarrow \mathcal{E}(B) \\ y \longmapsto \Psi(y) \end{cases}$$

Where $\Psi(y)$ is defined by
If $y \in \triangle_1$:

$$\Psi(y) := \begin{cases} (Qg^1(.,y)) \; o \; \pi_1 \;\; on \;\; X_1 \\ (K_D^{\mu_2} Qg^1(.,y)) \; o \; \pi_2 \;\; on \;\; X_2 \end{cases}$$

If $y \in \triangle_2$:

$$\Psi(y) := \begin{cases} (QK_D^{\mu_1} g^2(.,y)) \; o \; \pi_1 \;\; on \;\; X_1 \\ (Tg^2(.,y)) \; o \; \pi_2 \;\; on \;\; X_2 \end{cases}$$

The mapping Ψ is bijective because $g^1(x,.)$ and $g^2(x,.)$ separate $\triangle_1$ and $\triangle_2$ respectively. Ψ and its inverse Ψ^{-1} are continuous because g^1 and g^2 are continuous on $\triangle \times D$. Then there exist, from Choquet's representation ([11]), a unique measure ν supported by $\triangle_1 \cup \triangle_2$ such that:

$$\forall v \in B, \;\; v = \int_{\triangle_1 \cup \triangle_2} \Psi(y) d\nu(y).$$

Let ν_j, $j = 1, 2$ be the restriction of the measure ν on $\triangle_j$. Then, v may be written on X_1 as

$$v = \int_{\triangle_1} (Qg^1(.,y)) \; o \; \pi_1 d\nu_1(y) + \int_{\triangle_2} (QK_D^{\mu_1} g^2(.,y)) \; o \; \pi_1 d\nu_2(y)$$

and on X_2 as

$$v = \int_{\triangle_1} (K_D^{\mu_2} Qg^1(.,y)) \; o \; \pi_2 d\nu_1(y) + \int_{\triangle_2} (Tg^2(.,y)) \; o \; \pi_2 d\nu_2(y).$$

■

4. Restricted mean value property

Let D be a domain in $\mathbb{R}^d$, $d \geq 1$, and r be a numerical positive function on D such that the closed ball $\bar{B}(x, r(x))$ of center x and radius $r(x)$ is contained in D, for any $x \in D$. A numerical function f on D is called r-median if $\lambda_{x,r(x)}(f) = f(x)$, for any $x \in D$. Where λ is the Lebesgue measure on $\mathbb{R}^d$ and

$$\lambda_{x,r(x)} := (\lambda(B(x, r(x))))^{-1} . \chi_{B(x,r(x))} . \lambda.$$

If f is a harmonic function on D, then f is r-median. Several authors are interested to the converse question : If f is r-median, under what conditions, f is harmonic ? For a survey of the history we can see[2,8] .

In this section, we study the restricted mean value property for solutions of the system (S) for $L_j = \Delta$; $j = 1,\ 2$ using the Cornea - Vesely approach

Let $\tilde{r}$ be a positive function on $X = X_1 \cup X_2$ such that $\overline{B}(x, \tilde{r} \circ i_j(x)) \subset D$, for any $x \in D$. For a measurable function f defined on X. We say that f satisfy the restricted mean value property if $\mu_{z,\tilde{r}(z)}(f) = f(z)$. Where

$$\mu_{z,\tilde{r}(z)}(f) := \frac{d}{\tilde{r}(z)^d} \int_0^{\tilde{r}(z)} s^{d-1} S_{i_j(B(x,s))}(f)(z) ds, \text{ for } z = (x, j);\ j = 1, 2;\ x \in D$$

In other words a function f defined on X satisfies the restricted mean value property if and only if

$$\begin{aligned} f \circ i_1(x) &= \frac{d}{\tilde{r} \circ i_1(x)^d} \int_0^{\tilde{r} \circ i_1(x)} s^{d-1} H_{B(x,s)} f \circ i_1(x) ds \\ &+ \frac{d}{\tilde{r} \circ i_1(x))^d} \int_0^{\tilde{r} \circ i_1(x)} s^{d-1} K^{\mu_1}_{B(x,s)} f \circ i_2(x) ds, \end{aligned}$$

$$\begin{aligned} f \circ i_2(x) &= \frac{d}{\tilde{r} \circ i_2(x)^d} \int_0^{\tilde{r} \circ i_2(x)} s^{d-1} H_{B(x,s)} f \circ i_2(x) ds \\ &+ \frac{d}{\tilde{r} \circ i_2(x)^d} \int_0^{\tilde{r} \circ i_2(x)} s^{d-1} K^{\mu_2}_{B(x,s)} f \circ i_1(x) ds. \end{aligned}$$

for any $x \in D$. Where $H_{B(x,s)}$ is the classical harmonic kernel associated to the ball $B(x, s)$.

Remark 4.1. In an obvious way; for any couple of functions (f_1, f_2) on D, we can define the restricted mean value property for such couple if we replace $f \circ i_j$ by f_j and $\tilde{r}$ by r in the previous equalities.

In the following , we assume as in[7] that $r > 0$ and defined on D. Moreover, there exists $\varepsilon > 0$ and a lipschitzian function ρ on D, with Lipschitz constant 1, such that: $\rho < dist(., \partial D)$ and $\varepsilon\rho \le r \le (1 - \varepsilon)\rho$. For any measurable and positive function g on D, we consider the kernels

$$Ng(x) = \frac{d}{r(x)^d} \int_0^{r(x)} s^{d-1} H_{B(x,s)} g(x) ds.$$

and

$$R_j g(x) = \frac{d}{r(x)^d} \int_0^{r(x)} s^{d-1} K^{\mu_j}_{B(x,s)} g(x) ds;\ \ j = 1, 2$$

We consider also the kernel T defined on X by:

$$Tf := \begin{cases} (N(f \circ i_1) + R_1(f \circ i_2)) \circ \pi_1, \text{ on } X_1 \\ (N(f \circ i_2) + R_2(f \circ i_1)) \circ \pi_2, \text{ on } X_2 \end{cases}$$

Remark 4.2. 1) The function $\tilde{r}$ will be defined on each X_j by $r \circ \pi_j$
2) The restricted mean value property for a function f on X is equivalent to $Tf = f$.
3) We have,

$$R_j g(x) = \begin{cases} K^{\mu_j}_D g(x) - N(K^{\mu_j}_D g)(x) & for\ \ K^{\mu_j}_D g < +\infty \\ +\infty & for\ \ K^{\mu_j}_D g = +\infty. \end{cases}$$

Moreover, by definition of R_j, we have $R_j g \geq 0$, for any positive measurable function g. Hence the function $K_D^{\mu_j} g$ is a $N-$ supermedian function. Therefore, from [7, theorem 1.2.(f)], $K_D^{\mu_j} g$ is either locally λ-integrable or identically ∞.

Definition 4.1. A measurable positive function f on X is called T-supermedian if $T(f) \leq f$. A positive function on X such that $f \circ i_j;\ j = 1,2$ are locally $\lambda-$ integrable is called $T-$invariant if $Tf = f$.

Theorem 4.1 (2). *Let s be a T-supermedian function on X. Then $s \circ i_j$ is either identically ∞ or locally λ-integrable.*

Definition 4.2. We consider, for $x, y \in D$

$$\phi(x,y) = (\lambda(B(x,r(x))))^{-1} . \chi_{B(x,r(x))}(y)$$
$$\phi^1(x,y) = \phi(x,y)$$
$$\phi^{n+1}(x,y) = \int_D \phi^n(x,\xi)\phi(\xi,y)d\lambda(\xi)$$

Hence we have, for any measurable and positive function f on D ,

$$N^n f(x) = \int_D \phi^n(x,y) f(y) d\lambda(y).$$

Let $G_N = \sum_{n=0}^{\infty} N^n$ be the potential kernel associated to N. We have the following classical results: $N.G_N = G_N.N$, $N.G_N + I = G_N$. Note that if $f :\ D \longrightarrow R$ be a measurable function on D, we put $Nf := Nf^+ - Nf^-$ if $\{Nf^+ = +\infty\} \cap \{Nf^- = +\infty\}$ is empty.

Lemma 4.1 (2). *Let h be a T-invariant function on X. Then, for $j,k \in \{1,2\},\ j \neq k$, the function*

$$f_j = \begin{cases} h \circ i_j - G_N R_j h \circ i_k, & \text{if } h \circ i_j < +\infty \\ +\infty & , \text{if } h \circ i_j = +\infty \end{cases}$$

is positive and locally $\lambda-$ integrable on $\{h \circ i_j < \infty\}$.

Proposition 4.1. *If h is a harmonic function on X such that $h \circ i_j;\ j = 1,2$ are locally λ-integrable, then for any $U \in \mathcal{U}$ and for any $z \in X$ such that $\overline{B}(\pi(z), r(\pi(z)) \subset \pi(U)$, we have $Th(z) = h(z)$, for any $z \in X$*

Proof. Let $j,k \in \{1,2\},\ j \neq k$ and $U \in \mathcal{U}$. We have $h \circ i_j = H_{\pi(U)} h \circ i_j + K_{\pi(U)}^{\mu_j} h \circ i_k$. Hence $h \circ i_j - K_{\pi(U)}^{\mu_j} h \circ i_k$ is harmonic on $\pi(U)$. So, from [7, lemma 4.12],

$$N(h \circ i_j - K_{\pi(U)}^{\mu_j} h \circ i_k)(x) = h \circ i_j(x) - K_{\pi(U)}^{\mu_j} h \circ i_k(x),$$

for any $x \in D$ such that $\overline{B}(x, r(x)) \subset \pi(U)$. Hence

$$h \circ i_j(x) = N h \circ i_j(x) - N K_{\pi(U)}^{\mu_j} h \circ i_k(x) + K_{\pi(U)}^{\mu_j} h \circ i_k(x),$$

for any $x \in D$ such that $\overline{B}(x, r(x)) \subset \pi(U)$. It remains to prove that

$$R_j h \circ i_k(x) = K_{\pi(U)}^{\mu_j} h \circ i_k(x) - N(K_{\pi(U)}^{\mu_j} h \circ i_k)(x)$$

for any $x \in D$ such that $\overline{B}(x, r(x)) \subset \pi(U)$. From the definition of R_j, we have

$$R_j h \circ i_k(x) = K_D^{\mu_j} h \circ i_k(x) - N(K_D^{\mu_j}(h \circ i_k)(x)).$$

Or

$$K_D^{\mu_j} h \circ i_k(x) = K_{\pi(U)}^{\mu_j} h \circ i_k(x) + H_{\pi(U)}(K_D^{\mu_j} h \circ i_k)(x).$$

Hence

$$(1)\ \ R_j h \circ i_k(x) = K_{\pi(U)}^{\mu_j} h \circ i_k(x) + H_{\pi(U)}(K_D^{\mu_j} h \circ i_k)(x) - N(K_{\pi(U)}^{\mu_j} h \circ i_k)(x)$$
$$-N(H_{\pi(U)} K_D^{\mu_j} h \circ i_k)(x).$$

Since $H_{\pi(U)}(K_D^{\mu_j} h_k)$ is harmonic on $\pi(U)$, then from [7, lemma 4.12],

$$(2)\ \ N(H_{\pi(U)} K_D^{\mu_j} h \circ i_k)(x) = H_{\pi(U)}(K_D^{\mu_j} h \circ i_k)(x).$$

So, from (2) and (1) we have the result. ■

Theorem 4.2. *Any T-invariant function h on X such that $K_D^{\mu_j} h \circ i_k < \infty; j \neq k$; $j, k \in \{1, 2\}$, is harmonic with respect to $(\mathcal{S})_{U \in \mathcal{U}}$.*

Proof. On the one hand, $G_N(K_D^{\mu_j}(h \circ i_k)) < +\infty$. Indeed, since $G_N N + I = G_N$, then,

$$G_N N(K_D^{\mu_j}(h \circ i_k)) + (K_D^{\mu_j}(h \circ i_k)) = G_N(K_D^{\mu_j}(h \circ i_k)).\ \ (\star)$$

from the remark 4.1.3, and [7, lemma 1.1], we have, $N(K_D^{\mu_j}(h \circ i_k))$ is locally bounded. Thus, from the [7, lemma 1.5], $G_N N(K_D^{\mu_j}(h \circ i_k))$ is locally bounded. Hence, from ($\star$) and the hypothesis, we have $G_N(K_D^{\mu_j}(h \circ i_k)) < +\infty$.
On the other hand, since $K_D^{\mu_j}(h \circ i_k) < +\infty$, it results from the remark 4.1.3 that

$$R_j h \circ i_k = K_D^{\mu_j} h \circ i_k - N(K_D^{\mu_j}(h \circ i_k)),$$

and

$$G_N R_j h \circ i_k + G_N N(K_D^{\mu_j} h \circ i_k) = G_N(K_D^{\mu_j}(h \circ i_k)).$$

Therefore,

$$G_N R_j h \circ i_k + G_N N(K_D^{\mu_j} h \circ i_k) + K_D^{\mu_j} h \circ i_k = G_N(K_D^{\mu_j}(h \circ i_k)) + K_D^{\mu_j} h \circ i_k.$$

Using again the relation $G_N = G_N N + I$, we have,

$$G_N R_j h \circ i_k + G_N(K_D^{\mu_j} h \circ i_k) = G_N(K_D^{\mu_j}(h \circ i_k)) + K_D^{\mu_j} h \circ i_k.$$

Hence

$$G_N R_j h \circ i_k = K_D^{\mu_j} h \circ i_k$$

Since $N(h \circ i_j - G_N R_j h \circ i_k) = N(\lim_{n \to \infty} N^n h \circ i_j) = \lim_{n \to \infty} N^{n+1} h \circ i_j$, then $h \circ i_j - G_N R_j h \circ i_k$ is N-invariant. Hence, it results from [7, theorem 4.14] that,

$$H_{\pi(U)}(h \circ i_j - G_N R_j h \circ i_k) = h \circ i_j - G_N R_j h \circ i_k$$

Hence,

$$h \circ i_j - K_D^{\mu_j}(h \circ i_k) = H_{\pi(U)}(h \circ i_j - K_D^{\mu_j}(h \circ i_k)).$$

Therefore,

$$h \circ i_j = H_{\pi(U)} h \circ i_j + K_D^{\mu_j}(h \circ i_k) - H_{\pi(U)} K_D^{\mu_j}(h \circ i_k)$$

So,

$$h \circ i_j = H_{\pi(U)} h \circ i_j + K_{\pi(U)}^{\mu_j} h \circ i_k \quad U \in \mathcal{U}.$$

Thus h is harmonic with respect to $(\mathcal{S}_U)_{U \in \mathcal{U}}$. ■

Corollary 4.1. *Let D be a Green domain in $\mathbb{R}^d$, $d \geq 2$, and $r : D \to \mathbb{R}_+^\star$ be a measurable function such that there exists positive real number ε and a lipshitizian function ρ on D, with Lipschitz constant 1, dominated by dist$(.,\partial D)$ and $\varepsilon\rho \leq r \leq (1-\varepsilon)\rho$. Then, for any measurable and positive couple of functions (h_1, h_2) on D satisfying the restricted mean value property; the function h defined on X by $h \circ i_j = h_j$; $j = 1, 2$ is harmonic with respect to $(\mathcal{S}_U)_{U \in \mathcal{U}}$.*

References

1. A. Benyaiche, *Mesures de représentation sur les espaces biharmoniques*, in *Potential Theory. ICPT 91, Ed: Emile M.J. Bertin, p: 171-178*, (Kluwer Academic Publishers, Netherlands, 1994)
2. A. Benyaiche and S. Ghiate , *Propriété de moyenne restriente associée à un système d'E.D.P.*, Rendiconti Accademia Nazionale delle Scienze detta dei XL. Memorie di Matematica e Applicazioni. 121, Vol. XXVII, fasc. 1, p: 125-143 (2003).
3. A. Benyaiche and S. Ghiate , *Martin boundary associated with a system of PDE.*, Comment. Math. Univ. Carolin. 47, 3, p: 399-425 (2006).
4. J. Bliedtner and W. Hansen , *Potential Theory-An analytic and Probabilistic Approach to balayage* (Universitext, Springer, 1986).
5. A. Boukricha , *Espaces biharmoniques*, in *Théorie du potentiel, Proceedings, Orsay 1983, Eds: G. Mokobodzki and D. Pinchon, p: 116-149, Lectures Notes in Mathematics. 1096* (Springer- Verlag, 1984).
6. M. Brelot , *Eléments de la théorie classique du potentiel,* 4th edn (Centre de Documentation Universitaire, Paris, 1969).
7. A. Cornea and V. Veselý , *Martin compactification for discrete potential theory and the mean value property*, Potential Analysis 4, p: 1-21 (1995).
8. W. Hansen , *Restricted mean value property and harmonic functions*, Potential Theory-ICPT 94, Eds: J. Král, J. Lukeś, I. Netuka and J. Veselý, (Walter de Gruyter and Co., Berlin. New York 1996).
9. W. Hansen, *Modification of balayage spaces by transitions with application to coupling of PDE's*, Nagoya Math. J.169, p: 77-118 (2003).
10. R. M. Hervé and M. Hervé , *Les fonctions surharmoniques associées à un opérateur elliptique du second ordre à coefficients discontinus*, Ann. Inst. Fourier, 19, 1, p: 305-359 (1968).
11. R. R. Phelps, *Lectures on Choquet's Theorem* (Van Nostrand, 1966).
12. E.P. Smyrnelis , *Axiomatique des fonctions biharmoniques, première section,* Ann. Inst. Fourier, tome 26, 1, p: 35-98 (1976).

AN IMPLICIT FUNCTION THEOREM FOR SOBOLEV MAPPINGS

I. V. ZHURAVLEV

Mathematical Department, Volgograd State University, Volgograd, Russia
igor.zhuravlev@volsu.ru

A nonsmooth variant of the implicit function theorem is proved.

Keywords: implicit function; normalized Jacobi matrix; Clarke derivative.

F. H. Clarke proved the inverse function theorem for Lipschitz mappings more then forty years ago [1, 2] and as a consequence, the implicit function theorem for the class of Lipschitz mappings was obtained [3]. Several years ago M. Cristea [4] proved the inverse function theorem for more wide class of mappings. We present an implicit function theorem for mappings with Sobolev's derivatives. Our method uses normalized Jacobi matrix.

Let $m, n \geq 1$ be integers. Denote by $M_{m,n}$ the linear space of the (m,n)-matrices with real elements. For an arbitrary matrix $A \in M_{m,n}$ we put $\| A \| = (tr(A \cdot A^T))^{1/2}$ (T is the transposition of matrices), $|A| = \sup\limits_{|x|=1} |Ax|$. Let D be a domain in $\mathbf{R}^{n+m}$ and $F : D \to \mathbf{R}^m$ be a continuous function of variables $x \in \mathbf{R}^n$, $y \in \mathbf{R}^m$. Assume that the function $F(x,y)$ belongs to the Sobolev's class $W^1_{1,loc}(D)$ [5, 6]. For the function $F(x,y)$ the formal Jacobi matrix $F'(x,y)$ is defined on some set $D^*, D^* \subset D$, almost everywhere in D. Everywhere below we suppose $|F'(x,y)| > 0$ a. e. in the set D. For a point $a \in D$ we denote by $N(F,a)$ the convex hull in $M_{m,n+m}$ of a set $\{B \in M_{m,n+m}$: there exist a sequence $z_p = (x_p, y_p) \in D^*$ $(p \in N)$ such that $z_p \to a$ and $\frac{F'(z_p)}{|F'(z_p)|} \to B\}$. The set $N(F,a)$ is an analogue of the Clarke derivative [1, 2] and an analogue of the normalized Jacobi matrix [7]. Let $N_y(F,a)$ is a projection of the set $N(F,a) \subset M_{m,n+m}$ onto $M_{m,m}$ (last m columns of a matrix $B \in N(F,a)$ are considered as a square $m-$matrix). Let $B^n(x,r)$ be the ball in $\mathbf{R}^n$ with center at the point x and radius $r > 0$.

Theorem 1. Let $x_0 \in \mathbf{R}^n$, $y_0 \in \mathbf{R}^m$ and $a = (x_0, y_0) \in \mathbf{R}^{n+m}$. Let D be a domain in $\mathbf{R}^{n+m}$, $a \in D$, and $F : D \to \mathbf{R}^m$ be a continuous function of the Sobolev class $W^1_{1,loc}(D)$.

Suppose that $\|F'(x,y)\| \geq C$ a.e. in the domain D for some constant $C > 0$ and $detA \neq 0$ for all $A \in N_y(F,a)$. Then there exist $\rho > 0$ and a unique continuous mapping

$$G : B^n(x_0, \rho) \to R^m, \; G(x_0) = y_0,$$

such that $F(x, G(x)) = F(x_0, y_0)$ for all $x \in B^n(x_0, \rho)$.

Proof. In the proof we used the theorem on the the radius of injectivity for mappings with bounded distortion [5, 8].

Let Ω be a domain in $\mathbf{R}^n$ and $P : \Omega \to \mathbf{R}$ is a function of the class $L_1(\Omega)$. We shall denote by P_h the mean function of the function P [5, 6]. Here h is the parameter (radius) of averaging.

In the sequel we shall need the following lemma.

Lemma 2. Let A be a mapping defined in $\mathbf{R}^n$ and taking values in the set of real (m, k)-matrices. Suppose that the functions a^i_j – the elements of the matrix A, – are locally integrable in $\mathbf{R}^n$. Let A_h be a matrix-valued function obtained by averaging the functions a^i_j. Then $|A_h(x)| \leq |A|_h(x)$ and $||A_h(x)|| \leq ||A||_h(x)$ for every $x \in \mathbf{R}^n$.

By condition $detA \neq 0$ for all $A \in N_y(F, a)$ and Clarke results [1, 2] there exists d, $0 < d < 1$, and $r > 0$ such that the following conditions are satisfied: for every $u \in \mathbf{R}^m$, $|u| = 1$ there exist $w \in \mathbf{R}^m$, $|w| = 1$, such that the inequality $\langle w, \frac{F'_y(z)}{|F'(z)|}u\rangle \geq d$, $z = (x, y)$, holds almost everywhere on the neighbourhood $D_r = B^n(x_0, r) \times B^m(y_0, r) \subset D$. We have $w = O^{-1}u$ for some orthogonal matrix $O \in M_{m,m}$. By the inequality $\langle u, O\frac{F'y(z)}{|F'(z)|}u\rangle \geq d$ we have

$$|OF'_y(z)u - I(z)u| \leq kI(z), \tag{1}$$

where $I(z) = \frac{|F'(z)|}{d}$, $k = \sqrt{1 - d^2} < 1$. Using Lemma, from this we get

$$|OF_h{}'_y(z)u - I_h(z)u| < kI_h(z),$$

where $z \in D_{r/2} = B^n(x_0, r/2) \times B^m(y_0, r/2)$ and $h < r/4$. The function $I(z)$ is greater than zero almost everywhere on the neighbourhood D_r. Hence, $I_h(z) > 0$ everywhere on $D_{r/2}$. By the inequality (1) we have estimate

$$|F_h{}'_y(z)| \leq I_h(z)(k + 1). \tag{2}$$

Moreover, it follows from (2) that $I_h(z)(1 - k) \leq |F_h{}'_y(z)u|$ and we deduce that

$$I_h(z)(1 - k) \leq |\det(F_h\, y'(z))|^{1/m}. \tag{3}$$

The inequality (3) shows that for any fixed $x \in B^n(x_0, r/2)$ the mapping $F_h(z)$ is a local homeomorphism on $B^m(y_0, r/2)$ with respect to $y \in B^m(y_0, r/2)$.

Now we consider the case $m \geq 3$.

From (2) and (3) we deduce that the distortion coefficient

$$Q(F_h(z)) = \sup_{B(y_0, r/2)} \frac{|F_h{}'_y(z)|^m}{|\det(F_h{}'_y(z))|}$$

of the mapping $F_h(z)$ is bounded from above by $(\frac{1+k}{1-k})^m$, which does not depend on the choice of $h < r/4$ and $x \in B^n(x_0, r/2)$. Since any one of the mappings

$F_h(z)$, $h < r/4$, $x \in B^n(x_0, r/2)$, is a local homeomorphism on $B^m(y_0, r/2)$, by the theorem on the radius of injectivity for mappings with bounded distortion [5, 8] a neighborhood $B^m(y_0, r_1)$ $(r_1 < r/2)$ of the point y_0 can be found such that each mapping $F_h(z)$, $h < r/4$, $x \in B^n(x_0, r/2)$, is a homeomorphism on $B^m(y_0, r_1)$. From (3) and $Q(F_h(z)) \leq (\frac{1+k}{1-k})^m$ passing to the limit as h tends to zero, we conclude that the mapping $F(z)$, $z \in D_{r/2}$, is not constant and is a quasiconformal homeomorphism on $B^m(y_0, r_1)$ for any fixed $x \in B^n(x_0, r/2)$.

Now we consider the case $m \leq 2$. Fix arbitrary point $x \in B^n(x_0, r/2)$. Let $z_1 = (x, y_1)$, $z_2 = (x, y_2) \in D_{r/2} = B^n(x_0, r/2) \times B^m(y_0, r/2)$ and $y_1 \neq y_2$. Set

$$u = \frac{z_1 - z_2}{|z_1 - z_2|}, \; z_t = z_1 + t(z_2 - z_1), \; t \in [0, \, 1].$$

Then by (2) we obtain

$$|OF_{h}{}'_y(z_t)(z_2 - z_1) - I_h(z_t)(z_2 - z_1)| < kI_h(z_t)|z_1 - z_2|.$$

Integrating we find

$$|O(F_h(z_2) - F_h(z_1)) - (z_2 - z_1)\int_0^1 I_h(z_t)dt| \leq k|z_2 - z_1|\int_0^1 I_h(z_t)dt$$

and

$$(1 - k)|z_2 - z_1|\int_0^1 I_h(z_t)dt \leq |F_h(z_2) - F_h(z_1)|.$$

By the inequality $\int_0^1 I_h(z_t)dt \geq \frac{C}{d}$ we conclude that

$$|F_h(x, y_2) - F_h(x, y_1)| \geq \beta|y_2 - y_1|, \tag{4}$$

where $\beta = \frac{C(1-k)}{d}$, $y_2, y_1 \in \overline{B^m(y_0, r/2)}$, $x \in B^n(x_0, r/2)$. If $|Y - F_h(x, y_0)| < \beta\frac{r}{4}$, then for arbitrary point $y_1 \in \partial B^m(y_0, r/2)$ we have

$$|Y - F_h(x, y_1)| \geq |F_h(x, y_1) - F_h(x, y_0)| - |Y - F_h(x, y_0)| > \beta\frac{r}{2} - \beta\frac{r}{4} = \beta\frac{r}{4}.$$

Hence $\min\limits_{y \in \overline{B^m(y_0, r/2)}} |Y - F_h(x, y)|^2$ attains at some point $y^* \in B^m(y_0, r/2)$ and

$$(|Y - F_h(x, y)|^2)'_y(y^*) = F_{h}{}'_y(x, y^*)(Y - F_h(x, y^*)) = 0.$$

Since $det(F_{h}{}'_y(x, y^*)) \neq 0$ we have $Y = F_h(x, y^*)$ and $B^m(F_h(x, y_0), \beta\frac{r}{4}) \subset F(x, B^m(y_0, r/2))$. From (4) passing to the limit as h tends to zero, we conclude that the mapping $F(z)$, $z \in D_{r/2}$, is a homeomorphism on $B^m(y_0, r/2)$ for any fixed $x \in B^n(x_0, r/2)$.

For sufficiently small r_0 we have $B^m(Y_0, r_0) \subset F(x, B^m(y_0, r_1))$ $(Y_0 = F(x_0, y_0)$ for all $x \in B^n(x_0, r_0)$ in both case $m \geq 3$ and $m \leq 2$.

Consider the mapping $\Phi : D_{r_0} \to R^{n+m}$ defined by

$$(x, y) \xrightarrow{\Phi} (X, Y) = (x, F(x, y)),$$

$(x, y) \in D_{r_0} = B^n(x_0, r_0) \times B^m(y_0, r_0)$. From above the map Φ is a homeomorphism and $\Phi(D_{r_0}) \supset B^n(x_0, r_0) \times B^m(Y_0, r_0)$.

The mapping Φ had been defined such that its inverse map has the form $x = X$, $y = g(X, Y)$. Next we observe that

$$(X, Y) = \Phi(\Phi^{-1}(X, Y)) = (X, F(X, g(X, Y))) \tag{5}$$

and $F(X, g(X, Y)) = Y$. We put $G(x) = g(X, Y_0)$. By (5) we now find $F(x, G(x)) = Y_0 = F(x_0, y_0)$ and $G(x_0) = g(x_0, Y_0) = g(X_0, Y_0) = y_0$.

Uniqueness of the map follows from the bijectivity of Φ. ■

References

1. Clarke F. H. *On the invers function theorem*, Pac. J. Math., Vol. 64, No 1, 1976, p. 97-102.
2. Clarke F. H. *Optimimization and nonsmooth analysis*, "Nauka", Moscow, 1988. (In Russian).
3. Hiriart-Urruty J. B. *Tangent cones, generalized gradients and mathematical programming in Banach spaces*, Math. Oper. Res., **4**, 1979, p. 78-97.
4. Cristea M. *A generalization of some theorems of F.H. Clarke and B.H. Pourciau*, Rev. Roumanie Math. Pures Appl., **50**, N 2, 2005, p. 137-152.
5. Reshetnyak Yu. G., *Space mappings with bounded distortion*, "Nauka", Novosibirsk, 1982, 278 pp. (In Russian).
6. Sobolev S. L. *Some applications of functional analysis in mathematical physics*, "Nauka", Moscow, 1988. (In Russian).
7. Zhuravlev I. V. *Sufficient conditions for local quasiconformality of mappings with bounded distortion*, Russian Acad. Sci. Sb. Math., Vol. 78, No. 2, 1994, p. 437-445. R. V.
8. Martio O., Rickman S., Väisälä J. *Topological and metric properties of quasiregular mappings*, Ann. Acad. Sci. Fenn. Ser. A I, No 488, 1971, p. 1-31.

A RELATION AMONG RAMANUJAN'S INTEGRAL FORMULA, SHANNON'S SAMPLING THEOREM AND PLANA'S SUMMATION FORMULA

KUNIO YOSHINO

Department of Mathematics,
Sophia University, Tokyo, Japan

1. Introduction

It is well known that Shannon's sampling theorem is very important in digital signal analysis. On the other hand there is a very curious so called Ramanujan's Integral formula. But unfortunately Ramanujan's Integral formula is not always correct. The aims of this paper are

(1) *to give an interpretation of Ramanujan's Integral formula,*

(2) *to prove Ramanujan's Integral formula,*

(3) *to clarify the relation between Ramanujan's Integral formula and Shannon's sampling theorem.*

We will do these things by using the theory of Fourier-Borel transform and Avanissian-Gay transform of analytic functionals. Finally we will reveal the meaning of Plana's summation foumula in the theory of analytic functionals. Especially we will obtain the relation between Cauchy-Hilbert transform and Avanissian-Gay transform of analytic functionals by using Plana's summation foumula.

2. Ramanujan's Integral formula

In this section we will consider following Ramanujan's integral formula.
Ramanujan's integral formula(Ramanujan[6])

$$(2-1) \qquad \int_0^\infty u^{a-1}\left\{\sum_{n=0}^{\infty} f(n)(-u)^n\right\} du = \frac{\pi f(-a)}{\sin(\pi a)}$$

Now we list up some examples of Ramanujan's integral formula.

Example 1. $f(z) = 1$

$$\int_0^\infty u^{a-1}\left\{\sum_{n=0}^\infty f(n)(-u)^n\right\}du = \int_0^\infty \frac{u^{a-1}}{1+u}du = \frac{\pi}{\sin(\pi a)}$$

Following example 2 is equals to the definition of Euler Γ - function.

Example 2. $f(z) = \dfrac{1}{\Gamma(1+z)}$

$$\int_0^\infty u^{a-1}\left\{\sum_{n=0}^\infty \frac{1}{n!}(-u)^n\right\}du = \int_0^\infty u^{a-1}e^{-u}du = \Gamma(a) = \frac{\pi}{\Gamma(1-a)\sin(\pi a)}$$

Following example 3 tells us the necessity of some additional conditions on function $f(z)$.

Example 3. $f(z) = \sin(\pi z)$

In this case right hand side in (2-1) is $-\pi$, but left hand side of (2-1) is 0 . So Ramanujan's integral formula is not valid in this example 3. This means that we need some conditions on $f(z)$ in Ramanujan's integral formula.

3. Shannon's sampling theorem

The following formula is called Shannon's sampling theorem. It is very important in digital signal analysis.

Theorem 1 ([3] , [9])] Suppose that entire function $f(z)$ satisfies the following estimate : $\forall\varepsilon > 0, \exists C_\varepsilon > 0$ s.t.

$$(3-1) \quad |f(z)| \leq C_\varepsilon \exp(k|y| + \varepsilon|z|), \quad (\forall z = x + iy \in \mathbb{C}).$$

If $0 \leq k < \pi$, then we have

$$f(z) = \lim_{\delta\to 0}\sum_{n=-\infty}^{\infty}\frac{f(n)\sin\pi(z-n)}{\pi(z-n)}e^{-\delta|n|}.$$

Right hand side is sometime called cardinal series([3]).

4. Transformations of analytic functionals

4-1 Fourier Borel transform

Let K be a convex compact set in $\mathbb{C}^n$. $H(K)$ denotes the space of holomorphic functions defined near K. The element of the dual space $H'(K)$ is called analytic functional carried by K. Let T be an analytic functional carried by compact set K. We define Fourier-Borel transform $\widetilde{T}(z)$ of T as follows :

$$\widetilde{T}(z) = < T_t, e^{tz} >$$

$\widetilde{T}(z)$ is an entire function which satisfies the following estimate :

$\forall \varepsilon > 0, \exists C_\varepsilon > 0$ s.t.

$$(4-1) \quad |f(z)| \leq C_\varepsilon \exp(H_K(z) + \varepsilon|z|), \quad (\forall z = x + iy \in \mathbb{C}^n)$$

$Exp(\mathbb{C}^n : K)$ denotes the space of entire functions which satisfiy the estimate (4-1). Following theorem gives a characterization of Fourier-Borel image of analytic functionals.

Theorem 2 (Polya-Martineau [8]) *Fourier-Borel transform is a topological isomorphism from $H'(K)$ to $Exp(\mathbb{C}^n : K)$.*

4-2 Avanissian-Gay transform

Now we introduce Avanissian-Gay transform $G_T(w)$ of analytic functional T . It is called z-transform in digital signal analysis. In[2] V. Avanissian and R. Gay introduced Avanissian-Gay transform for analytic functionals with compact carrier. Later, in[8] M. Morimoto and K. Yoshino extended Avanissian-Gay transform for analytic functional with non-compact carrier. If K is contained in $\{t \in \mathbb{C} : |Im \ \ t| < \pi\}$, then we can define Avanissian-Gay transform $G_T(w)$ of $T \in H'(K)$ as follows:

$$G_T(w) = < T_t, \frac{1}{1 - we^t} > .$$

We have the following proposition.

Proposition 1 ([2] , [4] , [8])

Suppose that compact convex set K is contained in $\{t \in \mathbb{C} : |Im \ \ t| < \pi\}$ and $T \in H'(K)$.

(1) $G_T(w)$ is holomorphic in $\mathbb{C} \setminus exp(-K)$.

(2) $G_T(w) = \sum_{n=0}^{\infty} \tilde{T}(n) w^n$

(3) $\lim_{|w| \to \infty} G_T(w) = 0$

(4) **(inversion formula (1))**

$$\tilde{T}(z) = \frac{1}{2\pi i} \int_C G_T(e^{-u}) e^{uz} du,$$

where C is a path surrounding K.

(inversion formula (2))

$$\tilde{T}(z) = \frac{-1}{2\pi i} \int_{C'} G_T(w) w^{-z-1} dw,$$

where C' is a path surrounding $exp(-K)$.

For the details of Avanissian-Gay transform for analytic functional with non-compact carrier, we refer the reader to[8] .

5. A proof of Ramanujan's integral formula

Let $f(z)$ be a holomorphic function defined on a right half plane and satisfies the following estimate :

$\forall\varepsilon > 0, \forall\varepsilon' > 0, \exists C_{\varepsilon,\varepsilon'} > 0$ s.t.

$$(5-1) \quad |f(z)| \leq C_{\varepsilon,\varepsilon'} \exp(ax + b|y| + \varepsilon|z|), \quad (\forall x \geq \varepsilon', z = x + iy \in \mathbb{C})$$

Theorem 3 Suppose that holomorphic function $f(z)$ satisfies (5-1). If $0 \leq k < \pi$, then Ramanujan's Integral formula always valids in the following sense.

$$\frac{\pi f(-z)}{\sin(\pi z)} = \int_0^\infty G_T(-u)u^{z-1}du$$

where $\widetilde{T}(z) = f(z)$.

(Proof) There exists an analytic functional T carried by $\{t = u + iv : u \geq a, |v| \leq b\}$ such that $\widetilde{T}(z) = f(z)$ ([7]). By inversion formula (4) in Proposition 1, we have

$$f(z) = \widetilde{T}(z) = \frac{1}{2\pi i}\int_C G_T(e^{-u})e^{uz}du$$

where $C = a - \varepsilon + i[-b-\varepsilon, b+\varepsilon] \cup -i(b+\varepsilon) + [a-\varepsilon, \infty) \cup i(b+\varepsilon) + [a-\varepsilon, \infty)$.
Deforming the integral path C to $\pi i + (-\infty, \infty) \cup -\pi i + (-\infty, \infty)$,

$$f(z) = \frac{1}{2\pi i}\int_{+\infty}^{-\infty} G_T(-e^{-u})e^{\pi iz}du + \frac{1}{2\pi i}\int_{-\infty}^{+\infty} G_T(-e^{-u})e^{-\pi iz}du$$

$$= \frac{-\sin\pi z}{\pi}\int_{-\infty}^{\infty} G_T(-e^{-u})e^{uz}du$$

Putting $x = e^{-u}$, we obtain Ramanujan's integral formula.

6. A relation between Ramanujan's Integral formula and Shannon's sampling theorem

In section 5, we derived Ramanujan's formula from the inversion formula.
Now our conclusion is as follows :

Conclusion *Suppose that* $f(z) \in Exp(\mathbb{C}; K)$ *and* $K = [-ik, ik]$.
If $0 \leq k < \pi$, *then*

we can derive Ramanujan's integral formula and Shannon's sampling theorem by deforming the integral path in the inversion formula (4) in Proposition 1.

7. Plana's summation formula

Following formula is known as Plana's summation formula.
Plana's summation foumula [5] Suppose that $K \subset [-a,a]+i[-b,b]$ and $f \in Exp(\mathbb{C}:K)$.
If $0 \leq b < 2\pi$ and $|Im \quad s| + b < 2\pi$, then we have

$$\sum_{n=0}^{\infty} f(n)e^{-sn} = \frac{1}{2}f(0) + \int_0^{\infty} f(x)e^{-sx}dx + \imath\int_0^{\infty} \frac{f(\imath x)e^{-\imath sx} - f(-\imath x)e^{\imath sx}}{e^{2\pi x}-1}dx.$$

For the details of Plana's summation formula, we refer the reader to [5] and [10]. By using Plana's summation formula, we can show the following theorem.

Theorem 4. Put $K = [-a,a]+i[-b,b]$. Suppose that $f(z)$ satisfies following conditions (1) and (2).
(1) $f(z) \in Exp(\mathbb{C}:K)$
(2) $f(\mathbb{Z}) = 0$,
If $\pi \leq b < 2\pi$, then we have

$$f(z) = \sin \pi z \widetilde{S}(z),$$
where $S \in H'([-a,a]+i[\pi-b, \pi+b])$

(Proof) See [10] .

For another proofs of theorem 4, we refer the reader to[3] and[4] .

8. The meaning of Plana's summation formula in the theory of analytic functionals

Cauchy-Hilbert transform $CH(T)(w)$ of $T \in H'(K)$ is defined by

$$CH(T)(w) = < T_t, \frac{1}{t-w} > .$$

Proposition 2 ([8] , [2] , [4]) Suppose that compact convex set K is contained in $\{t \in \mathbb{C} : |Im \quad t| < \pi\}$. Then we have

$$(1) \quad CH(T)(w) = \int_0^{\infty} \widetilde{T}(z)e^{-wz}dz,$$

$$(2) \quad < T, g > = \frac{1}{2\pi i}\int_C CH(T)(t)g(t)dt, \quad (\forall g \in H(K)).$$

$$(3) < T, g > = \frac{1}{2\pi i}\int_C G_T(e^{-t})g(t)dt, \quad (\forall g \in H(K))$$

where C is the contour surrouding K.

By proposition 2 we have

$$\int_C (CH(T)(t) - G_T(e^{-t}))g(t)dt = 0, \quad (\forall g \in H(K))$$

Since we have following isomorphism

$$H'(K) \cong \frac{H(W \setminus K)}{H(W)},$$

$(K \subset W,\ W$: open) we can conclude that

$$G_T(e^{-t}) - CH(T)(t) \in H(W)$$

Now problem is :

What is $G_T(e^{-t}) - CH(T)(t)$?

Plana's summation formula gives an answer to this problem :

$$G_T(e^{-t}) - CH(T)(t)$$

$$= \frac{1}{2}\widetilde{T}(0) + \imath \int_0^\infty \frac{\widetilde{T}(\imath x)e^{-\imath tx} - \widetilde{T}(-\imath x)e^{\imath tx}}{e^{2\pi x} - 1} dx \in H(W).$$

where $W = \{t \in \mathbb{C} : |Im\ \ t| < \pi\}$

This is the meaning of Plana's summation formula in the theory of analytic functionals.

References

1. M. Andersson : *Topics in Complex Analysis*, Springer Verlag(1996)
2. V. Avanissian and R. Gay : *Sur une transformations des fonctionnle analytiques et ses applications aux fonctions entieres a plusieurs variables*, Bull.Soc.Math.France 103(1975) 341-384
3. R.P. Boas : *Entire Functions*, Academic Press(1982)
4. C. Berenstein and R. Gay : *Complex Analysis and Special Topics in Harmonic Analysis*, Springer Verlag(1995)
5. Erdelyi,Magnus, Oberhettinger and Tricomi : *Higher Transcendental Functions Vol. I*, Mcgrawhill(1953)
6. G.H. Hardy : *Ramanujan*, Chelsea, New York(1978)

7. M. Morimoto : *Analytic functionals with non-compact carriers*, Tokyo J.Math. 1(1978) 72-103.
8. M. Morimoto and K. Yoshino : *Uniqueness theorem for holomorphic functions of exponential type*, Hokkaido Math.J. 7(1978) 259-270.
9. K. Yoshino : *Liouville type theorem for entire functions of exponential type*, Complex Variables, 5(1985) 21-51
10. K. Yoshino and M. Suwa : *Plana's summation formula for entire functions of exponential type and its applications*, preprint,

ASYMPTOTIC EXPANSIONS OF THE SOLUTIONS TO THE HEAT EQUATIONS WITH GENERALIZED FUNCTIONS INITIAL VALUE

KUNIO YOSHINO and YASUYUKI OKA

Department of Mathematics,
Sophia University, Tokyo, 102-8554, Japan

We will drive the asymptotic expansions of solutions of the heat equation with generalized functions initial data.

Keywords: Heat Equation,Asymptotic expansion,Tempered Distributions, Distributions of exponential growth and Fourier Hyperfunctions

1. Introduction

In[4] , T.Matsuzawa characterized tempered distributions as the initial value of the solutions to the heat equation.

Theorem 1. For $u \in \mathcal{S}'$, $U(x,t) = (u * E)(x,t)$ satisfies

$$\left(\frac{\partial}{\partial t} - \Delta\right) U(x,t) = 0,$$

$$U(x,t) \in \mathcal{C}^\infty(\mathbb{R}^d \times (0,\infty)),$$

$$\exists M \ , \ \exists N \ , \ |U(x,t)| \leq Ct^{-N}(1+|x|)^M \ , \ (x \in \mathbb{R}^d \ , \ 0 < t < 1).$$

Moreover,

$$U(x,t) \to u, \ (t \to 0) \ in \ \mathcal{S}'(\mathbb{R}^d),$$

$$i.e. \ < u, \varphi >= \lim_{t \to 0} \int_{\mathbb{R}^d} U(x,t)\varphi(x)dx \ , \ (\forall \varphi \in \mathcal{S}(\mathbb{R}^d)), \qquad (\sharp)$$

where $E(x,t) = (4\pi t)^{-\frac{d}{2}} e^{-\frac{x^2}{4t}} \ , \ (x \in \mathbb{R}^d \ , \ 0 < t < +\infty)$.

We analyze ($\sharp$) more precisely. Namely, we will drive the asymptotic expansions of solutions of the heat equation with generalized functions initial data in this paper.

Main Theorem 1. Let $U(x,t) \in C^\infty(\mathbb{R}^d \times (0,\infty))$ satisfy the following conditions:

(i) $\left(\frac{\partial}{\partial t}-\Delta\right)U(x,t)=0$,
(ii) $\exists C>0,\ \exists\nu\geq 0$ *and* $\exists k\geq 0$ *s.t.* $|U(x,t)|\leq Ct^{-\nu}(1+|x|)^{k}$,
$(x\in\mathbb{R}^{d},\ 0<t<1)$,

Then $U(x,t)$ has the following asymptotic expansion :

$$U(x,t)\sim\sum_{k=0}^{\infty}\frac{t^{k}}{k!}\Delta_{x}^{k}u\ ,\ \left(u\in\mathcal{S}'(\mathbb{R}^{d})\ such\ that\ u=\lim_{t\to 0}U(x,t)\right)$$

i.e.

$$\lim_{t\to 0+}|<U(x,t)\ ,\ \varphi>-\sum_{k=0}^{\frac{N}{2}}\frac{t^{k}}{k!}<\Delta_{x}^{k}u\ ,\ \varphi>|t^{-\frac{N}{2}}=0\ ,\ (\varphi\in\mathcal{S}(\mathbb{R}^{d})),$$

where $\Delta_{x}=\partial_{x_1}^{2}+\cdots+\partial_{x_d}^{2}$.

Corollary 1. For $u\in\mathcal{S}'$, Put $U(x,t)=(u*E)(x,t)$, then $<u,\varphi>=\lim_{t\to 0}\int_{\mathbb{R}^{d}}U(x,t)\varphi(x)dx(\forall\varphi\in\mathcal{S})$.

We obtain the similar consequences with regard to distributions of exponential growth and Fourier Hyperfunctions by using the results of Suwa[5] and K.W.Kim *et al*[3] respectively.

2. Preliminaries

We use the multi-index notations such as

$$|\alpha|=\alpha_{1}+\cdots+\alpha_{d}\ \ for\ \alpha=(\alpha_{1},\cdots,\alpha_{d})\in\mathbb{Z}_{+}^{d}\ ,$$

where $\mathbb{Z}_{+}$ is the set of nonnegative integers and

$$\partial^{\alpha}=\partial_{1}^{\alpha_1}\cdots\partial_{d}^{\alpha_d}\ \ ,\ \ \partial_{j}=\frac{\partial}{\partial x_{j}}\ ,\ (j=1,\cdots,d)\ .$$

Defination 1. We put $E(x,t)=(4\pi t)^{-\frac{d}{2}}e^{-\frac{x^2}{4t}}$, $(x\in\mathbb{R}^{d}\ ,\ 0<t<+\infty)$,
$E(x,t)$ is called *heat kernel* and have the following properties :

$$\cdot\ E(x,t)\in\mathcal{S}(\mathbb{R}_{x}^{d})\ ,\ (t>0)\ ,$$

$$\cdot\ \int_{\mathbb{R}^{d}}E(x,t)dx=1\ ,\ (t>0)\ ,$$

$$\cdot\ \left(\frac{\partial}{\partial t}-\Delta\right)E(x,t)=0\ ,\ (x\in\mathbb{R}^{d}\ ,\ 0<t<\infty)\ ,$$

$$where\ \Delta=\frac{\partial^{2}}{\partial x_{1}^{2}}+\cdots+\frac{\partial^{2}}{\partial x_{d}^{2}}\ .$$

Defination 2. $\mathcal{S}(\mathbb{R}^d) = \{\varphi \in \mathcal{C}^\infty(\mathbb{R}^d) \mid \forall\alpha,\ \forall\beta \in \mathbb{Z}_+^d\ ,\ \sup_{x\in\mathbb{R}^d}|x^\alpha\partial^\beta\varphi(x)| < \infty\}$.
Defination 3. *$\mathcal{S}'(\mathbb{R}^d)$ denotes the dual space of $\mathcal{S}(\mathbb{R}^d)$.*
Proposition 1. Let u be a linear form from $\mathcal{S}$ to $\mathbb{C}$. Then the following statements (i) and (ii) are equivalent.

(i) $u \in \mathcal{S}'(\mathbb{R}^d)$,

(ii) $\exists C > 0\ s.t.\ | < u, \varphi > | \le C\sum_{|\beta|+k\le m}\sup_{x\in\mathbb{R}^d}(1+|x|^2)^k|\partial^\beta\varphi(x)|\ ,\ (\forall\varphi \in \mathcal{S}(\mathbb{R}^d))$.

3. Asymptotic expansions of the solutions to the heat equations with the tempered distributions initial data

Theorem 2. Let $U(x,t) \in C^\infty(\mathbb{R}^d \times (0,\infty))$ satisfy the following conditions :

(i) $\left(\frac{\partial}{\partial t} - \Delta\right)U(x,t) = 0$,

(ii)$\exists C > 0,\ \exists\nu \ge 0$ *and* $\exists k \ge 0\ s.t.\ |U(x,t)| \le Ct^{-\nu}(1+|x|)^k$,
$(x \in \mathbb{R}^d,\ 0 < t < 1)$,

Then $U(x,t)$ has the following asymptotic expansion :

$$U(x,t) \sim \sum_{k=0}^{\infty}\frac{t^k}{k!}\Delta_x^k u\ ,\ \left(u \in \mathcal{S}'(\mathbb{R}^d) \text{ such that } u = \lim_{t\to 0}U(x,t)\right).$$

i.e.

$$\lim_{t\to 0+} | < U(x,t)\ ,\ \varphi > - \sum_{k=0}^{\frac{N}{2}}\frac{t^k}{k!} < \Delta_x^k u\ ,\ \varphi > |t^{-\frac{N}{2}} = 0\ \ ,\ (\varphi \in \mathcal{S}(\mathbb{R}^d)),$$

where $\Delta_x = \partial_{x_1}^2 + \cdots + \partial_{x_d}^2$.

Lemma 1. Let φ be in $\mathcal{S}(\mathbb{R}^d)$ and $t > 0$. Then

$$\int_{\mathbb{R}^d} e^{-z^2}z^\alpha\left\{\int_0^1(1-\theta)^N\partial_y^\alpha\varphi(y+\sqrt{4t}z\theta)\ d\theta\right\}dz$$

is in $\mathcal{S}(\mathbb{R}^d)$.

Proof:

$$(1+|y|^2)^\gamma|\partial_y^\beta\int_{\mathbb{R}^d} e^{-z^2}z^\alpha\left\{\int_0^1(1-\theta)^N\partial_y^\alpha\varphi(y+\sqrt{4t}z\theta)\ d\theta\right\}dz|$$

$$= (1+|y|^2)^\gamma|\int_{\mathbb{R}^d} e^{-z^2}z^\alpha\left\{\int_0^1(1-\theta)^N\partial_y^{\alpha+\beta}\varphi(y+\sqrt{4t}z\theta)\ d\theta\right\}dz|$$

$$\le (1+|y|^2)^\gamma\int_{\mathbb{R}^d} e^{-z^2}|z|^\alpha\left\{\int_0^1|1-\theta|^N\partial_y^{\alpha+\beta}|\varphi(y+\sqrt{4t}z\theta)|\ d\theta\right\}dz. \qquad (3.1)$$

Since $\varphi(y) \in \mathcal{S}(\mathbb{R}^d)$, we have the following estimate,

$$(1+|y|^2)^\gamma |\partial_y^{\alpha+\beta}\varphi(y)| \leq C_{\gamma,\beta}.$$

So we obtain

$$|\partial_y^{\alpha+\beta}\varphi(y+\sqrt{4t}z\theta)| \leq C_{\gamma,\beta}(1+|y+\sqrt{4t}z\theta|^2)^{-\gamma}.$$

Thus,

$$\begin{aligned}
(3.1) \leq & C_{\gamma,\beta}(1+|y|^2)^\gamma \int_{\mathbb{R}^d} e^{-z^2}|z|^\alpha(1+|y+\sqrt{4t}z\theta|^2)^{-\gamma}dz. \\
\leq & C_{\gamma,\beta}\cdot 2^\gamma \int_{\mathbb{R}^d} e^{-z^2}|z|^\alpha(1+|\sqrt{4t}z\theta|^2)^\gamma dz \text{ , (by Peetre's inequality)} \\
\leq & C_{\gamma,\beta}\cdot 2^\gamma \int_{\mathbb{R}^d} e^{-z^2}|z|^\alpha(1+|2z|^2)^\gamma dz \\
= & C_{\gamma,\beta}\cdot 2^\gamma \sum_{\delta\leq\gamma}\binom{\gamma}{\delta}\int_{\mathbb{R}^d} e^{-z^2}|z|^\alpha|2z|^{2(\gamma-\delta)}dz \\
\leq & C_{\gamma,\beta}\cdot \sum_{\delta\leq\gamma}\binom{\gamma}{\delta}\left(\frac{1}{2}\right)^d \prod_{j=1}^{d}\Gamma\left(\frac{1}{2}(\alpha_j+2(\gamma_j-\delta_j)+1\right).
\end{aligned}$$

Hence we have

$$\int_{\mathbb{R}^d} e^{-z^2}z^\alpha \left\{\int_0^1 (1-\theta)^N \partial_y^\alpha\varphi(y+\sqrt{4t}z\theta)\, d\theta\right\} dz \ \in \mathcal{S}(\mathbb{R}^d).$$

This completes the proof of Lemma 1 . □

Proof of Theorem 2 : By Matsuzawa's result[4] , There exists $u \in \mathcal{S}'(\mathbb{R}^d)$ such that $U(x,t) = (u * E)(x,t)$. For any $\varphi \in \mathcal{S}(\mathbb{R}^d)$,

$$\begin{aligned}
< U(x,t) , \varphi > &= < (u*E)(x,t) , \varphi > \\
&= < u_y , E(x,t)*\varphi > \\
&= < u_y , \int_{\mathbb{R}^d} \pi^{-\frac{d}{2}} e^{-z^2}\varphi(y+\sqrt{4t}z)dz > .
\end{aligned}$$

By Taylor's formula,

$$\int_{\mathbb{R}^d} \pi^{-\frac{d}{2}} e^{-z^2} \varphi(y+\sqrt{4t}z)dz = \int_{\mathbb{R}^d} \pi^{-\frac{d}{2}} e^{-z^2} \Bigg\{ \sum_{|\alpha|\le N} \frac{(\sqrt{4t})^{|\alpha|}}{\alpha!} z^\alpha \partial_y^\alpha \varphi(y)$$

$$+ (\sqrt{4t})^{N+1} \sum_{|\alpha|=N+1} \frac{(N+1)z^\alpha}{\alpha!} \int_0^1 (1-\theta)^N \partial_y^\alpha \varphi(y+\sqrt{4t}z\theta)\, d\theta \Bigg\} dz$$

$$= \pi^{-\frac{d}{2}} \sum_{|\alpha|\le N} \frac{(\sqrt{4t})^{|\alpha|}}{\alpha!} \partial_y^\alpha \varphi(y) \int_{\mathbb{R}^d} e^{-z^2} z^\alpha \, dz$$

$$+ \pi^{-\frac{d}{2}} (\sqrt{4t})^{N+1} \sum_{|\alpha|=N+1} \frac{(N+1)}{\alpha!} \int_{\mathbb{R}^d} e^{-z^2} z^\alpha$$

$$\times \left\{ \int_0^1 (1-\theta)^N \partial_y^\alpha \varphi(y+\sqrt{4t}z\theta)\, d\theta \right\} dz \,. \tag{3.2}$$

Then we obtain the following equality :

$$(3.2) = \sum_{k=0}^{\frac{N}{2}} \frac{t^k}{k!} \Delta^k \varphi(y) + \pi^{-\frac{d}{2}} (\sqrt{4t})^{N+1}$$

$$\times \sum_{|\alpha|=N+1} \frac{(N+1)}{\alpha!} \int_{\mathbb{R}^d} e^{-z^2} z^\alpha \left\{ \int_0^1 (1-\theta)^N \partial_y^\alpha \varphi(y+\sqrt{4t}z\theta)\, d\theta \right\} dz \,,$$

where $\Delta = \partial_{y_1}^2 + \cdots + \partial_{y_d}^2$.
Thus,

$$< U(x,t), \varphi > = < u_y, \sum_{k=0}^{\frac{N}{2}} \frac{t^k}{k!} \Delta^k \varphi(y) >$$

$$+ < u_y, \pi^{-\frac{d}{2}} (\sqrt{4t})^{N+1} \sum_{|\alpha|=N+1} \frac{(N+1)}{\alpha!} \int_{\mathbb{R}^d} e^{-z^2} z^\alpha$$

$$\times \left\{ \int_0^1 (1-\theta)^N \partial_y^\alpha \varphi(y+\sqrt{4t}z\theta)\, d\theta \right\} dz >$$

Therefore we have the following equality :

$$\left| < U(x,t), \varphi > - \sum_{k=0}^{\frac{N}{2}} \frac{t^k}{k!} < \Delta^k u, \varphi > \right| t^{-\frac{N}{2}}$$

$$= \Bigg| < u_y, \pi^{-\frac{d}{2}} (\sqrt{4t})^{N+1}$$

$$\times \sum_{|\alpha|=N+1} \frac{(N+1)}{\alpha!} \int_{\mathbb{R}^d} e^{-z^2} z^\alpha \left\{ \int_0^1 (1-\theta)^N \partial_y^\alpha \varphi(y+\sqrt{4t}z\theta)\, d\theta \right\} dz > \Bigg| t^{-\frac{N}{2}} \tag{3.3}$$

We obtain the following estimate by Proposition 1 and Lemma 1. For any $n \in \mathbb{N}$,

$$(3.3)\ \leq C_{\alpha,\beta,\gamma,d} t^{\frac{N+1}{2}} t^{-\frac{N}{2}} = C_{\alpha,\beta,\gamma,d} t^{\frac{1}{2}} \to 0\ (t \to 0).$$

This completes the proof of Theorem 2 □

4. Asymptotic expansions of the solutions to the heat equations with the distributions of exponential growth initial data

We introduce Gel'fand-Shilov space $\mathcal{S}_1(\mathbb{R}^d)$ and recall its properties.

Definition 4. ([1])
We define $\mathcal{S}_1(\mathbb{R}^d)$ as follows :

$$\mathcal{S}_1(\mathbb{R}^d) := \left\{ \varphi \in \mathcal{C}^\infty(\mathbb{R}^d) \mid \exists A \in \mathbb{R}^d_+,\ \forall\alpha,\ \forall\beta \in \mathbb{Z}^d_+, \right.$$
$$\left. \exists C_\beta \geq 0\ s.t.\ \sup_{x \in \mathbb{R}^d} |x^\alpha \partial^\beta \varphi(x)| \leq C_\beta A^\alpha \alpha^\alpha \right\},$$

and $\mathcal{S}_{1,A}(\mathbb{R}^d)$ is defined as follows :

$$\mathcal{S}_{1,A}(\mathbb{R}^d) := \left\{ \varphi \in \mathcal{C}^\infty(\mathbb{R}^d) \mid \exists A \in \mathbb{R}^d_+,\ n \in \mathbb{N}^d, \right.$$
$$\left. \forall\alpha,\ \forall\beta \in \mathbb{Z}^d_+\ s.t. \|\varphi\|_{\beta,n} = \sup_{\substack{x\in\mathbb{R}^d \\ \alpha \in \mathbb{Z}^d_+}} \frac{|x^\alpha \partial^\beta \varphi(x)|}{\left(A + \frac{1}{n}\right)^\alpha \alpha^\alpha} < +\infty \right\},$$

where $\left(A + \frac{1}{n}\right)^\alpha = \left(A_1 + \frac{1}{n_1}\right)^{\alpha_1} \left(A_2 + \frac{1}{n_2}\right)^{\alpha_2} \cdots \left(A_d + \frac{1}{n_d}\right)^{\alpha_d}$.

- $\mathcal{S}_1(\mathbb{R}^d) = \lim_{A \to \infty} \mathcal{S}_{1,A}(\mathbb{R}^d)$.

The topology in $\mathcal{S}_{1,A}(\mathbb{R}^d)$ defined by the above semi-norms makes $\mathcal{S}_{1,A}(\mathbb{R}^d)$ the *Fréchet* space and $S_1(\mathbb{R}^d)$ is LF-space.

Proposition 2. ([1])
For $\xi \in \mathbb{R}^d$, we have the following estimate,

$$\exp\left(-\frac{1}{e}\sum_{j=0}^{d} |\xi_j|\right) \leq \mu(\xi) \leq C \exp\left(-\frac{1}{e}\sum_{j=0}^{d} |\xi_j|\right),$$

where C is some constant and

$$\mu(\xi) = \inf_k \frac{k^k}{|\xi|^k} .$$

Proposition 3.([1])
Let φ be a function in $\mathcal{C}^\infty(\mathbb{R}^d)$. Then the following statements (i) and (ii) are equivalent .

(i) $\varphi \in \mathcal{S}_1(\mathbb{R}^d)$.

(ii) $\exists A \in \mathbb{R}_+^d$, $\forall \beta \in \mathbb{Z}_+^d$, $\exists C_\beta \geq 0$ s.t. $|\partial^\beta \varphi(x)| \leq C_\beta \exp\left(-\sum_{j=1}^{d} a_j |x_j| \right)$,

where $a_j = \frac{1}{eA_j}$.

Proposition 4.([1])
Let φ be a function in $\mathcal{C}^\infty(\mathbb{R}^d)$. Then the following conditions (i) and (ii) are equivalent .

(i) $\varphi \in \mathcal{S}_{1,A}(\mathbb{R}^d)$.

(ii) $\exists A \in \mathbb{R}_+^d$, $n \in \mathbb{N}^d$, $\forall \beta \in \mathbb{Z}_+^d$, $\exists C_\beta \geq 0$ s.t. $|\partial^\beta \varphi(x)| \leq C_\beta \exp\left(-\sum_{j=1}^{d} a'_j |x_j| \right)$, where $a'_j = \frac{1}{e(A_j + \frac{1}{n_j})}$.

Definition 5. $\mathcal{S}'_1(\mathbb{R}^d)$ denotes the dual space of $\mathcal{S}_1(\mathbb{R}^d)$.
Proposition 5. Let u be a linear form from $\mathcal{S}_1$ to $\mathbb{C}$. Then the following statements (i) and (ii) are equivalent.

(i) $u \in \mathcal{S}'_1(\mathbb{R}^d)$,

(ii) $\forall A,\ \exists C > 0\ s.t.\ |< u, \varphi >| \leq C||\varphi||_{\beta,n}$, $(\forall \varphi \in \mathcal{S}_{1,A}(\mathbb{R}^d))$.

Remark : For the details of $\mathcal{S}'_1(\mathbb{R}^d)$, we refer the reader to[2] and.[5]

The following Theorem 3 is Main result in this section.
Theorem 3. Let $U(x,t) \in C^\infty(\mathbb{R}^d \times (0,\infty))$ satisfy the following conditions :

(i) $\left(\frac{\partial}{\partial t} - \Delta\right) U(x,t) = 0$,

(ii) $\forall \epsilon > 0$, $\exists N_\epsilon \geq 0$, $\exists C_\epsilon \geq 0$ s.t. $|U(x,t)| \leq C_\epsilon t^{-N_\epsilon} e^{\epsilon |x|}$, $(0 < t < 1$, $x \in \mathbb{R}^d)$.

Then $U(x,t)$ has the following asymptotic expansion :

$$U(x,t) \sim \sum_{k=0}^{\infty} \frac{t^k}{k!} \Delta_x^k u \ , \ \left(u \in \mathcal{S}'_1(\mathbb{R}^d) \ such\ that\ u = \lim_{t \to 0} U(x,t) \right) .$$

i.e.

$$\lim_{t\to 0+} | < U(x,t) \ , \ \varphi > - \sum_{k=0}^{\frac{N}{2}} \frac{t^k}{k!} < \Delta_x^k u \ , \ \varphi > | t^{-\frac{N}{2}} = 0 \quad , \ (\varphi \in \mathcal{S}_1(\mathbb{R}^d)),$$

where $\Delta_x = \partial_{x_1}^2 + \cdots + \partial_{x_d}^2$.

We need the following Lemma 2 for the proof of Theorem 3.

Lemma 2.(6) Let φ be in $\mathcal{S}_{1,A}(\mathbb{R}^d)$ and $t > 0$. Then

$$\int_{\mathbb{R}^d} e^{-z^2} z^\alpha \left\{ \int_0^1 (1-\theta)^N \partial_y^\alpha \varphi(y + \sqrt{4t}z\theta) \, d\theta \right\} dz$$

is in $\mathcal{S}_{1,A}(\mathbb{R}^d)$.

Proof of Theorem 3 : By Suwa's result[5], There exists $u \in \mathcal{S}_1'(\mathbb{R}^d)$ such that $U(x,t) = (u * E)(x,t)$. For any $\varphi \in \mathcal{S}_1(\mathbb{R}^d)$,

$$\begin{aligned} < U(x,t) \ , \ \varphi > &= < (u * E)(x,t) \ , \ \varphi > \\ &= < u_y \ , \ E(x,t) * \varphi > \\ &= < u_y \ , \ \int_{\mathbb{R}^d} \pi^{-\frac{d}{2}} e^{-z^2} \varphi(y + \sqrt{4t}z) dz > . \end{aligned}$$

By a similar calculation to Theorem 2, we have the following equality :

$$\begin{aligned} & \left| < U(x,t), \ \varphi > - \sum_{k=0}^{\frac{N}{2}} \frac{t^k}{k!} < \Delta^k u, \varphi > \right| t^{-\frac{N}{2}} \\ &= \left| < u_y, \ \pi^{-\frac{d}{2}} (\sqrt{4t})^{N+1} \sum_{|\alpha|=N+1} \frac{(N+1)}{\alpha!} \int_{\mathbb{R}^d} e^{-z^2} z^\alpha \right. \\ & \quad \left. \times \left\{ \int_0^1 (1-\theta)^N \partial_y^\alpha \varphi(y + \sqrt{4t}z\theta) \, d\theta \right\} dz > \right| t^{-\frac{N}{2}} \end{aligned} \tag{4.1}$$

We obtain the following estimate by Proposition 5 and Lemma 4. For any $n \in \mathbb{N}$,

$$(4.1) \ \leq C_{\gamma,N,A,n} t^{\frac{N+1}{2}} t^{-\frac{N}{2}} = C_{\gamma,N,A,n} t^{\frac{1}{2}} \to 0 \ (t \to 0).$$

This completes the proof of Theorem 5 □

Remark 1. We obtain the similar result for Fourier hyperfunctions in[6]

References

1. I.M.Gel'fand and G.E.Shilov, Generalized Functions, Volume 2, Space of Fundamental and Generalized Functions, Academy of Sciences Moscow, U.S.S.R, (1958) .
2. M.Hasumi, Note on the n-dimensional tempered ultra-distributions, Tohoku.Math.J.13,(1961),94-104.
3. K.W.Kim, S.Y.Chung and D.Kim,Fourier hyperfunctions as the boundary values of smooth solutions of the heat equation,
4. T.Matsuzawa, A calculus approach to the hyperfunctions III, Nagoya Math. J., 118 (1990), 133-153.
5. M.Suwa, Distributions of exponential growth with support in a proper convex cone, Publ, RIMS, Kyoto Univ. 40 (2004),no.2, 565-603.
6. K.Yoshino and Y.Oka, Asymptotic expansions of solutions to heat equations with Generalized Functions initial value, preprint.

ON THE EXISTENCE OF HARMONIC DIFFERENTIAL FORMS WITH PRESCRIBED SINGULARITIES

EUGENIA MALINNIKOVA

Department of Mathematics,
Norwegian University of Science and Technology
7491, Trondheim, Norway
eugeniamath.ntnu.no

In this note we obtain analogues of the Mittag–Leffler and the Weierstrass theorems for harmonic differential forms on noncompact Riemannian manifolds.

Keywords: Separation of singularities, harmonic differential forms, Mittag-Leffler theorem, Hodge decomposition

1. Introduction

A differential q-form ϕ, defined and smooth on an open subset of a smooth oriented Riemannian manifold, is called harmonic if

$$d\phi = 0 \quad \text{and} \quad \delta\phi = 0,$$

where d is the exterior differential operator and δ is its adjoint[1,2] . Harmonic differential forms on Riemannian manifolds can be considered as a generalization of analytic functions of one complex variable or analytic differentials on Riemann surfaces[3] . In this note we proof versions of the Mittag–Leffler and Weierstrass theorems for harmonic forms.

Construction of harmonic forms with prescribed singularities turns out to be a classical problem, it was discussed for example in the book by B.Rodin and L.Sario[4] . We will give another approach to this problem and prove a version of the Mittag–Leffler theorem in Section 3.

In Section 2 we formulate necessary and sufficient conditions for separation of singularities (in the sense of N.Aronszajn[5]). In several complex variable separation of singularities corresponds to the additive Cousin problem. This scheme of proofing Mittag–Leffler theorem was applied to harmonic functions on manifolds by T.Bagby and P.Blanchet[6] , and by P.Gauthier[7] ; and to harmonic differential forms on open subsets of $\mathbb{R}^n$ by the author[8] . In Sections 2 and 3 the results of Ref. 8 are generalized to forms on Riemannian manifolds. We give a new proof of Lemma 1 based on the Hodge decomposition. Further we formulate the result on separation on singularities for a given form (see Theorem 2), the last corollary repeats our result for forms on open subspace of $\mathbb{R}^n$. Finally, Theorem 2 is used to obtain a version of the Mittag–Leffler theorem.

In the last section a generalization of the Weierstrass theorem is given. In particular, we construct a harmonic form that vanishes on a prescribed discrete set of points, the form can be chosen to be both exact and coexact.

One of the main tools we employ for solving both problems is approximation of harmonic forms by elementary ones (as in the classical theorem by Runge), we adjust the methods used in Ref.9. We use also an abstract result on simultaneous approximation and interpolation due to H.Yamabe.[10] For harmonic and analytic functions simultaneous approximation and interpolation can be found in the works by L.Walsh and his coauthors.

2. Separation of singularities for harmonic differential forms

In this article M denotes a smooth (of class C^∞) connected oriented noncompact Riemannian manifold of dimension $n \geq 3$. We use the Hodge–de Rham decompositions for forms (and currents) on M

$$T = H_1 T + H_2 T + HT$$

and refer the reader to de Rham's book[1] .

Lemma 1. Let Ω be an open subset of M and let $1 \leq p \leq n$. For any $\alpha \in C_p^\infty(\Omega)$ there exists $\beta \in C_{p-1}^\infty(\Omega)$ such that $\delta d\beta = \delta\alpha$.

Proof. Let $\{K_j\}_j$ be an exhaustion of Ω by compact subsets such that $\Omega \setminus K_j$ has no connected components relatively compact in Ω. We take $\psi_j \in \mathcal{D}(\Omega)$ such that $0 \leq \psi_j \leq 1$ and $\psi_j \equiv 1$ on a neighborhood of K_j, $\psi_0 = 0$. Then $1 = \sum_j(\psi_j - \psi_{j-1}) = \sum_j \phi_j$, where $\phi_j \in \mathcal{D}(\Omega)$ and $\phi_j \equiv 0$ on a neighborhood of K_{j-1}.

We define $\kappa_j = H_1(\phi_j\alpha)$. Then $\kappa_j \in C_p^\infty(M)$ and κ_j is exact. Further,

$$\delta\kappa_j = \delta H_1(\phi_j\alpha) = \delta(\phi_j\alpha - H_2(\phi_j\alpha) - H(\phi_j\alpha)) = \delta(\phi_j\alpha).$$

Now we consider κ_j in a neighborhood of K_{j-1} where $\phi_j = 0$. We claim that κ_j can be uniformly on K_{j-1} approximated by exact in Ω forms that are coclosed on Ω.

First, suppose that Ω is relatively compact in M. We will show that κ_j can be approximated uniformly on K_{j-1} by forms of the form $H_1(c)$, where $c \cap \bar{\Omega} = \emptyset$. Let T be a current that is orthogonal to all such forms with $T \subset K_{j-1}$. So we have

$$0 = \{T, H_1(c)\} = \{H_2(T), c\},$$

for all c whose support does not intersect Ω. (We used that $T \cap c = \emptyset$ to move H_1, H_2 above, see details in the book by de Rham.[1]) Clearly $H_2(T)$ is a harmonic form off K_{j-1}. Last identity implies that $H_2(T)$ vanishes with all it's derivatives at each point of $M \setminus \bar{\Omega}$. But $M \setminus \bar{\Omega}$ intersects all components of $M \setminus K_{j-1}$, thus $H_2(T) = 0$ on $M \setminus K_{j-1}$. So we have

$$\{T, \kappa_j\} = \{T, H_1(\phi_j\alpha)\} = \{H_2 T, \phi_j\alpha\} = 0.$$

In general, let $\Omega = \cup_l \Omega_l$ and each Ω_l is relatively compact in Ω, $\Omega_l \subset\subset \Omega_{l+1}$. As we have seen κ_j can be approximated on K_{j-1} by $v_{j,1} = H_1(c_{j,1})$ where $(c_{j,1}) \cap \bar{\Omega}_1) = \emptyset$. We then approximate $v_{j,1}$ on $\bar{\Omega}_1$ by a form $H_1(c_{j,2})$ with $(c_{j,2}) \cap \bar{\Omega}_2 = \emptyset$. And continue the procedure. We get a sequence of forms $v_{j,l}$ that are exact and coclosed on Ω_l. This sequence can be done convergent uniformly on compact subsets of Ω. Then it follows from the de Rham theorems that the limit form w_j is exact and coclosed on Ω. It provides an`approximation to κ_j on K_{j-1},

$$\sup_{K_{j-1}} |\kappa_j - w_j| < 2^{-j}. \tag{1}$$

The series $w = \sum_j (\kappa_j - w_j)$ is uniformly absolutely convergent on compact subsets of Ω, for each l the sum $\sum_{j>l} (\kappa_j - w_j)$ is a harmonic form on a neighborhood of K_l and we have

$$\delta w = \sum_{j=1}^{l} \delta \kappa_j = \delta(\sum_{j=1}^{l} \phi_j \alpha) = \delta \alpha$$

on a neighborhood of K_l. Thus $\delta w = \delta \alpha$.

Our last step is to show that w is an exact form on Ω. For any finite p-cycle Γ in Ω we get

$$\int_\Gamma w = \lim_{n\to\infty} \sum_{j=1}^{\infty} \int_\Gamma (\kappa_j - w_j) = 0.$$

We applied dominated convergence theorem, see (1), and then used that forms κ_j and w_j are exact in Ω. Then, by the de Rham theorem, $w = d\beta$ for some $\beta \in C^\infty_{p-1}(\Omega)$.

We denote by $H_q(\Omega)$ the singular homology spaces of an open subset Ω of M (we consider smooth singular q-chains with real coefficients, see textbook 2 for details). There is a natural mapping

$$s_q : H_q(\Omega_1 \cap \Omega_2) \to H_q(\Omega_1) \oplus H_q(\Omega_2),$$

defined by $s_q(\Gamma) = (\Gamma, -\Gamma)$.

Let Ω be an open subset of M. We denote by $H^q(\Omega)$ the (de Rham) q-cohomology spaces, $H^q(\Omega)$ is the factor space of all (smooth) closed q-forms on Ω over the subspace of exact q-forms. De Rham cohomology spaces are dual to singular homology spaces. The dual operator to the operator s_q is defined as

$$a_q : H^q(\Omega_1) \oplus H^q(\Omega_2) \to H^q(\Omega_1 \cap \Omega_2),$$

where $a_q([w_1],[w_2]) = [(w_1 - w_2)|_{\Omega_1 \cap \Omega_2}]$, here we mean that the difference $w_1 - w_2$ is restricted to the intersection of the open sets and then the equivalence class is taken, we omit the restriction sign in what follows. It is easy to see that $\ker(s_q) = \{0\}$ if and only if a_q is surjective. Now, using Lemma 1, we can prove the theorem on separation of singularities.

Theorem 2. Let Ω_1, Ω_2 be two open subsets of M, and $\Omega = \Omega_1 \cap \Omega_2$. Then a harmonic q-form $v \in C_q^\infty(\Omega)$ can be written as $v = v_1 - v_2$, where $v_j \in C_q^\infty(\Omega_j)$ are harmonic q-forms if and only if $[v] \in \mathrm{Im}(a_q)$ and $[*v] \in \mathrm{Im}(a_{n-q})$.

Proof. The necessity of the condition is clear, if $v = v_1 - v_2$, then clearly $[v] = a_q([v_1],[v_2])$ and $[*v] = a_{n-q}([v_1],[v_2])$.

Suppose now that $[v] = a_q([v_1],[v_2])$. We have $v = v_1 - v_2 + dh$ in Ω. Moreover, using Lemma 1, we can find f_j such that $\delta df_j = \delta v_j$ in Ω_j then $w_j = u_j - df_j$ is both closed and coclosed and $v = w_1 - w_2 + dg$. Now $[*(w_1 - w_2)] = a_{n-q}([*w_1],[*w_2])$ and $*dg = *v - *(w_1 - w_2) \in \mathrm{Im}(a_{n-q})$. It is enough to show that dg can be written as the difference of two forms each harmonic on the corresponding domain Ω_j.

We can always write $g = g_1 - g_2$, where $g_j \in C^\infty(\Omega_j)$, then $*dg = *dg_1 - *dg_2$. Since $*dg \in \mathrm{Im}(a_{n-q})$, we have $*dg = \eta_1 - \eta_2$ for some forms η_1 and η_2 closed in Ω_1 and Ω_2 respectively. (Really we have $*dg = s_1 - s_2 + dr = (s_1 + dr_1) - (s_2 + dr_2)$, where $r = r_1 - r_2$ is a C^∞-decomposition.)

Now we consider

$$\kappa = \begin{cases} *\eta_1 - dg_1 \text{ on } \Omega_1 \\ *\eta_2 - dg_2 \text{ on } \Omega_2 \end{cases}$$

Clearly, κ is well-defined on $\Omega_1 \cup \Omega_2$. Using Lemma 1 once again, we find $\tau \in C_{q-1}^\infty$ such that $\delta d\tau = \delta\kappa$. Then $dg = (dg_1 + d\tau) - (dg_2 + d\tau)$ is the desired decomposition.

Corollary 3. A harmonic q-form $v \in C_q^\infty(\Omega)$ can be written as $v = v_1 - v_2$, where $v_j \in C_q^\infty(\Omega_j)$ are harmonic q-forms if and only if there exist q-forms u_1, u_2 such that $u = u_1 - u_2$, $u_j \in C_q^\infty(\Omega_j)$, and

$$\alpha = \begin{cases} du_1 \text{ on } \Omega_1 \\ du_2 \text{ on } \Omega_2 \end{cases} \quad \text{and} \quad \beta = \begin{cases} d*u_1 \text{ on } \Omega_1 \\ d*u_2 \text{ on } \Omega_2 \end{cases}$$

are exact forms on $\Omega_1 \cup \Omega_2$.

Proof. The corollary follows immediately from the theorem and the exactness of the Mayer-Vietoris exact sequence of the pair:

$$...H^p(\Omega_1) \oplus H^p(\Omega_2) \xrightarrow{a_p} H^p(\Omega_1 \cap \Omega_2) \xrightarrow{d_p} H^{p+1}(\Omega_1 \cup \Omega_2)....$$

We use that $\mathrm{Im}(a_p) = \mathrm{Ker}(d_p)$ and apply this for $p = q, n-q$. The statement of the corollary does not depend on the choice of the decomposition $v = u_1 - u_2$. Detailed discussion of the mapping d_p can be found in the book by R.Bott and L.Tu.[11]

Similar to the result[8] for harmonic forms on the open subsets of ^{n}, we formulate necessary and sufficient condition for separation of singularities, using mappings s_p:

Corollary 4. Let Ω_1, Ω_2 be two open subsets of M, and $\Omega = \Omega_1 \cap \Omega_2$. Then every harmonic q-form $v \in C_q^\infty(\Omega)$ can be written as $v = v_1 - v_2$, where $v_j \in C_q^\infty(\Omega_j)$ are harmonic q-forms if and only if $\ker s_q = \{0\}$ and $\ker s_{n-q} = \{0\}$.

3. Mittag – Leffler theorem for harmonic differential forms

As we have seen earlier,[8] even for the case of the Euclidean space there are some obstacles for construction of harmonic forms with prescribed (massive) singularities. On the manifold we will divide the problem into two parts. First, given a compact set e and a form u harmonic on $\omega \setminus e$ for some neighborhood of ω of e, we want to construct a form v harmonic on $M \setminus e$ that has same singularities on e, i.e. such that $v - u$ has harmonic extension to a neighborhood of e. This problem can be reformulated as separation of singularities problem. Let $\Omega_1 = M \setminus e$, $\Omega_2 = \omega$. We have a harmonic form on $\Omega_1 \cap \Omega_2$ and want to decompose it into the sum of two forms, one harmonic in Ω_1 and another harmonic in Ω_2. As we have seen in the previous section it can be done if and only if

$$s_q : H_q(\omega \setminus e) \to H_q(M \setminus e) \oplus H_q(\omega)$$

has trivial kernel for $q = p, n-p$. For example, a sufficient condition is

$$j_{q+1} : H_{q+1}(M \setminus e) \to H_{q+1}(M)$$

is surjective for $q = p, n-p$. (To see it one can write the exact homology sequence of the pair $(M \setminus e, \omega)$.) The last condition is satisfied if e is a subset of a coordinate chat of M.

Another part of the classical Mittag – Leffler theorem addresses the question of sewing together a sequence of singularities to one harmonic form. Here the situation is exactly the same as in the theory of functions of one complex variable.

Theorem 5. Let $e = \cup_j e_j$ be the union of a sequence of compact pairwise disjoint subsets of M. Assume that each compact subset of M intersects only finitely many sets of the sequence. Let h_j be a sequence of p-forms such that h_j is defined and harmonic on $M \setminus e_j$. Then there exists a q-form h harmonic on $M \setminus e$ such that $h - h_j$ has harmonic extension in a neighborhood of e_j.

Proof. Let σ_j be a relatively compact neighborhood of e_j such that σ_j is a sequence of pairwise disjoint subsets of M and each compact subset of M intersects only finitely many sets of the sequence. Further for each j we chose an open set ω_j such that $e_j \subset \omega_j \subset\subset \sigma_j$. We define v on $\omega = \cup_j \omega_j$ by $v = h_j$ on ω_j. We want to show that v can be written as the difference of a form harmonic on $M \setminus e$ and a form harmonic on ω.

Using the corollary proved after Theorem 2, for each j we get $h_j = u_j - r_j$, where $u_j \in C^\infty(M \setminus e_j)$, $r_j \in C^\infty(\omega_j)$ and

$$\alpha_j = \begin{cases} du_j \text{ on } M \setminus e_j \\ dr_j \text{ on } \omega_j \end{cases} \quad \text{and} \quad \beta_j = \begin{cases} d * u_j \text{ on } M \setminus e_j \\ d * r_j \text{ on } \omega_j \end{cases}$$

is exact and coexact on M. Moreover, we can choose u_j with compact support in σ_j. Then $v = \sum_j u_j - \sum_j r_j$ and both $\alpha = \alpha_j$ and $\beta = \beta_j$ are exact. To see it, we apply the de Rham theorem, all periods of α and β vanish.

4. Weierstrass's theorem for harmonic differential forms

We will show that there exists a harmonic form that interpolates given sequence of germs $\{u_j\}$ at a discrete set of points $\{p_j\}$ on M. We say that a smooth differential form ϕ has zero of order m at point p if in some coordinate chart all coefficients of ϕ vanish with all their partial derivatives up to order m.

The following result is simple after it is formulated [Yamabe[10]] Let Y be a dense linear subspace of the normed linear space X. Then for any $L_1, ..., L_n \in X^*$, any $x \in X$ and every $\epsilon > 0$ there exists $y \in Y$ such that $||x - y|| < \epsilon$ and $L_i(x) = L_i(y), i = 1, ..., n$.

We will use this result for simultaneous harmonic approximation and interpolation. It implies the following

Lemma 6. Let K be a compact subset of M such that $M \setminus K$ has no relatively compact connected components. Let further $\{p_1, ..., p_N\}$ be a finite sequence of points of K^o and $\{m_1, ..., m_N\}$ be a sequence of positive integers. If u is a q-form that is both exact and coexact on an open set containing K and if $\epsilon > 0$, then there exists a q-form U exact and coexact on M and such that $|u - U| < \epsilon$ on K, $U - u$ has zero of order m_j at point p_j,$j = 1, ..., N$.

Proof. Using coordinate chats we can rewrite the condition $U - u$ has zero of order m_j at p_j as finite number of conditions $L_{j,s}(U) = L_{j,s}(u)$, where $L_{j,s}$ are bounded linear functionals on the space $H_q^{c,e}(K)$ of forms exact and coexact in a neighborhood of K. We equip this space with the uniform norm.

Applying the methods we used in Section 1, we see that q-forms that are exact and coexact on M are dense in $H_q^{c,e}(K)$. Thus the statement of the lemma follows from the claim formulated above.

Theoem 7. Let $\{p_j\}_j$ be a discrete sequence of points in M and for each j let u_j be a q-form harmonic near p_j. Then there exists a harmonic q-form u on M such that $u - u_j$ has zero of order m_j at p_j for each j.

Proof. There exists an exhaustion of M by compact subsets, $M = \cup_{j=1}^{\infty} K_j$ such that $M \setminus K_j$ has no relatively compact in connected components. For simplicity (and without loss of generality) we may assume that $p_j \in K_j^0 \setminus K_{j-1}$. Poincaré's lemma implies that u_1 is exact and coexact near p_1. Then, applying Lemma ??, we get a q-form U_1 that is exact and coexact in M and such that $U_1 - u_1$ has zero of order m_1 at p_1. Suppose that we have constructed a q-forms U_{j-1} on M that is exact and coexact, and such that $U_{j-1} - u_l$ has zero of order m_l at p_l for $l = 1, ..., j - 1$.

Since $p_j \notin K_{j-1}$ we have $d_j = (p_j, K_{j-1}) > 0$. Let $E_j = K_{j-1} \cup B(p_j, r_j)$, where $r_j < d_j/2$ and r_j is such that u_j is harmonic in $B(p_j, 2r_j)$. We consider the following form near E_j

$$w_j(x) = \begin{cases} U_{j-1}(x) & \text{near } K_{j-1}, \\ u_j(x) & x \in B(p_j, r_j). \end{cases}$$

Then w_j is exact and coexact near E_j, there are no compact components of $M \setminus E_j$. Then using the lemma we can find a q-form U_j exact and coexact on M and such that $U_j - w_j$ vanishes to the order s_l at p_l for $l = 1, ..., j$. In addition we can have $||U_j - w_j||_{E_j} < \epsilon 2^{-j}$. Then

$$||U_j - U_{j-1}||_{K_{j-1}} < \epsilon 2^{-j}.$$

Thus sequence $\{U_j\}$ converges uniformly on compact subsets of M to some harmonic q-form that solves the interpolation problem.

Acknowledgments

This note is based on the talk given by the author at ICCAPT 2006 and it is my pleasure to thank the organizers of the conference. Thanks go also to A.Nicolau and J.Ortega-Cerda, whose question initiated author's interest in the interpolation problem.

The author was supported by the Research Council of Norway, project no 160192/V30, "PDE and Harmonic Analysis"

References

1. G. de Rham, *Differentiable manifolds. Forms, currents, harmonic forms.* (Springer-Verlag, 1984).
2. W. F. Warner, *Foundations of Differential Manifolds and Lie Groups*, (Springer-Verlag, 1983).
3. L. Ahlfors and L. Sario, *Riemann surfaces*, (Princeton University Press, 1960).
4. B. Rodin and L. Sario, *Principal functions*, (D.Van Nostrand company, inc., 1968).
5. N. Aronszajn, *Acta Math.* **65** (1935), 1–156.
6. T. Bagby and P. Blanchet, *J. Anal. Math.* **62** (1994), 47–76.
7. P. M. Gauthier, *Can. J. Math.*, **50**(3), (1998), 547–562.
8. E. Malinnikova, *Separation of singularities and Mittag-Leffler theorems for harmonic differential forms* (preprint, 2006).
9. E. Malinnikova, *St.Petersburg Math J.*, **11** (2000), no.4, 625–641.
10. H. Yamabe, *Osaka Math. J.* **2**, (1950), 15–17.
11. R. Bott and L. W. Tu, *Differential forms in algebraic topology.* (Springer-Verlag, 1982).

APPROXIMATE PROPERTIES OF THE BIEBERBACH POLYNOMIALS ON THE COMPLEX DOMAINS

DANIYAL M. ISRAFILOV
Balikesir University, Department of Mathematics
Balikesir, 10100, Turkey
mdaniyal@balikesir.edu.tr

Let G be a finite smooth Jordan domain of the complex plane $\mathbb{C}$, $z_0 \in G$, and let

$$\varphi_p(z) := \int_{z_0}^{z} \left[\varphi_0'(\zeta)\right]^{2/p} d\zeta, \quad z \in G, \quad p > 0,$$

where $w = \varphi(z)$ is the conformal mapping of G onto $\mathbb{D}(0, r_0) := \{w : \mid w \mid < r_0\}$ with $\varphi(z_0) = 0, \varphi'(z_0) = 1$. Let also $\pi_{n,p}(z)$, $n = 1, 2, \dots$, be the generalized Bieberbach polynomials for the pair (G, z_0). We investigate the uniform convergence of these polynomials on $\overline{G}$ and obtain an estimation for the quantity

$$\| \varphi_p - \pi_{n,p} \|_{\overline{G}} := \max_{z \in \overline{G}} |\varphi_p(z) - \pi_n(z)|$$

in term of the integral modulus of continuity of φ_p'.

Keywords: Generalized Bieberbach polynomials; Conformal mapping; Smooth boundaries; Uniform convergence.

1. Introduction and main results

Let G be a finite domain in the complex plane $\mathbb{C}$, bounded by a rectifiable Jordan curve L, and let $G^- := ExtL$. Further let $\mathbb{T} : = \{w \in \mathbb{C} : |w| = 1\}$, $\mathbb{D} := Int\mathbb{T}$ and $\mathbb{D}^- := Ext\mathbb{T}$. By the Riemann conformal mapping theorem, there exists a unique conformal mapping $w = \varphi(z)$ of G onto $\mathbb{D}(0, r_0) := \{w : \mid w \mid < r_0\}$ with the normalization $\varphi(z_0) = 0$, $\varphi'(z_0) = 1$. The inverse mapping of φ we denote by ψ.

Let also φ_0 be the conformal mapping of G^- onto $\mathbb{D}^-$ normalized by

$$\varphi_0(\infty) = \infty, \lim_{z \to \infty} \varphi_0(z)/z > 0,$$

and let ψ_0 be its inverse mapping. For an arbitrary function f given on G we set

$$\| f \|_{L_p(G)}^p := \int\int_G \mid f(z) \mid^p d\sigma_z, \quad p > 0.$$

If the function f has a continuous extension to $\overline{G}$ we use also the uniform norm:

$$\| f \|_{\overline{G}} := \sup\{\mid f(z) \mid, \quad z \in \overline{G}\}.$$

It is well known (see [12, p. 433]) that the function

$$\varphi_p(z) := \int_{z_0}^{z} [\varphi_0'(\zeta)]^{2/p} d\zeta, \quad z \in G, \quad p > 0$$

minimizes the integral $\| f' \|^p_{Lp(G)}$ $(p > 0)$ in the class of all functions analytic in G with the normalization $f(z_0) = 0$, $f'(z_0) = 1$. In the literature there are results concerning the region of exponents p which lead to the univalence of φ_p.In this work we study the approximation of φ_p by the extremal polynomials defined below. In fact this problem is a particular case of a more general one, formulated in [14, pp. 318-319, problems 1, 2] and is important for the approximately construction of the conformal mappings.

Let Π_n be the class of all polynomials p_n of degree at most n satisfying the conditions $p_n(z_0) = 0$, $p_n'(z_0) = 1$. Then we can prove that the integral $\| \varphi_p' - p_n' \|^p_{Lp(G)}$ $(p > 1)$ is minimized in Π_n by an unique polynomial $\pi_{n,p}$. We call (see also:[5,6]) these extremal polynomials $\pi_{n,p}$ the *generalized Bieberbach polynomials* for the pair (G, z_0). In case of $p = 2$ they are the usual Bieberbach polynomials π_n, $n = 1, 2, ...$ The approximation problems for $\varphi_2 = \varphi$ in closed domains with various boundary conditions, where approximation is conducted by the usual Bieberbach polynomials were intensively studied in[2,4,7–11,13] .

Similar problems for φ_p $(p > 1)$ using the generalized Bieberbach polynomials were investigated in[1,5,6] (see also,[8]). In the above cited works the rate of convergence to zero of the quantity

$$\| \varphi_p - \pi_{n,p} \|_{\overline{G}}$$

as n tends to ∞ , has been estimated by means of the geometric properties of G.

One of the interesting problem in this direction is the problem connected with a conjecture due to S. N. Mergelyan, who in[11] showed that the Bieberbach polynomials satisfy

$$\| \varphi_0 - \pi_n \|_{\overline{G}} \leq \frac{const}{n^{\frac{1}{2}-\varepsilon}} \tag{1}$$

for every $\varepsilon > 0$, whenever L is a smooth Jordan curve and stated it as a conjecture that the exponent $\frac{1}{2} - \varepsilon$ in (1) could be replaced by $1 - \varepsilon$.

In[7] it has been possible for us to obtain some improvement of the above cited Mergelyan's estimation (1). For its formulation, it is necessary to give some definitions as follows.

We denote by $L^p(L)$ and $E^p(G)$ the set of all measurable complex valued functions such that $| f |^p$ is Lebesgue integrable with respect to arclength, and the Smirnov class of analytic functions in G respectively. For a weight function ω given on L, and $p > 1$ we also set

$$L^p(L,\omega) := \{f \in L^1(L) : | f |^p \omega \in L^1(L)\},$$

$$E^p(G,\omega) := \{f \in E^1(G) : f \in L^p(L,\omega)\}.$$

Let $g \in L^p(\mathbb{T},\omega)$ and let $g_h(w)$ be the mean value function for g defined as:

$$g_h(w) := \frac{1}{2h}\int_{-h}^{h} g(we^{it})dt, \qquad 0 < h < \pi, \qquad w \in T.$$

The function $\Omega_{p,\omega}(g,\cdot) : [0,\infty) \to [0,\infty)$ defined by

$$\Omega_{p,\omega}(g,\delta) := \sup\{\| g - g_h \|_{L^p(T,\omega)}, h \leq \delta\}, \qquad 1 < p < \infty$$

is called the integral modulus of continuity in $L^p(\mathbb{T},\omega)$ for g.

The improvement obtained in[7] can be formulated in the following way: If the boundary L of the domain G is a smooth Jordan curve, then

$$\| \varphi_0 - \pi_n \|_{\overline{G}} \leq const \left(\frac{\ln n}{n}\right)^{\frac{1}{2}} \Omega_{2,|\psi_0'|}\left(\varphi'[\psi_0(w)](\psi_0'(w))^{\frac{1}{2}}, \frac{1}{n}\right), \tag{2}$$

for $n \geq 2$, where $\Omega_{2,|\psi_0'|}\left(\cdot,\frac{1}{n}\right)$ is the integral modulus of continuity in space $L^2(\mathbb{T}, |\psi_0'|)$ for $\varphi'[\psi_0(w)]\cdot(\psi_0'(w))^{\frac{1}{2}}$. From this result in particular it follows that if G is a finite domain with a smooth Jordan boundary, then

$$\| \varphi - \pi_n \|_{\overline{G}} \leq const \left(\frac{\ln n}{n}\right)^{\frac{1}{2}}, \qquad n \geq 2, \tag{3}$$

which improves the estimation (1).

Later in[8], under some restrictions of the boundary L, the improvement (2) and hence (3) has been extended to the finite Radon domains.

In this work, developing the idea used in[7-9] we shall obtain an estimation for the approximation of φ_p, $1 < p < \infty$, by means of the generalized Bieberbach polynomials $\pi_{n,p}$. This estimation in case of $p = 2$ appears simpler than (2).

We begin with the following definition.

Definition 1 *Let G be a domain with a smooth boundary L, $1 < p, r < \infty$ and let $\Phi_p := (\varphi_p' \circ \psi_0)$ on $\mathbb{T}$. The function*

$$\omega_r^*(\varphi_p',\delta) := \sup_{|h|\leq\delta} \| \Phi_p(we^{ih}) - \Phi_p(w) \|_{L^r(\mathbb{T})} =: \omega_r(\Phi_p,\delta),$$

is called the generalized integral modulus of continuity for φ_p'.

Our main result is the following.

Theorem 1 *Let G be a finite domain with a smooth boundary L and let $p > 1$. Then*

$$\| \varphi_p - \pi_{n,p} \|_{\overline{G}} \leq \begin{cases} c_1\sqrt{\frac{\ln n}{n}}\,\omega_{2+\varepsilon}^*(\varphi_2',\frac{1}{n}), & p = 2; \\ c_2 n^{-1/p}\omega_{p+\varepsilon}^*(\varphi_p',\frac{1}{n}), & p > 2; \\ c_3 n^{1/p-1+\varepsilon}\omega_{p+\varepsilon}^*(\varphi_p',\frac{1}{n}), & 1 < p < 2, \end{cases}$$

for every $\varepsilon > 0$ *and with the constants* $c_i = c_i(\varepsilon) > 0$, $i = 1, 2, 3$.

In spite of the fact that the function φ_p is defined on G, it has[6] a continuous extension to $\overline{G}$. Therefore, the uniform norm in the above inequality is well defined.

From this theorem in case of $p = 2$ we have the following result.

Corollary 1 *Let* G *be a domain with a smooth boundary* L. *Then*

$$\| \varphi - \pi_n \|_{\overline{G}} \leq c_1 \left(\frac{\ln n}{n} \right)^{\frac{1}{2}} \omega^*_{2+\varepsilon}(\varphi', \frac{1}{n}) \qquad n \geq 1$$

for every $\varepsilon > 0$ *and with a constant* $c_1 = c_1(\varepsilon) > 0$.

As it follows from Definition 1 the modulus of continuity $\omega^*_{2+\varepsilon}(\varphi', \frac{1}{n})$ is simpler than $\Omega_{2,|\psi_0'|}\left((\varphi' \circ \psi_0) (\psi_0')^{\frac{1}{2}}, \frac{1}{n} \right)$ defined above.

2. Auxiliary results

We shall use c, c_1, c_2, ... to denote constants (in general, different in different ralations) depending only on numbers that are not important for the questions of interest.

One of the important step in the proof of the main result is the following theorem, given here without proof.

Theorem 2 *Let* G *be a domain with a smooth boundary* L, $1 < p, r < \infty$ *and let*

$$S_n\left(\varphi_p', z\right) := \sum_{k=0}^{n} a_k\left(\varphi_p'\right) F_k(z), \qquad n = 0, 1, 2, ...$$

be *the* n^{th} *partial sums of the Faber series of* φ_p'. *Then*

$$\| \varphi_p' - S_n\left(\varphi_p', \cdot\right) \|_{L^r(L)} \leq c\, \omega_{r+\varepsilon}(\Phi_p, 1/n)$$

for every $\varepsilon > 0$ *and with a constant* $c = c(\varepsilon) > 0$.

For the function φ_p and a weight ω we set

$$\varepsilon_n\left(\varphi_p'\right)_r := \inf_{p_n} \| \varphi_p' - p_n \|_{L_r(G)}, \quad E_n^\circ\left(\varphi_p'\right)_r := \inf_{p_n} \| \varphi_p' - p_n \|_{L^r(L)},$$

$$E_n^\circ\left(\varphi_p', \omega\right)_r := \inf_{p_n} \| \varphi_p' - p_n \|_{L^r(L,\omega)},$$

where inf is taken over all polynomials p_n of degree at most n and

$$\| \varphi_p' - p_n \|_{L^r(L,\omega)} := \left(\int_L | \varphi_p'(z) - p_n(z) |^r \omega(z) \, | dz | \right)^{1/r}.$$

Corollary 2 *Let G be a domain with a smooth boundary L and $r>1$. Then*

$$E_n^{\circ}\left(\varphi_p', \frac{1}{|\varphi_0'|}\right)_r \leq c\omega_{r+\varepsilon}^*(\varphi_p', \frac{1}{n})$$

for every $\varepsilon > 0$ and with $c = c(\varepsilon) > 0$.

Proof. Since

$$E_n^{\circ}\left(\varphi_p', \frac{1}{|\varphi_0'|}\right)_r = \inf_{p_n} \| \varphi_p' - p_n \|_{L^r(L,1/|\varphi_0'|)}$$

$$\leq \| \varphi_p' - S_n \|_{L^r(L,1/|\varphi_0'|)}$$

and $1/\varphi_0' \in L_r(L)$, for every $r \geq 1$, by Hölder's inequality we have

$$E_n^{\circ}\left(\varphi_p', \frac{1}{|\varphi_0'|}\right)_r \leq \| \varphi_p' - S_n \|_{L^r(L,1/|\varphi_0'|)} \leq c_4 \| \varphi_p' - S_n \|_{L^{rp_0}(L)}$$

for any $p_0 > 1$. Now, applying Theorem 2 (in case of $r := rp_0$) we conclude that

$$E_n^{\circ}\left(\varphi_p', \frac{1}{|\varphi_0'|}\right)_r \leq c_5\omega_{rp_0+\varepsilon}^*(\varphi_p', \frac{1}{n}).$$

Choosing the number p_0 sufficiently close to 1 we finally get

$$E_n^{\circ}\left(\varphi_p', \frac{1}{|\varphi_0'|}\right)_r \leq c\omega_{r+\varepsilon}^*(\varphi_p', \frac{1}{n})$$

with $c = c(\varepsilon)$. ■

The proof of the following theorem is similar to that of Theorem 11 from[3] .

Theorem 3 *Let G be a domain with a smooth boundary L and let $p > 1$. Then*

$$\varepsilon_n\left(\varphi_p'\right)_p \leq cn^{-1/p}E_n^{\circ}\left(\varphi_p', \frac{1}{|\varphi_0'|}\right)_p$$

with a constant $c > 0$.

The approximation properties of the polynomials $\pi_{n,p}'$, $n = 1, 2, ...$, in the space $L_p(G)$ are given in the following lemmas.

Lemma 1 *Let G be a domain with a smooth boundary L and let $p > 1$. Then*

$$\| \varphi_p' - \pi_{n,p}' \|_{L_p(G)} \leq cn^{-1/p}\omega_{p+\varepsilon}^*(\varphi_p', \frac{1}{n}), \quad n = 1, 2, ...$$

for every $\varepsilon > 0$ and with $c = c(\varepsilon) > 0$.

Proof. For the polynomials $q_{n,p}(z)$ best approximating φ_p' in $L_p(G)$ we set

$$Q_{n,p}(z) := \int_{z_0}^{z} q_{n,p}(t)\,dt, \qquad t_{n,p}(z) := Q_{n,p}(z) + [1 - q_{n,p}(z_0)](z - z_0).$$

Then $t_{n,p}(z_0) = 0, \quad t'_{n,p}(z_0) = 1$ and hence by Theorem 3 we obtain

$$\left\| \varphi'_p - t'_{n,p} \right\|_{L_p(G)} = \left\| \varphi'_p - q_{n,p} - 1 + q_{n,p}(z_0) \right\|_{L_p(G)}$$

$$\leq \left\| \varphi'_p - q_{n,p} \right\|_{L_p(G)} + \left\| 1 - q_{n,p}(z_0) \right\|_{L_p(G)}$$

$$\leq cn^{-1/p} E_n^{\circ} \left(\varphi'_p, \frac{1}{|\varphi'_0|} \right)_p + \left\| \varphi'_p(z_0) - q_{n,p}(z_0) \right\|_{L_p(G)}.$$

This relation by the inequality

$$|f(z_0)| \leq c \|f\|_{L_p(G)},$$

which holds for every analytic function f with $\|f\|_{L^p(G)} < \infty$, and by the extremal property of the polynomials $\pi_{n,p}$ implies that

$$\left\| \varphi'_p - \pi'_{n,p} \right\|_{L_p(G)} \leq \left\| \varphi'_p - t'_{n,p} \right\|_{L_p(G)} \leq cn^{-1/p} E_n^{\circ} \left(\varphi'_p, \frac{1}{|\varphi'_0|} \right)_p.$$

Now applying Corollary 2 (in case of $r = p$) we obtain the statment of Lemma 1. ■

The proof of the following lemma can be found in[2] (for $p = 2$) and[6] (in case of $1 < p < \infty$).

Lemma 2 *Let G be a finite domain with a smooth boundary L and let p_n be any polynomial of degree $\leq n$ with $p_n(z_0) = 0$. Then*

$$\| p_n \|_{\overline{G}} \leq \begin{cases} c\sqrt{\log n} \, \| p'_n \|_{L_p(G)}, & p = 2; \\ c \, \| p'_n \|_{L_p(G)}, & p > 2; \\ c_6 n^{\frac{2}{p} - 1 + \varepsilon} \, \| p'_n \|_{L_p(G)}, & 1 < p < 2 \end{cases}$$

for every $\varepsilon > 0$, and with $c > 0$ and $c_6 = c_6(\varepsilon) > 0$.

The following lemma was proved in [2, Lemma 15] in case of $p = 2$ treating the Bieberbach polynomials π_n, $n = 1, 2, \ldots$. The general case $p \in (1, \infty)$, concerning the extremal polynomials $\pi_{n,p}$, $n = 1, 2, \ldots$, goes similarly.

Lemma 3 *Let G be a finite simple connected domain, and let p_n be a polynomial of degree $\leq n$ satisfying the condition $p_n(z_0) = 0$. Assume that*

$$\| p_n \|_{\overline{G}} \leq c_7 \alpha_n \| p'_n \|_{L_p(G)}$$

and

$$\| \varphi'_p - \pi'_{n,p} \|_{L_p(G)} \leq c_8 \beta_n,$$

with some positive constants c_7 and c_8, *where*

$$\{\alpha_n\}\nearrow, \{\beta_n\}\searrow \quad and \quad \{\gamma_n := \alpha_n \cdot \beta_n\}\searrow .$$

If in addition, there exists a sequence of indexes $\{n_k\}$ *such that*

$$\gamma_{n_{k+1}} \leq \varepsilon\gamma_{n_k}, \quad \alpha_{n_{k+1}} \leq c_8\alpha_{n_k}, \qquad k = 1, 2, ...$$

for some $\varepsilon \in (0,1)$ *and a constant* $c \geq 1$, *then*

$$\| \varphi_p - \pi_{n,p} \|_{\overline{G}} \leq c\gamma_n.$$

3. Proof of Main Result

Proof of Theorem 1. We apply the traditional method (see, for example[2] in case of $p = 2$), its modification in case of $1 < p < \infty$ given in[6] , and also the idea developed in the papers[7–9] .

The proof goes following to the procedure drawing by Lemma 3 with an suitable choice of α_n from Lemma 2, of β_n from Lemma 1, and defining $n_k := 2^k$ in Lemma 3.

References

1. F. G. Abdullayev, : *Uniform convergence of the generalized Bieberbach polynomials in domains of complex plane,* In: Approximation Theory and its Applications, Pr. Inst. Mat. Nats. Akad. Nauk Ukr., Kiev (1999), 5-18.
2. V. V. Andrievskii, : *Uniform convergence of Bieberbach polynomials in domains with piecewise-quasiconformal boundary* (Russian), In: Theory of Mappings and Approximation of Functions, 3-18. Kiev: Naukova Dumka (1983).
3. E. M. Dyn'kin, : *The rate of polynomial approximation in the complex domain.* In Complex analysis and spectral theory (Leningrad, 1979/1980), pp. 90-142. Berlin: Springer 1981.
4. D. Gaier, : *On the convergence of the Bieberbach polynomials in regions with corners.* Constr. Approx. **4** (1988), 289-305.
5. D. M. Israfilov, : *On the approximation properties of the extremal polynomials,* Dep. VINITI, No:5461 (1981) (in Russian).
6. D. M. Israfilov,:*Uniform convergence of some extremal polynomials in domains in domains with quasi conformal boundary.* East Journal on Approx., **4** (1998), 527-539.
7. D. M. Israfilov,:*Approximation by* $p-$ *Faber -Laurent rational functions in the weighted Lebesgue space* $L^p(L, \omega)$ *and the Bieberbach polynomials,* Constr. Approx. **17** (2001), 335-351.
8. D. M. Israfilov,:*Bergman type kernel function and approximation properties of some extremal polynomials on quasidisks,* East Journal on Approx. vol **9, 2** (2003), 157-173.
9. D. M. Israfilov,:*Uniform convergence of the Bieberbach polynomials in closed smooth domains of bounded boundary rotation,* Journal of Approximation Theory, 125 (2003), 116-130.
10. M. V. Keldych,: *Sur l'approximation en moyenne quadratique des fonctions analytiques,* Math. Sb., **5 (47)** (1939), 391-401.
11. S. N. Mergelyan,: *Certain questions of the constructive theory of functions* (Russian), Trudy Math. Inst. Steklov, **37** (1951), 1-91.

12. I. I. Privalov,: Introduction to the Theory of Functions of a Complex Variable, Nauka, Moscow, (1984).
13. P. K. Suetin,: *Polynomials ortogonal over a region and Bieberbach polynomials,* Proc. Steklov Inst. Math., Providence, RI: American Mathematical Society, **100** (1974).
14. J. L. Walsh,: Interpolation and Approximation by Rational Functions in the Complex Domain, Third edition, Rhode island, (1960).

HARMONIC TRANSFINITE DIAMETER AND CHEBYSHEV CONSTANTS

N. SKIBA

Rostov State University of Transportation,
Rostov-na-Donu, Russia
skibanimail.ru

VYACHESLAV ZAKHARYUTA

Sabanci University, Orhanli,
34956 Tuzla/Istanbul, Turkey
zahasabanciuniv.edu

For a given compact set K in $\mathbb{R}^{p+2}$ the harmonic transfinite diameter $d^h(K)$, directional harmonic Chebyshev constants $\tau^h(K,\theta)$ and the main harmonic constant $\tau^h(K)$ are introduced by analogy with the corresponding characteristics for sets in $\mathbb{C}^n$ (Leja, Zakharyuta). It is proved that always $d^h(K) \leq \tau^h(K)$. Unlike the case of seneral complex variables (considered in[11]), the equality $d^h(K) = \tau^h(K)$ is proved under some restrictions on K. For the spheroid Φ_σ, obtained by rotating of the ellipse with the half-axes $\sinh\sigma$ and $\cosh\sigma$ around the main axis, the following formula is obtained:

$$d^h(K) = \tau^h(K) = \frac{\sinh\sigma}{\sqrt{e}} \left(\coth\frac{\sigma}{2}\right)^{\frac{\cosh\sigma}{2}}$$

1. Introduction

In[11] the second author introduced the directional Chebyshev constants $\tau(K,\theta)$ of a compact set $K \subset \mathbb{C}^n$ and proved that the main Chebyshev constant $\tau(K)$, defined as the integral geometric mean of the directional constants, coincides with the Fekete-Leja transfinite diameter $d(K)$ introduced in;[6] this fact may be considered as a natural generalization of the classical result of Fekete-Szegö (see, e.g.,[4]). The existence of the usual (instead of upper) limit in the definitions of the directional Chebyshev constants and of the transfinite diameter in multidimensional case is based on some *arithmetic multiplicative* properties of the monomial basis, ordered in a natural way. This was the reason for the successful applications of ideas from[11] in studying of capacities in arithmetic geometry (see, e.g.,[7,8])

Here, by analogy with the classical approach in $\mathbb{C}^n$,[6,9] we define the *harmonic transfinite diameter* $d^h(K)$ of a compact set K in $\mathbb{R}^{p+2}$, $p \geqslant 0$. Following,[11] we introduce a natural notion of the *main harmonic Chebyshev constant* $\tau^h(K)$ defined as the integral geometric mean of *directional Chebyshev constants* $\tau^h(K,\theta)$ of K. We discuss some interconnections among these characteristics in general case and show, in particular, that $d^h(K) = \tau^h(K) = \frac{\sinh\sigma}{\sqrt{e}}\left(\coth\frac{\sigma}{2}\right)^{\frac{\cosh\sigma}{2}}$ for the prolate spheroid $K := \left\{(t_1,t_2,t_3)\in\mathbb{R}^3 : \frac{t_1}{\cosh^2\sigma} + \frac{t_2^2+t_3^2}{\sinh^2\sigma} \le 1\right\}$, $0<\sigma<\infty$. The last computation is based on the asymptotics of associated Legendre functions suggested in[13] .

Since the natural basis of harmonic polynomials (see below) has, contrary to the monomial basis in $\mathbb{C}^n$, no explicit arithmetic properties the results for harmonic characteristics are considerably weaker than those from[11] . Nevertheless the equality $d^h(K) = \tau^h(K)$ for the spheroidal case encourages our hopes that there might be some hidden recurrent "arithmetic" properties of the basis (3) which may confirm the following

Conjecture 1.1. *The equality $d^h(K) = \tau^h(K)$ holds for each compact set K in $\mathbb{R}^{p+2}$.*

2. Preliminaries

2.1. *Notation*

By $\mathbb{R}^{p+2}$ we denote the $(p+2)$-dimensional Euclidean space, $p \ge 1$; $\mathbb{B}_r(a)$ stands for the open ball in $\mathbb{R}^{p+2}$ centered at a with the radius r; $\mathbb{B}_r := \mathbb{B}_r(0)$; $\mathbb{S} = \mathbb{S}^{p+1}$ is the unit sphere in $\mathbb{R}^{p+2}$. For every set F and each function $x : F \to \mathbb{C}$ we denote $|x|_F := \sup\{|x(t)| : t \in F\}$.The sign $\hookrightarrow$ means a continuous linear embedding. The open p-dimensional simplex

$$\{(\theta_1,\theta_2,\ldots,\theta_p)\in\mathbb{R}^p : 1>\theta_1>\theta_2>\ldots>\theta_{p-1}>|\theta_p|\ge 0\} \tag{1}$$

is denoted by Σ. Notice that, in particular, $\Sigma := \{\theta_1 : -1 < \theta_1 \leqslant 1\}$ for $p = 1$.

If K is a compact set in $\mathbb{R}^{p+2}$, then $h(K)$ stands for the space of all harmonic germs on K endowed with the standard inductive limit topology (for more details about spaces of harmonic functions see, e.g.,[12]) .

Denote by $h_2(\mathbb{B})$ the harmonic L_2-Hardy space in the unit ball $\mathbb{B}$, i.e. the Hilbert space obtained as a closure of the set of all harmonic polynomials in the space $L_2(\mathbb{S},\sigma)$ with the Lebesgue measure σ on the sphere $\mathbb{S}$.

2.2. *Harmonic polynomials*

For each $n \in \mathbb{Z}_+$ consider the standard system of homogeneous harmonic polynomials of degree n of the variables $t = (t_1,\ldots,t_p,t_{p+1},t_{p+2}) \in \mathbb{R}^{p+2}$(see, e.g.,[3] , 11.2). Namely, let $\mathcal{M}_n$ be the set of all vectors $m = (m_1,\ldots,m_p)$ with integer coefficients such that

$$n = m_0 \geqslant m_1 \geqslant \ldots \geqslant |m_p| \geqslant 0, \tag{2}$$

and r_k defined by the formula $r_k = \sqrt{t_{k+1}^2 + t_{k+2}^2 + \cdots + t_{p+2}^2}$, where $k = 0, 1, \ldots, p$. Then the functions

$$H(n, m; t) =: \left(\frac{t_{p+1}}{r_p} + i\frac{t_{p+2}}{r_p}\right)^{m_p} r_p^{m_p} \prod_{k=0}^{p-1} r_k^{m_k - m_{k+1}} C_{m_k - m_{k+1}}^{m_{k+1} + \frac{p}{2} - \frac{k}{2}} \left(\frac{t_{k+1}}{r_k}\right) \quad (3)$$

with $m \in \mathcal{M}_n$, form the complete system of $s(n) = s(n,p) = (2n+p)\dfrac{(n+p-1)!}{p!n!}$ linearly independent harmonic polynomials of degree n; here $C_n^\nu(x)$ is the Gegenbauer (ultraspheric) polynomial of degree n and order ν (see,[3]).

Definition 2.1. The *standard (complete orthonormal in the space $h_2(\mathbb{B})$) system of harmonic polynomials* $e_i = e_i(t)$, $i \in \mathbb{N}$, is obtained from the system (3) by normalizing and enumerating in the strict lexicographic order $\prec$ with respect to the indices $m_0, m_1, \ldots, m_{p-1}, m_p$ (therewith, the last index m_p is ordered as usual: $0, -1, 1, \ldots, -m_p, m_p$):

$$\widetilde{H}(m_0, m; t) = \frac{H(m_0, m; t)}{\sqrt{N(m_0, m)}}, \quad e_i(t) = \widetilde{H}(m_0(i), m(i); t), \; i \in \mathbb{N}, \quad (4)$$

where $(m_0(i), m(i))$ is the enumeration, determined by the order $\prec$ and the normalizing numbers $N(m_0, m)$ are defined in[3] , 11.3(5)).

The chosen order provides, in particular, that the degree $n(i) = m_0(i)$ of the polynomial e_i is a *non-decreasing function of* i. We denote by $\mathcal{H}_i$ the set of all harmonic polynomials represented in the form: $p(t) = \sum\limits_{j=1}^{i} c_j\, e_j(t)$.

For the case $p = 1$, in the spherical coordinates (r, ψ, φ), we have

$$\widetilde{H}(n, m; t) = \left(\frac{(2n+1)(n-|m|)!}{2\pi(n+|m|)!}\right)^{1/2} r^n\, P_n^{|m|}(\cos\psi)\, e^{im\varphi}, \quad (5)$$

where P_n^m are the associated Legendre functions.

3. Harmonic characteristics of compacta

3.1. *Harmonic transfinite diameter*

Let K be a compact set in $\mathbb{R}^{p+2}$. Given $\xi_j \in K$, $j = 1, 2, \ldots, i$, we define the Vandermondian: $V^{(h)}(\xi_1, \xi_2, \ldots, \xi_i) = \det\left(e_\mu(\xi_\nu)\right)_{\mu,\nu=1}^{i}$, where $\{e_i\}_{i\in\mathbb{N}}$ is the standard complete orthonormal system of harmonic polynomials (4). Introduce the numbers

$$V_i^h = V_i^h(K) := \sup\left\{|V^h(\xi_1, \xi_2, \ldots, \xi_i)| : \xi_j \in K, \; j = 1, 2, \ldots, i\right\}, \; i \in \mathbb{N}. \quad (6)$$

One can easily prove that there exist extremal points $\xi_1^{(i)}, \xi_2^{(i)}, \ldots, \xi_i^{(i)}$ (in general, not uniquely defined) for which the supremum in (6) is attained.

Definition 3.1. *The harmonic transfinite diameter of* K is determined by the formula $d^h(K) = \limsup\limits_{n\to\infty} [V_{r(n)}^h]^{\frac{1}{l(n)}}$, where $l(n) = \sum\limits_{\nu=1}^{n} \nu \cdot s(\nu)$ and $r(n) = \sum\limits_{\nu=0}^{n} s(\nu)$ is the number of all harmonic polynomials (3) of degree $\leqslant n$.

Following[9] we say that a compact set K is *unisolvent* if there is a sequence $\{\xi_i\}_1^\infty \subset K$ such that $V^h(\xi_1, \xi_2, \ldots, \xi_n) \neq 0$ for each $n \in \mathbb{N}$. Notice that if K is not unisolvent then there is i_0 such that $V_i^h(K) = 0$ for all $i \geq i_0$.

3.2. *Harmonic Chebyshev constants*

Given a compact set $K \subset \mathbb{R}^{p+2}$, consider for each $i \in \mathbb{N}$ the minimal deviation of harmonic polynomial $p \in \mathcal{H}_i$, with the leading coefficient equal 1, from the origin in $C(K)$:

$$M_i = M_i(K) =: \inf\{|p|_K : p(t) = e_i(t) + \sum_{j=1}^{i-1} c_j e_j(t)\}. \tag{7}$$

It is a common place (see, e.g., [1]) that there exists a Chebyshev harmonic polynomial

$$T_i(t) = e_i(t) + \sum_{j=1}^{i-1} c_j^{(i)} e_j(t)\} \tag{8}$$

(maybe not unique) such that $M_i = |T_i|_K$. The numbers (6) can be estimated through the numbers (7) in the following way.

Lemma 3.1. *Let $K \subset R^{p+2}$ be a compact set such that $V_{i-1}^h = V_{i-1}^h(K) \neq 0$. Then*

$$M_i \leq \frac{V_i^h}{V_{i-1}^h} \leq i M_i. \tag{9}$$

In particular, if K is unisolvent, then (9) holds for all $i = 2, 3, \ldots$.

Proof. The polynomial $P_i(t) = \frac{V^h\left(\xi_1^{(i-1)}, \xi_2^{(i-1)}, \ldots, \xi_{i-1}^{(i-1)}, t\right)}{V^h\left(\xi_1^{(i-1)}, \xi_2^{(i-1)}, \ldots, \xi_{i-1}^{(i-1)}\right)}$, with the extremal points $\xi_1^{(i-1)}, \xi_2^{(i-1)}, \ldots, \xi_{i-1}^{(i-1)}$ providing $V^h\left(\xi_1^{(i-1)}, \xi_2^{(i-1)}, \ldots, \xi_{i-1}^{(i-1)}\right) = V_{i-1}^h$, belongs to $\mathcal{H}_i$ and has the leading coefficient equal 1. Hence $M_i \leq |P_i|_K$. On the other hand, since $V^h\left(\xi_1^{(i-1)}, \xi_2^{(i-1)}, \ldots, \xi_{i-1}^{(i-1)}, t\right) \leq V_i^h$, $t \in K$, we have $|P_i|_K \leq \frac{V_i^h}{V_{i-1}^h}$. Thus the left hand side of (9) is proved. Let us prove the right hand side. Taking the extremal points $\xi_1^{(i)}, \xi_2^{(i)}, \ldots, \xi_i^{(i)}$, it is easily seen that the right hand side of the equality $V_i^h = \det[e_\mu(\xi_\nu^{(i)})]_{\mu,\nu=1}^i$ is not changed if we substitute the last row of $\det\left(e_i(\xi_\nu^{(i)})\right)_{\nu=1}^i$ by the row $\left(T_i\left(\xi_\nu^{(i)}\right)\right)_{\nu=1}^i$, where T_i is a Chebyshev polynomial (8). Therefore, expanding the resulting determinant along the last row, we obtain the desired estimate:

$$V_i^h \leq \sum_{\nu=1}^{i} \left|T_i\left(\xi_\nu^{(i)}\right)\right| \left|V^h\left(\xi_1^{(i)}, \ldots, \xi_{\nu-1}^{(i)}, \xi_{\nu+1}^{(i)}, \ldots, \xi_i^{(i)}\right)\right| \leq i\, M_i\, V_{i-1}^h.$$

Definition 3.2. Denote $\tau_i = \tau_i^h(K) =: (M_i)^{1/n(i)}$ and $\theta(i) = (\theta_k(i)) =: \left(\frac{m_k(i)}{n(i)}\right)_{k=1}^p \in \Sigma$ with $m_k(i)$ and $n(i)$ from Definition 2.1. The number

$$\tau^h(K,\theta) = \limsup_{\theta(i)\to\theta} \tau_i, \quad \theta = (\theta_k) \in \Sigma, \tag{10}$$

is the *harmonic Chebyshev constant of the compactum K for the direction θ* .

Lemma 3.2. *The function $\tau^h(K,\theta)$ is upper semicontinuous on Σ.*

Proof. Given $\theta^{(0)} \in \Sigma$, choose a sequence $\theta^{(j)} \in \Sigma$, $j \in \mathbb{N}$, so that $\theta^{(j)} \to \theta^{(0)}$, $\theta^{(j)} \neq \theta^{(0)}$ and $\limsup_{\theta\to\theta^{(0)}} \tau^h(K,\theta) = \lim_{j\to\infty} \tau^h(K,\theta^{(j)})$. In accordance with Definition 3.1, for each $j \in \mathbb{N}$ we find $i(j)$ such that

$$\tau^h(K,\theta^{(j)}) - 1/j \leq \tau_{i(j)}, \; \left|\theta(i(j)) - \theta^{(j)}\right| < 1/j.$$

Then, because of $\theta(i(j)) \to \theta^{(0)}$, we obtain the required estimate:

$$\limsup_{\theta\to\theta^{(0)}} \tau^h(K,\theta) = \lim_{j\to\infty} \tau^h(K,\theta^{(j)}) \leq \limsup_{j\to\infty} \tau_{i(j)} \leq \tau^h(K,\theta^{(0)}).$$

Since the function $\ln \tau^h(K,\theta)$ is bounded from above and, by Lemma 3.2, measurable with respect to the induced Lebesgue measure λ on Σ, we are ready to give the following

Definition 3.3. The *(main) harmonic Chebyshev constant of a compactum $K \subset \mathbb{R}^{p+2}$* is defined as the geometric mean of the directional Chebyshev constants (10), namely,

$$\tau^h(K) = \exp\left\{\frac{1}{\lambda(\Sigma)} \int_\Sigma \ln \tau^h(K,\theta)\, d\lambda(\theta)\right\}; \tag{11}$$

it is assumed here that $\tau^h(K) = 0$ if the integral equals $-\infty$.

It is easy to see that

$$d^h(\mathbb{B}_r) = \tau^h(\mathbb{B}_r) = \tau^h(\mathbb{B}_r,\theta) \equiv r, \quad \theta \in \Sigma. \tag{12}$$

Lemma 3.3. *Let K be a compact set in $\mathbb{R}^{p+2}$, $\tau_i = \tau_i^h(K)$ and*

$$\widetilde{\tau}_\nu =: \left(\prod_{n(i)=\nu} \tau_i\right)^{1/s(\nu)}. \tag{13}$$

Then $\limsup_{\nu\to\infty} \widetilde{\tau}_\nu \leq \tau^h(K)$.

Proof. Introduce the rectangles

$$\Delta_i =: \{\theta = (\theta_k) \in \mathbb{R}^p : \theta_k(i) \leq \theta_k < \theta_k(i) + 1/n(i),\; k = 1,\ldots,p\},\; i \in \mathbb{N}$$

and denote by S_ν the union of all Δ_i with $r(\nu-1) < i \le r(\nu)$. For each $n \in \mathbb{N}$ define the function on the set $S_\nu \supset \Sigma$:

$$\tau_\nu(K,\theta) = \{ \ \tau_i \quad \text{if} \quad \theta \in \Delta_i, \ \ r(\nu-1) < i \le r(\nu)\,.$$

It is clear that

$$\ln \widetilde{\tau}_\nu = \frac{1}{\lambda(S_\nu)} \int_{S_\nu} \ln \tau_\nu(K,\theta)\, d\lambda(\theta) \tag{14}$$

and $\limsup\limits_{\nu\to\infty} \tau_\nu(K,\theta) \le \tau^h(K,\theta)$ if $\theta \in \Sigma$. Therefore, taking into account that $\limsup\limits_{\nu\to\infty} \int\limits_{S_\nu \setminus \Sigma} \ln \tau_\nu(K,\theta)\, d\lambda(\theta) \le 0$ and using the Fatou Theorem, we obtain

$$\begin{aligned} \limsup_{\nu\to\infty} \widetilde{\tau}_\nu &\le \limsup_{\nu\to\infty} \frac{1}{\lambda(S_\nu)} \int_{S_\nu} \ln \tau_\nu(K,\theta)\, d\lambda(\theta) \\ &\le \frac{1}{\lambda(\Sigma)} \int\limits_{\Sigma} \ln \tau^h(K,\theta)\, d\lambda(\theta)\} = \tau^h(K). \end{aligned}$$

3.3. *Relation between the characteristics*

Theorem 3.1. *The estimate $d^h(K) \le \tau^h(K)$ holds for any compact set $K \subset \mathbb{R}^{p+2}$.*

This theorem follows from Lemma 3.3 combined with the following

Lemma 3.4. *For any unisolvent compact set $K \subset \mathbb{R}^{p+2}$ the following is true:*

$$\liminf_{n\to\infty} \widetilde{\tau}_\nu \le d^h(K) \le \limsup_{n\to\infty} \widetilde{\tau}_\nu. \tag{15}$$

Proof. First we derive from the estimate (9) that

$$\prod_{\nu=1}^{n} (\widetilde{\tau}_\nu)^{\nu s(\nu)} \le \prod_{i=1}^{r(n)} (\tau_i)^{n(i)} \le V^h_{r(n)} \le r(n)! \prod_{i=1}^{r(n)} (\tau_i)^{n(i)} \le r(n)! \prod_{\nu=1}^{n} (\widetilde{\tau}_\nu)^{\nu s(\nu)},$$

where $\widetilde{\tau}_\nu$ is defined in (13). Then consider the sequence

$$c_j =: \begin{cases} \ln \widetilde{\tau}_0 = 0 & \text{if} \quad j = 1 \\ \ln \widetilde{\tau}_\nu & \text{if} \quad r(\nu-1) < j \le r(\nu) \end{cases}.$$

Taking into account that $\lim \frac{\ln r(n)!}{l(n)} = 1$ and the upper limit by Cesaro does not exceed the usual upper limit, we obtain, by Lemma 3.3, the estimates:

$$\limsup_{n\to\infty} \frac{\ln V^h_{r(n)}}{l(n)} \le \limsup_{n\to\infty} \frac{1}{l(n)} \sum_{j=1}^{l(n)} c_j \le \limsup_{j\to\infty} c_j \le \limsup_{\nu\to\infty} \ln \widetilde{\tau}_\nu. \tag{16}$$

The left part of (15) can be obtained in an analogous way.

Unlike the case of $\mathbb{C}^N$-characteristics, we can prove the coincidence of the harmonic characteristics $d^h(K)$ and $\tau^h(K)$ only under certain quite restrictive assumptions about K. Fortunately, these conditions can be checked in some important concrete cases (see, section 4).

Theorem 3.2. *Suppose that in (10) the usual limit exists for each $\theta \in \Sigma$ and $\liminf_{i\to\infty} \tau_i > 0$. Then $d^h(K) = \tau^h(K)$.*

Proof. It is clear that the existence of the usual limit in (10) implies that $\tau_\nu(K,\theta) \to \tau(K,\theta)$ for all $\theta \in \Sigma$. Then, due to the second condition, we pass to the limit under the integral in (14) and, applying (15), obtain the desired equality.

Remark 3.1. The second condition in Theorem 3.2 holds if 0 is an interior point of K.

4. Prolate spheroids

Consider the following one-parameter family of compact sets enclosed by confocal prolate spheroids with the focuses $(\pm 1, 0, 0)$:

$$\Phi_\sigma = \{t = (t_1, t_2, t_3) \in \mathbb{R}^3 : \frac{t_1}{\cosh^2\sigma} + \frac{t_2^2 + t_3^2}{\sinh^2\sigma} \le 1\},\ 0 < \sigma < \infty. \tag{17}$$

By $(\eta, \vartheta, \varphi)$ we denote the spheroidal coordinates corresponding to the family (17), which are connected with the Cartesian coordinates $t = (t_1, t_2, t_3)$ as follows:

$$t_1 = \cosh\eta\cos\vartheta,\ t_2 = \sinh\eta\sin\vartheta\cos\varphi,\ t_3 = \sinh\eta\sin\vartheta\sin\varphi,$$

with $0 \leqslant \eta < \infty$, $0 \leqslant \vartheta \leqslant \pi$, $-\pi < \varphi \leqslant \pi$. Let G_σ be the Hilbert space defined as the completion of the set of all complex valued harmonic polynomials with respect to the norm

$$\|x\|_{G_\sigma} = \left(\int_{-\pi}^{\pi} d\varphi \int_0^{\pi} |\tilde{x}(\sigma, \vartheta, \varphi)|^2 \sin\vartheta d\vartheta\right)^{1/2},$$

where $\tilde{x}(\eta, \vartheta, \varphi) = x(t)$. The system of harmonic polynomials

$$u_{n,m}(t) = \left(\frac{(2n+1)(n-|m|)!}{2\pi(n+|m|)!}\right)^{1/2} P_n^{|m|}(\cosh\eta)\, P_n^{|m|}(\cos\vartheta)e^{im\varphi}, \tag{18}$$

with $n \in \mathbb{Z}_+$, $m \in \mathbb{Z}$, $-n \leqslant m \leqslant n$, forms an orthogonal basis in each Hilbert space G_σ, $0 < \sigma < \infty$. The norms of polynomials (18) are expressed by the following formula (,[5] item 247):

$$\|u_{n,m}\|_{G_\sigma} = P_n^{|m|}(\cosh\sigma), \quad 0 < \sigma < \infty. \tag{19}$$

Lemma 4.1. *The system (18), after the same enumeration of indices $(n(i), m(i))$ as in Definition 2.1, can be represented in the form:*

$$f_i = u_{n(i),m(i)} = c_{ii}e_i + \sum_{j<i} c_{ij}\, e_j \tag{20}$$

with

$$c_{ii} = \frac{(2n\,(i))!}{2^{n(i)} n\,(i)!(n\,(i) - m\,(i))!}\,.$$

Proof. We apply the result (proved in[2] and mentioned implicitly long ago, e.g., in,[5] chapter X) about the representation

$$P_n^m(\cosh\eta)\, P_n^m(\cos\vartheta) = \sum_{l\le n} k_{n,m,l}\; r^l P_l^m(\cos\psi).$$

Then the system f_i is represented in the form (20) with $c_{ii} = k_{n(i),m(i),n(i)}$. So, applying the expression for $k_{n,m,n}$ from,[2] we obtain the desired representation.

Lemma 4.2. *Let* $K = \Phi_\sigma$, $\sigma > 0$. *Then for any* $\theta \in \Sigma = (-1,1)$ *there exists a usual limit in (10) and the equality*

$$\tau^{(h)}(\Phi_\sigma, \theta) = \frac{(1+|\theta|)^{|\theta|}}{2} F(\sigma, |\theta|), \tag{21}$$

holds, where

$$F(\sigma,\gamma) = \frac{(\sinh\sigma)^\gamma(\cosh\sigma + \sqrt{\gamma^2 + (\sinh\sigma)^2})}{(\gamma\cosh\sigma + \sqrt{\gamma^2+(\sinh\sigma)^2})^\gamma}, \quad 0 < \gamma < 1. \tag{22}$$

Proof. For any harmonic polynomial p we have the estimates

$$\frac{\|p\|_{G_\sigma}}{\sqrt{\lambda\,(\Phi_\sigma)}} \le |p| \le C\,(\varepsilon)\,\|p\|_{G_{\sigma+\varepsilon}}, \quad \varepsilon > 0.$$

Therefore, using the extremal property of partial sums of expansions by an orthogonal system, we obtain the estimates

$$\frac{\|f_i\|_{G_\sigma}}{c_{ii}\sqrt{\lambda\,(\Phi_\sigma)}} \le (\tau_i)^{n(i)} \le \frac{C\,(\varepsilon)\,\|f_i\|_{G_{\sigma+\varepsilon}}}{c_{ii}} \tag{23}$$

It was proved in[13] that

$$\limsup_{\frac{m}{n}\to\gamma} \left(\frac{P_n^m(\cosh\sigma)}{n^m}\right)^{1/n} = \frac{(1+\gamma)^\gamma\, F(\sigma,\gamma)}{e^\gamma\,(1-\gamma)^{1-\gamma}}, \quad 0<\gamma<1, \ \sigma>0,$$

with $F(\sigma,\gamma)$ defined above. From here, taking into account Lemma 4.1, we obtain that

$$\begin{aligned}\limsup_{\frac{m(i)}{n(i)}\to\gamma} \left(\left\|\frac{f_i}{c_{ii}}\right\|_{G_\sigma}\right)^{1/n(i)} &= \limsup_{\frac{m}{n}\to\gamma}\left(\frac{P_n^m(\cosh\sigma)\cdot n^m \cdot n!(n-m)!2^n}{n^m(2n)!}\right)^{1/n} \\ &= \frac{(1+\gamma)^\gamma F(\sigma,\gamma) e^\gamma (1-\gamma)^{1-\gamma}}{e^\gamma(1-\gamma)^{1-\gamma}\cdot 2} = \frac{(1+\gamma)^\gamma}{2} F(\sigma,\gamma).\end{aligned}$$

Therefore, by the estimate (23) and continuity of the function $F(\sigma,\gamma)$, we come to the equality $\tau^{(h)}(\Phi_\sigma,\theta) = \frac{(1+|\theta|)^{|\theta|}}{2} F(\sigma,|\theta|)$ for any $\theta\in\Sigma = (-1,1)$.

Theorem 4.1. *For any spheroid (17) we have*

$$d^h(\Phi_\sigma) = \tau^h(\Phi_\sigma) = \frac{\sinh \sigma}{\sqrt{e}} \left(\coth \frac{\sigma}{2}\right)^{\frac{\cosh \sigma}{2}}. \tag{24}$$

Proof. By Lemma 4.2 the conditions of the Theorem 3.2 are met, hence $d^h(K) = \tau^h(K)$. Putting (21) into (10), we obtain

$$\tau^h(\Phi_\sigma) = \exp \frac{1}{2} \int_{-1}^{1} \tau^h(\Phi_\sigma, \theta)\, d\theta = \exp \int_0^1 \ln \left(\frac{(1+\gamma)^\gamma}{2} F(\sigma, \gamma) \right) d\gamma.$$

The direct calculation of the integral gives the formula (24).

5. Final remarks

The following questions, which have been solved positively for the corresponding characteristics in $\mathbb{C}^n$ ([11]) , are open for an arbitrary compact set K in $\mathbb{R}^{p+2}$.

(1) Does there exist a usual limit in (10)?
(2) Does the equality $d^h(K) = \tau^h(K)$ holds?
(3) Is the function $\tau^h(K, \theta)$ continuous on Σ?
(4) Are the characteristics $d^h(K)$ and $\tau^h(K)$ invariant under translations? In particular, would the equality (12) hold if $\mathbb{B}_r$ be changed to $\mathbb{B}_r(a)$?

References

1. N. I. Akhiezer, *Lectures on the theory of approximation*, 2nd revised ed., Nauka, Moscow, 1965 (English transl. of the 1st ed., New York. 1956).
2. H. A. Buchdahl, N. P. Buchdahl and P. J. Stiles, *J.Phys. Gen.* **10,** 1833 (1977).
3. A. Erdelyi, *Higher Transcendental Functions,* Vol. 2. McGraw-Hill, New York, 1953.
4. G. M. Goluzin, *Geometric Theory of Functions of a Complex Variable,* Translations of Mathematical Monographs, **26**, 1969.
5. E. W. Hobson, *The Theory of Spherical and Ellipsoidal Harmonics,* Cambridge Univ. Press., Cambridge, 1931.
6. F. Leja, *Colloq. Math.* **7,** 153 (1959).
7. R. Rumely and Chi Fong Lau, *Math. Z.,* **533** (1994).
8. R. Rumely, Chi Fong Lau and R. Varley, *Existence of the sectional capacity*, Memoirs of AMS **145 (690),** 2000.
9. J. Siciak, *Trans. Amer. Math. Soc.* **105,** 322 (1962).
10. N. Skiba and V. Zahariuta, *Israel Mathematical Conferences Proceedings* **15,** 357 (2001).
11. V. P. Zahariuta, *Math USSR Sbornik* **25,** 350 (1975).
12. V. Zahariuta, *Spaces of harmonic functions,* in *Lect. Notes in Pure and Appl. Math.*, Dekker, NY **150** (1993), pp. 497-522.
13. V. Zahariuta, *Indiana University Math. J.,* **50** (2001), 1047.

ON PROPERTIES OF MODULI OF SMOOTHNESS OF CONFORMAL MAPPINGS

OLENA W. KARUPU

Department of Higher and Numerical Mathematics, National Aviation University, Kyiv, 1 Komarov ave, Ukraine
karupu@ukr.net
www.nau.edu.ua

We consider some new estimates for the uniform, local and integral moduli of smoothness for the function realizing conformal mapping of the unit disk onto the simply connected domain.

Keywords: conformal mapping; modulus of smoothness; finite difference smoothnesses.

1. Introduction

Suppose G is a simply connected domain in the complex plane bounded by a smooth Jordan curve Γ, $\tau = \tau(s)$ is the angle between the tangent to Γ and the positive real axis, $s = s(w)$ is the arc length on Γ. Suppose $w = \varphi(z)$ is a homeomorphism of the closed unit disk $\overline{D} = \{z : |z| \leq 1\}$ onto the closure $\overline{G}$ of the domain G, conformal in the open unit disk D.

Kellog in 1912 [1] proved the theorem in which it had been established that if $\tau = \tau(s)$ satisfies Holder condition with index α, $0 < \alpha < 1$, then the derivative $\varphi'(e^{i\theta})$ satisfies Holder condition with the same index α. Afterwards this result was generalized in works by several authors.

S.E. Warshawski received generalization of Kellog's theorem for higher order derivatives of the function $w = \varphi(z)$ [2] and for domains with nonsmooth boundary [3]. J.L. Geronimus [4] and S. J. Alper [5] received results for general moduli of continuity satisfying some integral conditions. P. M. Tamrazov [6] obtained solid reinforcement of these results. Some close problems were investigated by E.P. Dolzenko [7] and V.A. Danilov [8].

Generalization of Kellog's theorem for the moduli of smoothness of the second order satisfying Holder condition was received by R.N. Kovalchuk [9]. L. I. Kolesnik obtained inversion of this result [10].

Results by S.E. Warshawski, R. N. Kovalchuk, L I. Kolesnik and author [11] for moduli of smoothness of the 3-d order were proved by means of method due to S.E. Warshawski based on the introduction of additional point. But the mentioned method contains a roughering step in the replacement of finite differences (and moduli of smoothness) of order $k(k \geq 2)$ by finite differences (and moduli of smoothness) of the 2-nd order. And as a result the less sharp inequalities obtained by means of this method do not posess any property important for applications and have essentially a restricted range of applications.

P. M. Tamrazov [12] solved the problem of estimating for finite difference smoothnesses of composite function.

These results gave posibility to receive generalizations and inversations of Kellog's type theorems for general moduli of smoothness of arbitrary order.

2. Estimates for uniform moduli of smoothness of arbitrary order

The localization $l = l(k, E, z, \delta)$ is called the rule which to every ordered collection (k, E, z, δ), that consists of nonnegative integer k, set $E \subset C$, point $z \in C$ and positive number δ, corresponds the unique set $l(k, E, z, \delta) \subset E^{k+1}$.

Let the finite function $w = f(z)$ be defined on a curve $\gamma \subset C$. Let $(z_0, ..., z_k)$ be the collection of the points on the curve γ and $[z_0, ..., z_k; f, z_0]$ be the finite difference of order k for the function $w = f(z)$. P. M. Tamrazov [6] defined on rectifiable curves the uniform curvilinear moduli of smoothness of order k for the function $w = f(z)$ as $\omega_{k,N,\gamma}((z), \delta)_p = \sup\limits_{w\in\gamma} \sup\limits_{(z_0,...,z_k)\in\gamma_{w,\delta}(N)} |[z_0, ..., z_k; f, z_0]|$, where $\gamma_{w,\delta}(N)$ is the set of collections $(z_0, ..., z_k)$ such that curvilinear (with respect to the curve γ) distances between points $z_0, ..., z_k \in \gamma$ satisfy the condition ${\rho(z_i, z_{i+1})}/{\rho(z_j, z_{j+1})} \leq N$, $(N \in [1, +\infty))$, and $\rho(z_i, w) \leq \delta$ $(i, j = 1, ..., k)$

Theorem 2.1 (O.W. Karupu, [13]). *Let modulus of smoothness $\omega_k(\tau(s), \delta)$ of order k $(k \in N)$ for the function $\tau(s)$ satisfy the condition $\omega_k(\tau(s), \delta) = O[\omega(\delta)]$ $(\delta \to 0)$ where $\omega(\delta)$ is normal majorant satisfying the condition $\int\limits_0^l \frac{\omega(t)}{t}dt < +\infty$.*

Then the modulus of smoothness $\omega_k(\varphi'(e^{i\theta}), \delta)$ of the same order k for the derivative $\varphi'(e^{i\theta})$ of the function $\varphi(z)$ on ∂D satisfies conditions

$$\omega_k(\arg\varphi'(e^{i\theta}), \delta) = O(\mu(\delta))\,(\delta \to 0),$$

$$\omega_k(\log\varphi'(e^{i\theta}), \delta) = O(\nu(\delta))\,(\delta \to 0),$$

$$\omega_k(\varphi'(e^{i\theta}), \delta) = O(\nu(\delta))\,(\delta \to 0),$$

where

$$\mu(\delta) = \omega_k(\tau(s), \delta) + \sum_{j=1}^{k-1}\sum_{r_1=1}^{j-1} ... \sum_{r_j=1}^{r_{j-1}-1} [\omega_k(\tau(s), \delta)]^{\frac{k-r_1}{k}} \times$$

$$\times \int_0^l \cdots \int_0^l \frac{[\omega_k(\tau(s), x_j)]^{\frac{r_j}{k}} \prod_{i=2}^{j} [\omega_k(\tau(s), x_{i-1})]^{\frac{r_{i-1}-r_i}{k}}}{\prod_{p=1}^{j} x_p \left(1 + \left({}^{x_p}/_{x_{p-1}}\right)^{r_p}\right)} dx_1 ... dx_j,$$

$$\nu(\delta) = \int_0^l \frac{\omega(x_1)}{x_1 \left(1 + \left({}^{x_1}/_{\delta}\right)^k\right)} dx_1 +$$

$$+ \sum_{j=1}^{k-1} \sum_{r_1=1}^{j-1} \cdots \sum_{r_j=1}^{r_{j-1}-1} \delta^{k-r_1} \int_0^l \cdots \int_0^l x_{j+1}^{r_{j-1}} \left(1 + \int_{x_{j+1}}^{l} \frac{\omega(y)}{y_j^{r_j+1}} dy\right) \left(1 + \left(\frac{x_{j+1}}{x_j}\right)^{r_{j+1}}\right)$$

$$\times \prod_{p=1}^{j} \left(1 + \int_{x_p}^{l} \frac{\omega(t_p)}{t_p^{r_{p-1}-r_p+1}} dt_p\right) \left(1 + \left(\frac{x_p}{x_{p-1}}\right)^{r_{p-1}}\right)^{-1} x_p^{r_{p-1}-r_p-1} dx_1 ... dx_{j+1}.$$

In partial case when modulus of smoothness $\omega_k(\tau(s), \delta)$ of order k for the function $\tau(s)$ satisfies Holder condition

$$\omega_k(\tau(s), \delta) = O(\delta^\alpha)\,(\delta \to 0), 0 < \alpha < k,$$

the modulus of smoothness $\omega_{k,\theta}(\varphi'(e^{i\theta}), \delta)$ of the same order k for the derivative $\varphi'(e^{i\theta})$ of the function $\varphi(z)$ on ∂D satisfies the condition

$$\omega_k(\varphi'(e^{i\theta}), \delta) = O(\delta^\alpha)\,(\delta \to 0)$$

with the same index α.

3. Estimates for local moduli of smoothness of arbitrary order

Let the finite function $w = f(z)$ be defined on the set $E \subset C$. Let $l(k, E, z, \delta)$ be the maximal Euclidian localization described by nonnegative integer number k, positive number δ and point $z \in C$.

Then local modulus of smoothness of order k for the function $w = f(z)$ is determined by formula

$$\omega_{k,E}^l(f, z, \delta) = \sup_{(z_0,...,z_k) \in l(k,E,z,\delta)} |[z_0, ..., z_k; f, z_0]|,$$

where $[z_0, ..., z_k; f, z_0]$ is the finite difference of order k for the function $w = f(z)$.

Let consider the noncentralized local arithmetic modulus of smoothness $\omega_{k,z}(f(z), \delta)$ of order k $(k \in N)$ for the function $w = f(z)$ on a curve γ, that is

$$\omega_{k,z}(f(z), \delta) = \sup_{(z_0,...,z_k) \in \gamma_{w,\delta}(N)} |[z_0, ..., z_k; f, z_0]|,$$

where $\gamma_{w,\delta}(N)$ is the set of collections $(z_0, ..., z_k)$ such that curvilinear (with respect to the curve γ) distances between points $z_0, ..., z_k \in \gamma$ satisfy the condition $\rho(z_i, z_{i+1})/\rho(z_j, z_{j+1}) \leq N$ $(N \in [1, +\infty))$, and $\rho(z_i, w) \leq \delta$ $(i, j = 1, ..., k)$.

Theorem 3.1 (O.W. Karupu, [14]). *If the local modulus of smoothness $\omega_k(\tau(s), \delta)$ of order k for the function $\tau(s)$ at the point $w_0 = w(s_0)$ on the curve Γ satisfies the condition*

$$\omega_{k,s_0}(\tau(s), \delta) = O(\delta^\alpha)\,(\delta \to 0),$$

then the local modulus of smoothness $\omega_{k,\theta}(\varphi'(e^{i\theta}), \delta)$ of the same order k for the derivative $\varphi'(e^{i\theta})$ of the function $\varphi(z)$ at the point $z_0 = e^{i\theta_0}$ on ∂D satisfies the condition

$$\omega_{k,\theta_0}(\varphi'(e^{i\theta}), \delta) = O(\delta^\alpha)\,(\delta \to 0).$$

Theorem 3.2 (O.W. Karupu, [14]). *If the local modulus of smoothness $\omega_{k,\theta}(\varphi'(e^{i\theta}), \delta)$ of order k for the derivative $\varphi'(e^{i\theta})$ of the function $\varphi(z)$ at the point $z_0 = e^{i\theta_0}$ on ∂D satisfies the condition*

$$\omega_{k,\theta_0}(\varphi'(e^{i\theta}), \delta) = O(\delta^\alpha)\,(\delta \to 0),$$

then the local modulus of smoothness of the same order k for the function $\tau(s)$ at the point $w_0 = w(s_0)$ on the curve Γ satisfies the condition

$$\omega_{k,s_0}(\tau(s), \delta) = O(\delta^\alpha)\,(\delta \to 0).$$

Theorem 3.3 (O.W. Karupu, [14]). *Let the local modulus of smoothness of order k for the derivative $\varphi'(e^{i\theta})$ of the function $\varphi(z)$ at the point $z_0 = e^{i\theta_0}$ on ∂D satisfy the condition*

$$\omega_{k,\theta_0}(\varphi'(e^{i\theta}), \delta) = O\left(\delta^k \log\frac{1}{\delta}\right)(\delta \to 0). \tag{1}$$

Then the local modulus of smoothness $\omega_k(\tau(s), \delta)$ of the same order k for the function $\tau(s)$ at the point $w_0 = w(s_0)$ on the curve Γ satisfies the condition

$$\omega_{k,s_0}(\tau(s), \delta) = O\left(\delta^k \log\frac{1}{\delta})\right)(\delta \to 0). \tag{2}$$

Proof. We will consider at first the case $k = 1$. After the wellknown Lindelof' theorem on ∂D the formula

$$\arg\varphi'(e^{i\theta}) = \tau \circ s(\theta) - \theta - \frac{\pi}{2} \tag{3}$$

takes place.

Consequently

$$\omega_{1,\theta_0}(\tau \circ s(\theta), \delta) = \omega_{1,\theta_0}(\arg\varphi'(e^{i\theta}), \delta) + C_1\delta$$

$$\leq C_2\omega_{1,\theta_0}(\arg\varphi'(e^{i\theta}), \delta) \leq C_3\delta\log\frac{1}{\delta}, \tag{4}$$

where constants C_1, C_2 and C_3 do not depend on δ.

So,

$$\omega_{1,\theta_0}(\tau \circ s(\theta), \delta) = O\left(\delta \log \frac{1}{\delta}\right) (\delta \to 0). \tag{5}$$

As

$$\omega_{1,s_0}(\tau(s), \delta) \leq C_4 \omega_{1,\theta_0}(\tau \circ s(\theta), b\delta),$$

where constants C_4 and b do not depend on δ, then

$$\omega_{1,s_0}(\tau(s), \delta) = O\left(\delta \log \frac{1}{\delta}\right) (\delta \to 0), \tag{6}$$

and theorem is proved for $k = 1$.

Let consider now the case $k \geq 2$.

Let condition of theorem be satisfied for $k \geq 2$. Then it follows from equality (3), that for all $m \geq 1$ the identity

$$\omega_{m,\theta_0}(\tau \circ s(e^{i\theta}), \delta) = \omega_{m,\theta_0}(\arg \varphi'(e^{i\theta}), \delta) \tag{7}$$

is true.

It follows from the supposition of the theorem, that

$$\omega_{k,\theta_0}(\arg \varphi'(e^{i\theta}), \delta) \leq b_1 \delta^k \log \frac{1}{\delta},$$

where constant b_1 does not depend on δ. So,

$$\omega_{k,\theta_0}(\tau \circ s(e^{i\theta}), \delta) = O(\delta^k \log \frac{1}{\delta}))(\delta \to 0). \tag{8}$$

Applying inequalities from [12], we receive the estimates of the local modulus of smoothness $\omega_{k,s_0}(\tau(s), \delta)$ via the local moduli of smoothness $\omega_{k,\theta_0}(\tau \circ s(e^{i\theta}), \delta)$ and $\omega_{k,\theta_0}(s(e^{i\theta}), \delta)$. We have

$$\omega_{k,s_0}(\tau(s), \delta) \leq \omega_{k,\theta_0}(\tau \circ s(e^{i\theta}), \delta) + b_2 \delta^{-k(k+1)/2} \sum_{j=1}^{k-1} \omega_{j,\theta_0}(\tau \circ s(e^{i\theta}), b\delta)$$

$$\times \sum_{\substack{r_1,\ldots,r_{k(k-1)/2} \geq 1 \\ r_1+\ldots+r_{k(k-1)/2}=k(k+1)/2-j}}^{k-1} \prod_{q=1}^{k(k-1)/2} \omega_{r_q,\theta_0}(s(e^{i\theta}), b\delta), \tag{9}$$

where constant b_2 does not depend on δ. We will also use inequalities:

$$\omega_{m,s_0}(\tau(s), \delta) \leq b_{(m)} \left(\delta^m + \delta^m \int_{\delta}^{l} \frac{\omega_{k,s_0}(\tau(s), t)}{t^{m+1}} dt\right) \leq B_{(m)} \cdot \delta^m, \tag{10}$$

where constants $b_{(m)}$, $B_{(m)}$, $(m = 1, ..., k-1)$, do not depend on δ.

As for the local moduli of smoothness of the function $\tau(s)$ the estimates, that are the analogues of Marchaud inequalities, take place, then it follows from the inequalities (9) and (10), that for all $j = 1, ..., k-1$ the estimates

$$\omega_{j,s_0}(\tau \circ s(e^{i\theta}), \delta) \leq b'_{(j)} \left(\delta^j + \delta^j \int_{\delta}^{l} \frac{\omega(t)}{t^{j+1}} dt \right) \leq B'_{(j)} \delta^j, \tag{11}$$

where constants $b'_{(j)}, B'_{(j)}$, $(j = 1, 2, ..., k-1)$ do not depend on δ, are true. So,

$$\omega_{k,s_0}(\tau, \delta) \leq C_5 \cdot \delta^k \log \frac{1}{\delta},$$

where constant C_5 does not depend on δ.

Theorem is proved.

4. Estimates for integral moduli of smoothness of arbitrary order

Let $\omega_{k,z}(f(z), \delta)$ be a noncentralized local arithmetic modulus of smoothness of order k $(k \in N)$ of the function $w = f(z)$ at a point z on the curve γ. Then the integral modulus of smoothness of order k for the function $w = f(z)$ on the curve γ is introduced by the formula

$$\omega_k((z), \delta)_p = \left\{ \int_{\gamma} [\omega_{k,z}(f(z), \delta)]^p d\lambda(z) \right\}^{1/p}, 1 \leq p < +\infty, k \in N,$$

where $\lambda = \lambda(z)$ is the linear Lebesgue's measure on the curve. These integral moduli is the special case of integral moduli of smoothness introduced by P. M. Tamrazov in 1977. He defined integral moduli of smoothness as averaging on arbitrary measure on the curve of the respective local moduli of smoothness. Difference between these moduli and traditional integral moduli of smoothness, introduced as the least upper bound of averaging absolute values of finite differences, is that the operators of averaging and taking of least upper bound are applied in reverse order.

Theorem 4.1 (O.W. Karupu, [16]). *Let $\tau(s) \in L_{pk}[0, l]$, $1 \leq p < +\infty$, $k \in N$. Let integral modulus $\omega_k(\tau(s), \delta)_{pk}$ of smoothness of order k for the function $\tau = \tau(s)$ satisfy condition*

$$\omega_k(\tau(s), \delta)_{pk} = O\left[\omega(\delta)\right](\delta \to 0),$$

where $\omega(\delta)$ is normal majorant satisfying the condition

$$\int_0^l \frac{\omega(t)}{t} dt < +\infty.$$

Then integral modulus of smoothness of order k of the derivative $\varphi'(e^{i\theta})$ for the function $\varphi(z)$ on ∂D satisfies condition:

$$\omega_k(\varphi'(e^{i\theta}),\delta)_p = O\left[\nu(\delta)\right](\delta\to 0),$$

where

$$\nu(\delta) = \int\limits_0^l \frac{\omega(x_1)}{x_1\left(1+\left(x_1/\delta\right)^k\right)}dx_1$$

$$+\sum_{j=1}^{k-1}\sum_{r_1=1}^{j-1}\dots\sum_{r_j=1}^{r_{j-1}-1}\delta^{k-r_1}\int\limits_0^l\dots\int\limits_0^l x_{j+1}^{r_{j-1}}\left(1+\int\limits_{x_{j+1}}^l\frac{\omega(y)}{y_j^{r_j+1}}dy\right)\left(1+\left(\frac{x_{j+1}}{x_j}\right)^{r_{j+1}}\right)$$

$$\times\prod_{p=1}^{j}\left(1+\int\limits_{x_p}^l\frac{\omega(t_p)}{t_p^{r_{p-1}-r_p+1}}dt_p\right)\left(1+\left(\frac{x_p}{x_{p-1}}\right)^{r_{p-1}}\right)^{-1}x_p^{r_{p-1}-r_p-1}dx_1\dots dx_{j+1}.$$

Theorem 4.2.

If integral modulus of smoothness of order k of the m-th derivative of the function $\tau=\tau(s)$ satisfies Holder condition with index α, where $m\in N$ $(m<k)$ and $0<\alpha<k$:

$$\omega_k(\tau^{(m)}(s),\delta)_p = O\left[\delta^\alpha\right](\delta\to 0),$$

then integral modulus of smoothness of order k of the $(m+1)$-th derivative of the function $\varphi(e^{i\theta})$ satisfies Holder condition with the same index α:

$$\omega_k(\varphi^{(m+1)}(e^{i\theta}),\delta)_{pk} = O\left(\delta^\alpha\right)(\delta\to 0),$$

If

$$\omega_k(\tau^{(m)}(s),\delta)_p = O\left(\delta^k\log\frac{1}{\delta}\right)(\delta\to 0),$$

then

$$\omega_k(\varphi^{(m+1)}(e^{i\theta}),\delta)_{pk} = O\left(\delta^k\log\frac{1}{\delta}\right)(\delta\to 0).$$

Proof of this theorem is similar to the proof of theorem 4.1.

References

1. O. D. Kellog, Harmonic functions and Green's integral, *Trans Amer Math. Soc.*, **13** (1912), 109 - 132.
2. S.E. Warshawski, Uber einen Satz von O. D. Kellog, *Nachr. Ges. Wiss.*, Gottingen, (1932), No 1, 73 - 86.

3. S.E. Warshawski, On differentiability at the boundary in conformal mapping, *Proc. Amer Math. Soc.*, **14** (1961), No 4, 614 - 620.
4. J.L. Geronimus, On some properties of functions continuous in closed disk, *Soviet mathematics Dokl.*, **98** (1954), No 6, 889 - 891 [In Russian].
5. S. J. Alper, On uniform approximation of functions of complex variable in closed domains, *Soviet mathematics Izv., Ser. Math.*, **19** (1955), No 6, 423 - 444 [In Russian].
6. P. M. Tamrazov, *Smoothnesses and polynomial appoximations*, Kiev: Naukova dumka (1975), 270 p. [In Russian].
7. E.P. Dolzenko, Smoothness of harmonic and analytic functions at boundary points of domain, *Soviet mathematics Izv., Ser. Math.*, **29** (1965), No 5, 1069 - 1084 [In Russian].
8. V.A. Danilov, On construction of the smooth quasiconformal automorphism of the plane transforming the given smooth curve into the circle, *Siberian. Math. J.*, **14** (1973, No 6, 1341 - 1345 [In Russian].
9. R. N. Kovalchuk, On some generalization of Kellog's theorem, *Ukr. Math. J.*, **17** (1965), No 4, 104 - 108 [In Russian].
10. L I. Kolesnik, Inversation of Kellog's type theorem, *Ukr. Math. J.*, **21** (1969), No 1, 104 - 108 [In Russian].
11. O. W. Karupu, *On finite difference smoothnesses of conformal mappings*, Kiev: Institute of mathematics of Ukrainian Academy of sciences, (1977), 12 p. [In Russian].
12. P. M. Tamrazov, *Finite difference identities and estimates for moduli of smoothness of composite functions*, Kiev: Institute of mathematics of Ukrainian Academy of sciences, (1977), 24 p[In Russian]).
13. O. W. Karupu, On moduli of smoothness of conformal mappings, *Ukr. Math. J.*, **30** (1978), No 4, 540 - 545 [In Russian].
14. O. W. Karupu, , Local moduli of smoothness of conformal homeomorphisms, in: *Metrical questions of functions theory*, Kiev:Naukova dumka, (1980), 49 - 53. [In Russian].
15. O. W. Karupu, On some properties of moduli of smoothness of conformal mappings, *Acad. Nauk Ukrain. Works Inst. Mat.*, **31**, Kiev, 2000, 237-243 [In Ukrainian].
16. O. W. Karupu, *Integral moduli of smoothness of conformal mappings*, Kiev: Kiev Air Force Institute Publishers (2000), 12 p. [In Ukrainian].
17. O. W. Karupu, On some chracteristics of conformal mapping, in: *Proceedings of the second world congress*, Ukraine, Kiev, 19 - 21 September, 2005, **5**, (2005), NAU Publishers, 586 - 591.

STRICT STABILITY CRITERIA OF PERTURBED SYSTEMS WITH RESPECT TO UNPERTURBED SYSTEMS IN TERMS OF INITIAL TIME DIFFERENCE

COŞKUN YAKAR

University Department, University Name,
Gebze Institute of Technology,Department of Mathematics,
Faculty of Sciences Gebze, Kocaeli, TURKEY 141-41400
cyakarpenta.gyte.edu.tr

In this paper we investigate the strict stability criteria between an unperturbed differential system and a perturbed differential system that have different initial positions and an initial time difference.

Keywords: strict stability, initial time difference, perturbed differential systems, Lyapunov's stability.

1. Introduction

The method of Lyapunov's stability has been very useful and fruitful techniques in the qualitative theory of differential equations since it is a practical tool in the investigation of the behavior of solutions of differential equations. It has been applied to investigate the relationship of unpertubed and perturbed systems in nonlinear differential systems [2, 4, 5, 6, 7, 8] the boundedness, stability properties of perturbed systems. Recently in [1, 5, 6, 7, 8], the study of initial value problems with an initial time difference has been initiated and the corresponding theory of differential inequalities has been investigated. As in all measurements, errors occur in measurements of starting time and the solutions of an unperturbed differential system may start at an initial time that is different than the starting time of the perturbed differential system. In real situations it may be impossible to have only a change in the space variable and not also in the initial time.

We investigate strict stability criteria between unperturbed systems with different initial conditions and perturbed systems with different initial conditions. We not only consider the change in initial position, but also an initial time difference. Previously, the investigation of initial value problems in differential equations has been restricted to a perturbation in the space variable only with the initial time unchanged.

2. Definition and Notation

Consider the differential systems

$$x' = f(t,x), x(t_0) = x_0 \text{ for } t \geq t_0, t_0 \in \mathbb{R}_+ \tag{2.1}$$

$$\widetilde{x}' = f(t-\eta,\widetilde{x}), \widetilde{x}(\tau_0) = x_0 \text{ for } t \geq \tau_0, \tau_0 \in \mathbb{R}_+ \tag{2.2}$$

and the perturbed differential system of (2.1)

$$y' = F(t,y), y(\tau_0) = y_0 \text{ for } t \geq \tau_0 \tag{2.3}$$

where $f, F \in C[\mathbb{R}_+ \times \mathbb{R}^n, \mathbb{R}^n]$ and $\eta = \tau_0 - t_0$. A special case of (2.3) is where $F(t,y) = f(t,y) + R(t,y)$ and $R(t,y)$ is the perturbation term. Assume that the existence and uniqueness of the solution $x(t) = x(t,t_0,x_0)$ of (2.1) for $t \geq t_0$ and $y(t) = y(t,\tau_0,y_0)$ of (2.3) for $t \geq t_0$.

Let us give the definition of the strict stability criteria with initial time difference.

Definition 2.1: The solution $y(t,\tau_0,y_0)$ of the system (2.3) through (τ_0, y_0) is said to be initial time difference strict stable with respect to the solution $x(t-\eta,t_0,x_0)$, where $x(t,t_0,x_0)$ is any solution of the system (2.1) for $t \geq \tau_0 \geq 0, t_0 \in \mathbb{R}_+$ and $\eta = \tau_0 - t_0$. If given any $\epsilon_1 > 0$ and $\tau_0 \in \mathbb{R}_+$ there exist $\delta_1 = \delta_1(\epsilon_1,\tau_0) > 0$ and $\delta_2 = \delta_2(\epsilon_1,\tau_0)$ such that

$$\|y(t,\tau_0,y_0) - x(t-\eta,t_0,x_0)\| < \epsilon_1 \text{ for } t \geq \tau_0$$

whenever $\|y_0 - x_0\| < \delta_1$ and $|\tau_0 - t_0| < \delta_2$ and, for $\delta_1^* < \delta_1$ and $\delta_2^* < \delta_2$ there exist $\epsilon_2 < \min\{\delta_1^*, \delta_2^*\}$ such that

$$\|y(t,\tau_0,y_0) - x(t-\eta,t_0,x_0)\| > \epsilon_2 \text{ for } t \geq \tau_0$$

whenever $\|y_0 - x_0\| > \delta_1^*$ and $|\tau_0 - t_0| > \delta_2^*$.

Definition 2.2: If δ_1, δ_2 and ϵ_2 in Definition 2.1 are independent of τ_0, then the solution $y(t,\tau_0,y_0)$ of the system (2.3) is initial time difference uniformly strict stable with respect to the solution $x(t-\eta,t_0,x_0)$ for $t \geq \tau_0$.

Definition 2.3: The solution $y(t,\tau_0,y_0)$ of the system (2.3) through (τ_0,y_0) is said to be initial time difference strictly attractive with respect to the solution $x(t-\eta,t_0,x_0)$, where $x(t,t_0,x_0)$ is any solution of the system (2.1) for $t \geq \tau_0 \geq 0, t_0 \in \mathbb{R}_+$ and $\eta = \tau_0 - t_0$. If given any $\alpha_1 > 0, \gamma_1 > 0, \epsilon_1 > 0$ and $\tau_0 \in \mathbb{R}_+$, for every $\alpha_2 < \alpha_1$ and $\gamma_2 < \gamma_1$, there exist $\epsilon_2 < \epsilon_1, T_1 = T_1(\epsilon_1,\tau_0)$ and $T_2 = T_2(\epsilon_1,\tau_0)$ such that

$$\|y(t,\tau_0,y_0)-x(t-\eta,t_0,x_0)\|<\epsilon_1 \text{ for } T_1+\tau_0\le t\le T_2+\tau_0$$

whenever $\|y_0-x_0\|<\alpha_1$ and $|\tau_0-t_0|<\gamma_1$
and

$$\|y(t,\tau_0,y_0)-x(t-\eta,t_0,x_0)\|>\epsilon_2 \text{ for } T_2+\tau_0\ge t\ge T_1+\tau_0$$

whenever $\|y_0-x_0\|>\alpha_2$ and $|\tau_0-t_0|>\gamma_2$.

If $T_1$and T_2 in Definition 2.3 are independent of τ_0, then the solution $y(t,\tau_0,y_0)$ of the system (2.3) is initial time difference uniformly strictly attractive stable with respect to the solution $x(t-\eta,t_0,x_0)$ for $t\ge\tau_0$.

Definition 2.4: The solution $y(t,\tau_0,y_0)$ of the system (2.3) through (τ_0,y_0) is said to be initial time difference strictly assimptotically stable with respect to the solution $x(t-\eta,t_0,x_0)$ if Definition 2.3 satisfies and the solution $y(t,\tau_0,y_0)$ of the system (2.3) through (τ_0,y_0) is initial time difference stable with respect to the solution $x(t-\eta,t_0,x_0)$.

If $T_1$and T_2 in Definition 2.3 are independent of τ_0, then the solution $y(t,\tau_0,y_0)$ of the system (2.3) is initial time difference uniformly strictly assimptotically stable with respect to the solution $x(t-\eta,t_0,x_0)$ for $t\ge\tau_0$.

Definition 2.5: For any function $V_n\in C[\mathbb{R}_+\times\mathbb{R}^n,\mathbb{R}_+]$, we define the Dini derivative of the function with respect to the solution of the system (2.1)

$$D^+V(t,x)=\lim_{h\longrightarrow 0^+}\sup\frac{1}{h}\left[V(t+h,x+hf(t,x))-V(t,x)\right]$$

and

$$D_-V(t,x)=\lim_{h\longrightarrow 0^-}\inf\frac{1}{h}\left[V(t+h,x+hf(t,x))-V(t,x)\right]$$

for $(t,x)\in\mathbb{R}_+\times\mathbb{R}^n$.

Definition 2.6: $\mathcal{K}$ is said to be the class $\mathcal{K}$ set of functions such that

$\mathcal{K}:=[a_1,b_1,c_1:a_1,b_1,c_1\in C[0,\rho]\longrightarrow\mathbb{R}_+$, increasing and $a_1(0)=0,b_1(0)=0,c_1(0)=0]$ and $S_\rho=\{z\in\mathbb{R}^n.:\|z\|\le\rho\}$.

3. Main Results

In this section we obtain the strict stability concepts with initial time difference parallel to the Lyapunov's results.

Theorem 3.1: Assume that

(A_1) for each $\eta, 0<\eta<\rho, V_n\in C[\mathbb{R}_+\times S_\rho,\mathbb{R}_+]$ and V_η is locally Lipschitzian in z and for $(t,z)\in\mathbb{R}_+\times S_\rho$ and $\|z\|\ge\eta$,

$$b_1(\|z\|) \le V_\eta(t,z) \le a_1(\|z\|), a_1, b_1 \in \mathcal{K}$$

$$D^+V_\eta(t,z) \le 0; \tag{3.1}$$

(A_2) for each $\theta, 0 < \theta < \rho, V_\theta \in C[\mathbb{R}_+ \times S_\rho, \mathbb{R}_+]$ and V_θ is locally Lipschitzian in z and for $(t,z) \in \mathbb{R}_+ \times S_\rho$ and $\|z\| \le \theta$,

$$b_2(\|z\|) \le V_\theta(t,z) \le a_2(\|z\|), a_2, b_2 \in \mathcal{K}$$

$$D^+V_\eta(t,z) \ge 0; \tag{3.2}$$

where $z(t) = y(t,\tau_0,y_0) - x(t-\eta,t_0,x_0)$ for $t \ge \tau_0, y(t,\tau_0,y_0)$ of the system (2.3) through (τ_0, y_0) and $x(t-\eta,t_0,x_0)$, where $x(t,t_0,x_0)$ is any solution of the system (2.1) for $t \ge \tau_0 \ge 0, t_0 \in \mathbb{R}_+$ and $\eta = \tau_0 - t_0$.

Then the solution $y(t,\tau_0,y_0)$ of the system (2.3) is the initial time difference uniformly strictly stable with respect to the system (2.1) for $t \ge \tau_0 \ge 0, t_0 \in \mathbb{R}_+$ and $\eta = \tau_0 - t_0$.

Proof of Theorem 3.1: Let $0 < \epsilon_1 < \rho$ and $\tau_0 \in \mathbb{R}_+$. Let us choose $\delta_1 = \delta_1(\epsilon_1, \tau_0) > 0$ such that

$$a_1(\delta_1) < b_1(\epsilon_1). \tag{3.3}$$

Then we claim that

$$\|y(t,\tau_0,y_0) - x(t-\eta,t_0,x_0)\| < \epsilon_1 \text{ for } t \ge \tau_0 \tag{3.4}$$

whenever $\|y_0 - x_0\| < \delta_1$ and $|\tau_0 - t_0| < \delta_2$.

If (2.7) is true, then there exist $t_1 > t_2 > \tau_0$ and the solution of (1.2) and (1.3) with $\|y_0 - x_0\| < \delta_1, |\tau_0 - t_0| < \delta_2$ satisfying

$$\left\|y(t_1) - \widetilde{x}(t_1)\right\| = \epsilon_1, \left\|y(t_2) - \widetilde{x}(t_2)\right\| = \delta_1 \text{ and } \delta_1 \le \left\|y(t) - \widetilde{x}(t)\right\| \le \epsilon_1 \text{ for } t \in [t_2, t_1].$$

Let us set $\eta = \delta_1$, we can obtain

$$b_1(\epsilon_1) = b_1(\left\|y(t_1) - \widetilde{x}(t_1)\right\|) \le V_\eta(t_1, y(t_1) - \widetilde{x}(t_1))$$
$$\le V_\eta(t_2, y(t_2) - \widetilde{x}(t_2)) \le a_1(\left\|y(t_2) - \widetilde{x}(t_2)\right\|) = a_1(\delta_1)$$

which contradicts with (3.3). Hence (3.4) is valid.

Now let $0 < \delta_1^* < \delta_1, 0 < \delta_2^* < \delta_2$ and $\epsilon_2 < \delta = \min\{\delta_1^*, \delta_2^*\}$ such that

$$a_2(\epsilon_2) < b_2(\delta). \tag{3.5}$$

Then we can prove that

$$\epsilon_2 < \|y(t,\tau_0,y_0) - x(t-\eta,t_0,x_0)\| < \epsilon_1 \text{ for } t \geq \tau_0 \tag{3.6}$$

whenever $\delta_1^* < \|y_0 - x_0\| < \delta_1$ and $\delta_2^* < |\tau_0 - t_0| < \delta_2$.

In fact, if (3.6) is not true, then there would exist $t_1 > t_2 > \tau_0$ and the solution of (1.2) and (1.3) with $\delta_1^* < \|y_0 - x_0\| < \delta_1, \delta_2^* < |\tau_0 - t_0| < \delta_2$ satisfying

$$\left\|y(t_1) - \widetilde{x}(t_1)\right\| = \epsilon_2, \left\|y(t_2) - \widetilde{x}(t_2)\right\| = \delta \text{ and } \left\|y(t) - \widetilde{x}(t)\right\| \leq \delta \text{ for } t \in [t_2, t_1]. \tag{3.7}$$

Let us set $\theta = \delta$ and using (A_2), we get

$$\begin{aligned} a_2(\epsilon_2) &= a_2(\left\|y(t_1) - \widetilde{x}(t_1)\right\|) \geq V_\theta(t_1, y(t_1) - \widetilde{x}(t_1)) \\ &\geq V_\theta(t_2, y(t_2) - \widetilde{x}(t_2)) \geq b_2(\left\|y(t_2) - \widetilde{x}(t_2)\right\|) = b_2(\delta) \end{aligned}$$

which contradicts with (3.5). Thus (3.6) is valid. Then the solution $y(t,\tau_0,y_0)$ of the system (2.3) through (τ_0, y_0) is initial time difference strictly uniformly stable with respect to the solution $x(t-\eta,t_0,x_0)$.

This completes the proof of Theorem 3.1.

Theorem 3.2: Assume that

(A_1) for each $\eta, 0 < \eta < \rho, V_\eta \in C[\mathbb{R}_+ \times S_\rho, \mathbb{R}_+]$ and V_η is locally Lipschitzian in z and for $(t,z) \in \mathbb{R}_+ \times S_\rho$ and $\|z\| \geq \eta$,

$$b_1(\|z\|) \leq V_\eta(t,z) \leq a_1(\|z\|), a_1, b_1 \in \mathcal{K},$$

$$D^+V_\eta(t,z) \leq -c_1(\|z\|)\ c_1 \in \mathcal{K}; \tag{3.8}$$

(A_2) for each $\theta, 0 < \theta < \rho, V_\theta \in C[\mathbb{R}_+ \times S_\rho, \mathbb{R}_+]$ and V_θ is locally Lipschitzian in z and for $(t,z) \in \mathbb{R}_+ \times S_\rho$ and $\|z\| \leq \theta$,

$$b_2(\|z\|) \leq V_\theta(t,z) \leq a_2(\|z\|), a_2, b_2 \in \mathcal{K},$$

$$D^+V_\theta(t,z) \geq -c_2(\|z\|)\ c_2 \in \mathcal{K}; \tag{3.9}$$

where $z(t) = y(t,\tau_0,y_0) - x(t-\eta,t_0,x_0)$ for $t \geq \tau_0, y(t,\tau_0,y_0)$ of the system (2.3) through (τ_0,y_0) and $x(t-\eta,t_0,x_0)$, where $x(t,t_0,x_0)$ is any solution of the system (2.1) for $t \geq \tau_0 \geq 0, t_0 \in \mathbb{R}_+$ and $\eta = \tau_0 - t_0$.

Then the solution $y(t,\tau_0,y_0)$ of the system (2.3) through (τ_0,y_0) is the initial time difference uniformly strictly asymptotically stable with respect to the solution of the system (2.1) for $t \geq \tau_0 \geq 0, t_0 \in \mathbb{R}_+$ and $\eta = \tau_0 - t_0$.

Proof of Theorem 3.2: We note that (3.9) implies (2.1). However, (3.8) does not yield (2.2). As a result of these, we obtain because of (3.8) only stability of perturbed systems with initial time difference with respect to $x(t-\eta,t_0,x_0)$ that is for given any $\epsilon_1 \leq \rho$ and $\tau_0 \in \mathbb{R}_+$ there exist $\delta_1 = \delta_1(\epsilon_1,\tau_0) > 0$ and $\delta_2 = \delta_2(\epsilon_1,\tau_0) > 0$ such that

$$\|y(t,\tau_0,y_0) - x(t-\eta,t_0,x_0)\| < \epsilon_1 \text{ for } t \geq \tau_0 \text{ whenever } \|y_0 - x_0\| < \delta_{10} \text{ and } |\tau_0 - t_0| < \delta_{20}. \tag{3.10}$$

To prove the conclusion of Theorem 3.2 we need to show that the solution $y(t,\tau_0,y_0)$ of the system (2.3) through (τ_0,y_0) is strictly uniformly attractive with respect to $x(t-\eta,t_0,x_0)$ for this purpose, let $\epsilon_1 = \rho$ and set $\delta_{10} = \delta_1(\rho)$ and $\delta_{20} = \delta_2(\rho)$ so that (3.10) yields $\|y(t,\tau_0,y_0) - x(t-\eta,t_0,x_0)\| < \rho$ for $t \geq \tau_0$ whenever $\|y_0 - x_0\| < \delta_{10}$ and $|\tau_0 - t_0| < \delta_{20}$.

Let $\|y_0 - x_0\| < \delta_{10}$ and $|\tau_0 - t_0| < \delta_{20}$. We show, using standard argument, that there exists a $t^* \in [\tau_0,\tau_0+T]$, where $T = T(\epsilon,\tau_0) > \frac{a_1(\delta_{10},\delta_{20})}{c_1(\delta_1,\delta_2)}$ where δ_{10} and δ_{20} are the numbers corresponding to ϵ_1 in (3.10) that is in stability of perturbed systems with initial time difference with respect to $x(t-\eta,t_0,x_0)$ such that $\|y(t^*,\tau_0,y_0) - x(t^*-\eta,t_0,x_0)\| < \delta_1$ for any solutions of the systems (2.3) and (2.2) with $\|y_0 - x_0\| < \delta_{10}$ and $|\tau_0 - t_0| < \delta_{20}$. If this is not true, we will have $\|y(t^*,\tau_0,y_0) - x(t^*-\eta,t_0,x_0)\| \geq \delta_1$ for $t \in [\tau_0,\tau_0+T]$. Then, $\eta = \delta_1$ and using (A_1) with (3.8), we have

$$\begin{aligned}
b_1(\delta_1) &= b_1(\left\|y(\tau_0+T) - \widetilde{x}(\tau_0+T)\right\|) \\
&\leq V_\eta(\tau_0+T, y(\tau_0+T) - \widetilde{x}(\tau_0+T)) \\
&\leq V_\eta(\tau_0, y_0 - x_0) - \int_{\tau_0}^{\tau_0+T} c_1(\left\|y(s) - \widetilde{x}(s)\right\|)ds \\
&\leq a_1(\delta_{10},\delta_{20}) - c_1(\delta_1,\delta_2)T = b_1(\delta_1)
\end{aligned}$$

in view of the choice of T. This contradiction implies that there exist a $t^* \in [\tau_0,\tau_0+T]$ satisfying $\|y(t^*,\tau_0,y_0) - x(t^*-\eta,t_0,x_0)\| < \delta_1$. Because of the stability $y(t,\tau_0,y_0)$ of perturbed systems with initial time difference with respect to $x(t-\eta,t_0,x_0)$, this yields that

$$\|y(t,\tau_0,y_0) - x(t-\eta,t_0,x_0)\| < \epsilon_1 \text{ for } t \geq \tau_0 + T \geq t^*$$

which implies that there exists a $\tau_0 < T_1 < T$ such that

$$\|y(\tau_0+T,\tau_0,y_0) - x(\tau_0+T-\eta,t_0,x_0)\| = \epsilon_1$$

Now, for any $\delta_{12}, 0 < \delta_{12} < \delta_{10}$ and $0 < \delta_{12} < \delta_{20}$ we can choose ϵ_2 such that $b_2(\epsilon_1) > a_2(\epsilon_2)$ and $0 < \epsilon_2 < \epsilon_1 < \delta_{12}$.

Suppose that $\delta_{12} < \|y_0 - x_0)\| < \min\{\delta_{10}, \delta_{20}\}$ and $\delta_{12} < |\tau_0 - t_0| < \min\{\delta_{10}, \delta_{20}\}$. Let us define $\tau = \frac{b_2(\epsilon_1)-a_2(\epsilon_2)}{c_2(\epsilon_1)}$, and $T_2 = T_1 + \tau$.

$$\begin{aligned} a_2(\left\|y(t) - \widetilde{x}(t)\right\|) &\geq V_\theta(t, y(t) - \widetilde{x}(t)) \\ &\geq V_\theta(\tau_0 + T_1, y(\tau_0 + T_1) - \widetilde{x}(\tau_0 + T_1)) \\ &\quad - \int_{\tau_0+T_1}^{t} c_2(\left\|y(s) - \widetilde{x}(s)\right\|)ds \\ &\geq b_2(\epsilon_1) - \int_{\tau_0+T_1}^{t} c_2(\left\|y(s) - \widetilde{x}(s)\right\|)ds \\ &\geq b_2(\epsilon_1) - c_2(\epsilon_1)[t - (\tau_0 + T_1)] \end{aligned}$$

Since, $t - (\tau_0 + T_1) \geq \tau$, it follows that

$$a_2(\left\|y(t) - \widetilde{x}(t)\right\|) \geq b_2(\epsilon_1) - c_1(\epsilon_1)[\frac{b_2(\epsilon_1) - a_2(\epsilon_2)}{c_2(\epsilon_1)}] = a_2(\epsilon_2).$$

This yields that

$$\|y(t, \tau_0, y_0) - x(t - \eta, t_0, x_0)\| \geq \epsilon_2 \text{ for } t \in [\tau_0 + T_1, \tau_0 + T_2]$$

and therefore,

$$\epsilon_2 < \|y(t, \tau_0, y_0) - x(t - \eta, t_0, x_0)\| < \epsilon_1 \text{ for } t \in [\tau_0 + T_1, \tau_0 + T_2].$$

This completes the proof. Then the solution $y(t, \tau_0, y_0)$ of the system (2.3) through (τ_0, y_0) is initial time difference uniformly strictly asymptotically stable with respect to the solution $x(t - \eta, t_0, x_0)$, where $x(t, t_0, x_0)$ is any solution of the system (2.1) for $t \geq \tau_0 \geq 0, t_0 \in \mathbb{R}_+$ and $\eta = \tau_0 - t_0$.

Before we prove the general result in terms of the comparison principle. Let us consider the uncoupled comparison differential systems,

$$\begin{cases} (i)\ u_1' = g_1(t, u_1), u_1(\tau_0) = u_{10} \geq 0 \\ (ii)\ u_2' = g_2(t, u_2), u_2(\tau_0) = u_{20} \geq 0 \end{cases} \text{where } g_1, g_2 \in C\left[\mathbb{R}_+^2, \mathbb{R}\right] \qquad (3.11)$$

The comparison system (3.11) is said to be strictly stable:
If given any $\epsilon_1 > 0$ and $t \geq \tau_0, \tau_0 \in \mathbb{R}_+$, there exist a $\delta_1 > 0$ such that

$$u_{10} \leq \delta_1 \text{ implies } u_1(t) < \epsilon_1 \text{ for } t \geq \tau_0$$

and for every $\delta_2 < \delta_1$ there exists an $\epsilon_2 > 0, 0 < \epsilon_2 < \delta_2$ such that

$$u_{20} \geq \delta_2 \text{ implies } \ u_2(t) < \epsilon_2 \text{ for } t \geq \tau_0.$$

Here, $u_1(t)$ and $u_2(t)$ are any solutions of (i) in (3.11) and (ii) in (3.11); respectively.

Following main result based on this definition that result is formulated in terms of comparison principle.

Theorem 3.3: Assume that

(A_1) for each $\eta, 0 < \eta < \rho, V_\eta \in C[\mathbb{R}_+ \times S_\rho, \mathbb{R}_+]$ and V_η is locally Lipschitzian in z and for $(t,z) \in \mathbb{R}_+ \times S_\rho$ and $\|z\| \geq \eta$,

$$b_1(\|z\|) \leq V_\eta(t,z) \leq a_1(\|z\|), a_1, b_1 \in \mathcal{K},$$

$$D^+V_\eta(t,z) \leq g_1(t, V_\eta(t,z)); \tag{3.12}$$

(A_2) for each $\theta, 0 < \theta < \rho, V_\theta \in C[\mathbb{R}_+ \times S_\rho, \mathbb{R}_+]$ and V_θ is locally Lipschitzian in z and for $(t,z) \in \mathbb{R}_+ \times S_\rho$ and $\|z\| \leq \theta$,

$$b_2(\|z\|) \leq V_\theta(t,z) \leq a_2(\|z\|), a_2, b_2 \in \mathcal{K},$$

$$D^+V_\theta(t,z) \geq g_2(t, V_\theta(t,z)); \tag{3.13}$$

where $g_2(t,u) \leq g_1(t,u), g_1, g_2 \in C[\mathbb{R}^2_+, \mathbb{R}], g_1(t,0) = g_2(t,0) = 0$ and $z(t) = y(t,\tau_0,y_0) - x(t-\eta,t_0,x_0)$ for $t \geq \tau_0, y(t,\tau_0,y_0)$ of the system (2.3) through (τ_0,y_0) and $x(t-\eta,t_0,x_0)$, where $x(t,t_0,x_0)$ is any solution of the system (2.1) for $t \geq \tau_0 \geq 0, t_0 \in \mathbb{R}_+$ and $\eta = \tau_0 - t_0$.

Then any strict stability concept of the comparison system implies the corresponding strict stability concept of the solution $y(t,\tau_0,y_0)$ of the system (2.3) through (τ_0,y_0) with respect to the solution $x(t-\eta,t_0,x_0)$ of the system (2.2) with initial time difference where $x(t,t_0,x_0)$ is any solution of the system (2.1) for $t \geq \tau_0 \geq 0,\ t_0 \in \mathbb{R}_+$.

Proof of Theorem 3.3: We will only prove the case of strict uniformly asymptotically stability. Suppose that the comparison differential systems in (3.11) is strictly uniformly asymptotically stable, then for any given $\epsilon_1,\ 0 < \epsilon_1 < \delta$, there exist a $\delta^* > 0$ such that $u_{10} \leq \delta^*$ implies that $u_1(t,\tau_0,u_{10}) < b_1(\epsilon_1)$ for $t \geq \tau_0$.

For this ϵ_1 we choose δ_1 and δ_{11}, such that $a_1(\delta_1^*) \leq \delta^*$ and $\delta_1^* < \epsilon_1$ where $\delta_1^* = \max\{\delta_1, \delta_{11}\}$, then we claim that

$$\|y_0 - x_0\| < \delta_1, |\tau_0 - t_0| < \delta_{11} \text{ imply that} \tag{3.14}$$

$$\|y(t,\tau_0,y_0) - x(t-\eta,t_0,x_0)\| < \epsilon_1 \text{ for } t \geq \tau_0.$$

If it is not true, then there exist t_1 and t_2, $t_2 > t_1 > \tau_0$ and a solution of

$$z' = \tilde{f}(t,z),\ z(\tau_0) = y_0 - x_0 \text{ for } t \geq \tau_0 \text{ with } |\tau_0 - t_0| < \delta_{11} \text{ and } \|y_0 - x_0\| < \delta_1$$

$$\|y(t,\tau_0,y_0) - x(t-\eta,t_0,x_0)\| < \delta_1^*,\ \ \|y(t,\tau_0,y_0) - x(t-\eta,t_0,x_0)\| = \epsilon_1 \text{ and}$$

$$\delta_1^* \leq \|y(t,\tau_0,y_0) - x(t-\eta,t_0,x_0)\| < \epsilon_1 \text{ for } [t_1,t_2].$$

Choosing $\eta = \delta_1^*$ and using the theory of differential inequalities we get

$$\begin{aligned} b_1(\epsilon_1) &= b_1(\|y(t_2,\tau_0,y_0) - x(t_2-\eta,t_0,x_0)\|) \\ &\leq V_\eta(t_2, y(t_2,\tau_0,y_0) - x(t_2-\eta,t_0,x_0)) \\ &\leq r(t_2,t_1,V_\eta(t_1,y(t_1,\tau_0,y_0) - x(t_1-\eta,t_0,x_0))) \\ &\leq r(t_2,t_1,a_1(\delta_1^*)) \\ &\leq r(t_2,t_1,\delta^*) \\ &< b_1(\epsilon_1) \end{aligned}$$

which is a contradiction. Here $r(t,t_0,u_{10})$ is the maximum solution of (3.11). Hence, (3.14) is true and we have uniformly stability with initial time difference. Now, we shall prove strictly uniformly attractive with initial time difference.

For any given δ_2, $\epsilon_2 > 0$, $\delta_2 < \delta^*$ we choose $\bar{\delta}_2$and $\bar{\epsilon}_2$such that $a_1(\delta_2) < \bar{\delta}_2$and $b_1(\epsilon_2) \geq \bar{\epsilon}$. For these $\bar{\delta}_2$ and $\bar{\epsilon}_2$, since (2.11) is strictly uniformly attractive, for any $\bar{\delta}_3 < \bar{\delta}_2$there exist $\bar{\epsilon}_3$ and T_1 and T_2 (we assume $T_2 < T_1$) such that $\bar{\delta}_3 < u_{10} = u_{20} < \bar{\delta}_2$ implies

$$r(t,\tau_0,u_{10}) \leq r(t,\tau_0,\bar{\delta}_2) < \bar{\epsilon_2}$$

$$\rho(t,\tau_0,u_{20}) \geq \rho(t,\tau_0,\bar{\delta}_3) > \bar{\epsilon_2}$$

where $r(t,\tau_0,u_{10})$ and $\rho(t,\tau_0,u_{20})$ is the maximal solution and minimal solution of (3.11) (i) and (3.11) (ii); respectively.

Now, for any δ_3, let $b_2(\delta_3) \geq \bar{\delta_3}$. We choose ϵ_3 such that $a_2(\epsilon_3) < \bar{\epsilon_3}$. Then by using comparison principle (3.11) (i) and (A_1), we have

$$
\begin{aligned}
b_1(\|y(t,\tau_0,y_0)-x(t-\eta,t_0,x_0)\|) &\le V_\eta(t,y(t,\tau_0,y_0)-x(t-\eta,t_0,x_0))\\
&\le r(t,\tau_0,V_\eta(\tau_0,y_0-x_0))\\
&\le r(t,\tau_0,a_1(\|y_0-x_0\|))\\
&\le r(t,\tau_0,\bar{\delta}_2)\\
&< \bar{\epsilon}_2 \le b_1(\epsilon_2)
\end{aligned}
$$

which implies that $\|y(t,\tau_0,y_0)-x(t-\eta,t_0,x_0)\| < \epsilon_2$ for $t \in [\tau_0+T_2, \tau_0+T_1]$.

Similarly, by using comparison principle in (3.11) (ii) and (A_2) we get

$$
\begin{aligned}
a_2(\|y(t,\tau_0,y_0)-x(t-\eta,t_0,x_0)\|) &\ge V_\theta(t,y(t,\tau_0,y_0)-x(t-\eta,t_0,x_0))\\
&\ge \rho(t,\tau_0,V_\theta(\tau_0,y_0-x_0))\\
&\ge r(t,\tau_0,b_2(\delta_3))\\
&\ge r(t,\tau_0,\bar{\delta}_3)\\
&> \bar{\epsilon}_3 \ge a_2(\epsilon_3)
\end{aligned}
$$

which implies that for $\|y(t,\tau_0,y_0)-x(t-\eta,t_0,x_0)\| < \epsilon_3$ for $t \in [\tau_0+T_2, \tau_0+T_1]$. Hence, the solution $y(t,\tau_0,y_0)$ of the system (2.3) through (τ_0,y_0) is strictly uniformly attractive with respect to the solution $x(t-\eta,t_0,x_0)$ is any solution of the system (2.1) for $t \ge \tau_0 \ge 0,\ t_0 \in \mathbb{R}_+$. The proof is complete.

References

1. Koksal, S. and Yakar, C., Generalized Quasilinearization Method with Initial Time Difference. (2002): Journal of Engineering Simulation 24 (5).
2. Lakshmikantham, V. and Leela, S. (1969): Differential and Integral Inequalities Vol. 1, Academic Press, New York.
3. Lakshmikantham, V. and Mohapatra, R. N. (2001): Strict Stability of Differential Equations, Nonlinear Analysis, Volume 46, Issue 7, Pages 915-921.
4. Rajalakshmi, S and Sivasundaram, S. (1993): Variational Lyapunov Second Method, Dyn. Sys. and Appl. 2, 485-490.
5. Shaw, M. D. and Yakar, C., Lipschitz Stability Criteria with Initial Time Difference. Journal of Applicable Analysis. (to appear)
6. Shaw, M. D. and Yakar, C. (1999): Generalized Variation of Parameters with Initial Time Difference and A Comparison Result in terms of Lyapunov-like Functions. International Journal of Nonlinear Differential Equations Theory-Methods and Applications. Vol. 5, No: 1&2, Jan-Dec. (86-108).
7. Shaw, M. D. and Yakar, C. (2000): Stability Criteria and Slowly Growing Motions with Initial Time Difference. Journal of Problems of Nonlinear Analysis in Engineering Systems. Volume 6, Issue 1(11) . (50-66).
8. Yakar, C. and Shaw, M.D. (2005): A Comparison Result and Lyapunov Stability Criteria with Initial Time Difference. Dynamics of Continuous, Discrete and Impulsive Systems. A: Mathematical Analysis. Volume 12, Number 6, Pages 731-741.

PIECEWISE CONTINUOUS RIEMANN BOUNDARY VALUE PROBLEM ON A CLOSED JORDAN RECTIFIABLE CURVE

YU.V. VASIL'EVA

Institute of Mathematics of the National Academy of Sciences of Ukraine,
3 Tereshchenkivska Street, Kiev, Ukraine
kudjavina@mail.ru

There are expanded classes of closed Jordan rectifiable curves and given functions in the theory of piecewise continuous Riemann boundary value problem and characteristic singular integral equation with Cauchy kernel that is connected with it.

Keywords: Cauchy type integral; Riemann boundary value problem; closed Jordan rectifiable curve; local centered module of smoothness.

1. Introduction

Let γ be a closed Jordan rectifiable curve in the complex plane $\mathbb{C}$, D^+ and D^- be, respectively, the interior and exterior domains bounded by γ, $0 \in D^+$, and let $T := \{a_1, a_2, \cdots, a_m\}$ be a fixed finite set of points of the curve γ.

By $H_T^{\pm}$ we denote the set of holomorphic in $D^{\pm}$ functions F (having also a limit at infinity in the case of the set H_T^-) which are extended continuously on $\gamma \setminus T$ and satisfy the estimation

$$|F(z)| \leq c \prod_{j=1}^{m} |z - a_j|^{-\nu_F} \qquad \forall\, z \in D^{\pm}, \tag{1}$$

where the constant c does not depend on z, and ν_F is a certain number from the interval $(0; 1)$, depending on the function F.

Consider the piecewise continuous Riemann boundary value problem: to find functions $\Phi^+ \in H_T^+$ and $\Phi^- \in H_T^-$ which satisfy the equality

$$\Phi^+(t) = G(t)\Phi^-(t) + g(t) \qquad \forall\, t \in \gamma \setminus T\,, \tag{2}$$

where G and g are given functions. If $g(t) \not\equiv 0$, then the problem is called nonhomogeneous, and if $g(t) \equiv 0$, then the problem is called homogeneous.

In this paper there are expanded classes of closed Jordan rectifiable curves and given functions in the theory of the piecewise continuous Riemann boundary value problem, which was studied earlier in [1 – 9].

2. Homogeneous problem

For $\varepsilon > 0$ and $X \subset \mathbb{C}$ denote $\gamma_\varepsilon(X) := \bigcup\limits_{x\in X} \{t \in \gamma : |t - x| \le \varepsilon\}$. If $X = \{x\}$, then $\gamma_\varepsilon(X) =: \gamma_\varepsilon(x)$.

All integrals over the curve γ are understood in terms of their principal value, i.e.

$$\int\limits_\gamma \varphi(t)\,dt := \lim_{\varepsilon\to 0} \int\limits_{\gamma\setminus\gamma_\varepsilon(X)} \varphi(t)\,dt\,,$$

where X is a finite set of discontinuity points for the function φ.

Introduce into consideration the Cauchy type integral

$$\widetilde{p}(z) := \frac{1}{2\pi i}\int\limits_\gamma \frac{p(t)}{t-z}\,dt\,, \qquad z \in \mathbb{C}\setminus\gamma\,. \tag{1}$$

If the function p is summable on γ or $p \in H_T := H_T^+ + H_T^-$, then the function $\widetilde{p}$ is holomorphic in $\mathbb{C}\backslash\gamma$.

For every point $a_j \in T$ define the numbers

$$\Delta_p(a_j) := \liminf_{z\to a_j,\, z\in\mathbb{C}\backslash\gamma} \frac{\operatorname{Re}\, \widetilde{p}(z)}{\ln|z-a_j|}\,,$$

$$\Delta^p(a_j) := \liminf_{r\to 0} \frac{1}{\ln r} \inf_{z\in\mathbb{C}\backslash\gamma:|z-a_j|=r} \operatorname{Re}\, \widetilde{p}(z)$$

and, moreover, suppose, that the following relation is fulfilled

$$\Delta^p(a_j) \le \Delta_p(a_j) + c \qquad \forall\, a_j \in T\,, \tag{2}$$

where c is a constant.

Relation (2) means, that for all points $a_j \in T$ there exist the equalities $\Delta^p(a_j) = \Delta_p(a_j) = +\infty$, or the equalities $\Delta^p(a_j) = \Delta_p(a_j) = -\infty$, or the numbers $\Delta^p(a_j)$ and $\Delta_p(a_j)$ are finite.

If the numbers $\Delta^p(a_j)$ and $\Delta_p(a_j)$ are finite for all $a_j \in T$, then the index æ of piecewise continuous Riemann boundary value problem is defined by

$$æ := \sum_{j=1}^m æ_j\,,$$

where

$$æ_j := \begin{cases} \Delta_p(a_j)\,, & \text{if } \Delta_p(a_j) \text{ is integer,} \\ [\Delta_p(a_j)] + 1\,, & \text{otherwise.} \end{cases}$$

If there exists $-\infty$ among the values $\Delta_p(a_j)$, then $æ = -\infty$. If $+\infty$ exists among the values $\Delta_p(a_j)$ but $-\infty$ does not, then $æ = +\infty$.

The following theorem is proved using the scheme, proposed in [4, p. 46], and generalizes Theorem 1 from [7].

Theorem 2.1. *Let γ be a closed Jordan rectifiable curve and G be a function of the form $G(t) = \exp(p(t))$, where $p \in H_T$ and, furthermore, relation (2) is fulfilled. Then:*

1) if the relations $-\infty \leq æ < 0$ are satisfied, then the homogeneous Riemann boundary value problem has only trivial solution;

2) if the equality $æ = +\infty$ is satisfied, then the homogeneous Riemann boundary value problem has an infinite set of linearly independent solutions;

3) if the relations $0 \leq æ < \infty$ are satisfied, then the homogeneous Riemann boundary value problem has $æ + 1$ linearly independent solutions and its general solution is given by the expression

$$\Phi^{\pm}(z) = \exp(\widetilde{p}(z))P_{æ}(z)\prod_{j=1}^{m}(z-a_j)^{-æ_j}, \qquad z \in D^{\pm},$$

where $P_{æ}$ is an arbitrary polynomial of degree at most $æ$.

3. Nonhomogeneous problem

The nonhomogeneous Riemann boundary value problem is considered in the case where the index is finite and for all points $a_j \in T$ the following numbers

$$\Delta_p^*(a_j) := \limsup_{z \to a_j,\, z \in \mathbb{C}\setminus\gamma} \frac{\operatorname{Re}\widetilde{p}(z)}{\ln|z-a_j|}$$

are finite.

We use the following metric characteristic of the curve γ (see [10]):

$$\theta(\varepsilon) := \sup_{z\in\gamma} \theta_z(\varepsilon),$$

where $\theta_z(\varepsilon) := \{t \in \gamma : |t-z| \leq \varepsilon\}$ and denotes the linear Lebesgue measure on γ.

For a function q given on $\gamma\setminus T$ and for a point $x \in \gamma\setminus T$ we introduce the local centered module of smoothness of the first order (see [7]):

$$\Omega_x(q,\gamma,\varepsilon) := \begin{cases} \sup\limits_{t\in\gamma\,:\,|t-x|=\varepsilon} |q(t)-q(x)|, & \text{if } \{t\in\gamma : |t-x|=\varepsilon\} \neq \emptyset, \\ 0, & \text{if } \{t\in\gamma : |t-x|=\varepsilon\} = \emptyset. \end{cases}$$

In general case, the function $\Omega_x(q,\gamma,\varepsilon)$ is not monotone with respect to ε as distinct from the classical module of continuity. Therefore, the function $\Omega_x(q,\gamma,\varepsilon)$ takes into account oscillations of the function q. Notice that a function q is continuous at a point x if and only if $\Omega_x(q,\gamma,\varepsilon) \to 0$ as $\varepsilon \to 0$.

By $\widetilde{p}^{+}(x), \widetilde{p}^{-}(x)$ we denote the limiting values of function (1) at a point $x \in \gamma \backslash T$ taken in the domains D^{+} and D^{-}, respectively.

The following theorem describes the solvability of the nonhomogeneous Riemann boundary value problem with a finite index under minimal assumptions on the coefficient G of the problem.

Theorem 3.1. *Let γ be a closed Jordan rectifiable curve satisfying the following condition*

$$\theta(\varepsilon) = O(\varepsilon^{\nu}), \qquad \varepsilon \to 0, \tag{1}$$

where $0 < \nu \leq 1$. Suppose that the function G is represented as $G(t) = \exp(p(t))$, where $p \in H_T$, and, furthermore, the Cauchy type integral $\widetilde{p}$ has limiting values $\widetilde{p}^{+}(x), \widetilde{p}^{-}(x)$ on $\gamma \backslash T$, and the numbers $\Delta_p(a_j)$, $\Delta_p^{*}(a_j)$ are finite for all $a_j \in T$. Suppose also that the function g is represented as $g = g^{+} + g^{-}$, where $g^{+} \in H_T^{+}$, and the function g^{-} is holomorphic in D^{-}, continuous on $\overline{D^{-}} \setminus T$ and satisfies the following condition

$$\sup_{x \in \gamma \backslash \gamma_\delta(T)} \int\limits_{[0,\varepsilon]} \frac{\Omega_x(g^{-}, \gamma, \eta)}{\eta} \, d\theta_x(\eta) \to 0, \qquad \varepsilon \to 0, \qquad \forall \delta > 0 \tag{2}$$

and estimations

$$|g^{-}(t)| \leq c \prod_{j=1}^{m} |t - a_j|^{\alpha_j} \qquad \forall t \in \gamma \setminus T, \qquad \alpha_j > \Delta_p^{*} - \Delta_p - \nu, \tag{3}$$

$$\int\limits_{[0,r_t]} \frac{\Omega_t(g^{-}, \gamma, \eta)}{\eta} \, d\theta_t(\eta) \leq c \prod_{j=1}^{m} |t - a_j|^{\beta_j} \qquad \forall t \in \gamma \setminus T, \qquad \beta_j > \Delta_p^{*} - \Delta_p - 1, \tag{4}$$

where the constant c does not depend on t.

Then for $æ \geq -1$ the nonhomogeneous Riemann boundary value problem is solvable, but when $æ < -1$ it is solvable if and only if the following $-æ-1$ conditions are satisfied:

$$\int\limits_{\gamma} \frac{g(t)}{\exp(\widetilde{p}^{+}(t))} \prod_{j=1}^{m} (t - a_j)^{æ_j} \, t^{s-1} \, dt = 0, \qquad s = 1, 2, \ldots, -æ - 1. \tag{5}$$

The general solution of the nonhomogeneous Riemann boundary value problem is given by the formula

$$\Phi^{\pm}(z) = \Phi_0(z) + \exp(\widetilde{p}(z)) \, P_{æ}(z) \prod_{j=1}^{m} (z - a_j)^{-æ_j}, \qquad z \in D^{\pm}, \tag{6}$$

where

$$\Phi_0(z) = \begin{cases} g^{+}(z) + \widehat{\Phi}(z), & \text{if} \quad z \in D^{+}, \\ \widehat{\Phi}(z), & \text{if} \quad z \in D^{-}, \end{cases}$$

$$\widehat{\Phi}(z) = \prod_{j=1}^{m} (z - a_j)^{-æ_j} \exp(\widetilde{p}(z)) \frac{1}{2\pi i} \int_{\gamma} \frac{g^{-}(t)}{\exp(\widetilde{p}^{+}(t)) \prod_{j=1}^{m} (t - a_j)^{-æ_j}} \frac{dt}{t - z},$$

and $P_æ$ *is an arbitrary polynomial of degree at most* $æ$ *when* $æ \geq 0$, *and* $P_æ(z) \equiv 0$ *when* $æ < 0$.

Proof. Taking into account the equalities $g(t) = g^{+}(t) + g^{-}(t)$ and $G(t) = \exp(\widetilde{p}^{+}(t) - \widetilde{p}^{-}(t))$, which are satisfied for all $t \in \gamma \setminus T$, rewrite the boundary condition (2) as

$$\frac{(\Phi^{+}(t) - g^{+}(t)) \prod_{j=1}^{m} (t - a_j)^{æ_j}}{\exp(\widetilde{p}^{+}(t))} = \frac{\Phi^{-}(t) \prod_{j=1}^{m} (t - a_j)^{æ_j}}{\exp(\widetilde{p}^{-}(t))} + \frac{g^{-}(t) \prod_{j=1}^{m} (t - a_j)^{æ_j}}{\exp(\widetilde{p}^{+}(t))}.$$

Define $F^{+}(t) := \exp(-\widetilde{p}^{+}(t)) \prod_{j=1}^{m} (t - a_j)^{æ_j}$. The function F^{+} satisfies the inequality

$$|F^{+}(t)| \leq c\,|t - a_j|^{æ_j - \Delta_p^{*}(a_j) - \varepsilon}, \tag{7}$$

for some $\varepsilon > 0$ and for all $t \in \gamma \setminus T$, which are sufficiently close to $a_j \in T$.

The consequence of inequalities (3) and (7) is the following estimation

$$|g^{-}(t)\, F^{+}(t)| \leq c\,|t - a_j|^{\alpha_j + æ_j - \Delta_p^{*}(a_j) - \varepsilon}, \tag{8}$$

where constants c in (7) and (8) are different, not depending on t. Moreover, for sufficiently small ε the inequality $\alpha_j + æ_j - \Delta_p^{*}(a_j) - \varepsilon > -\nu$ is fulfilled and, therefore, the function $h(t) := g^{-}(t)F^{+}(t)$ is summable on γ. Then, taking into account condition (2), conclude, that the integral $\widetilde{h}$ has limiting values on $\gamma \setminus T$ in the domains D^{+}, D^{-} and Sokhotski formulas are satisfied. Therefore, the function Φ_0 is satisfied the boundary condition (2).

Now estimate the function $\Phi_0(z)$ in a neighbourhood of a point a_j $(j = \overline{1, m})$. For this notice, that for $\varepsilon > 0$ in the sufficiently small neighbourhood of this point the following estimation is fulfilled

$$|\exp(\widetilde{p}(z))| \prod_{j=1}^{m} |z - a_j|^{-æ_j} \leq c\,|z - a_j|^{\Delta_p(a_j) - æ_j - \varepsilon}, \tag{9}$$

where the constant c does not depend on z. Also the following estimation is fulfilled

$$\left| \int_{\gamma} \frac{g^{-}(t)F^{+}(t)}{t - z}\, dt \right| \leq c \prod_{j=1}^{m} \max\{|z - a_j|^{\mu_j + \beta_j}, V_{\alpha_j + \mu_j + \nu}(|z - a_j|)\} \quad \forall\, z \in C \setminus \gamma, \tag{10}$$

where

$$V_{\alpha_j + \mu_j + \nu}(r) = \begin{cases} r^{\alpha_j + \mu_j + \nu - 1}, & \textit{if} \quad \alpha_j + \mu_j + \nu < 1; \\ \ln \frac{d}{r}, & \textit{if} \quad \alpha_j + \mu_j + \nu = 1; \\ 1, & \textit{if} \quad \alpha_j + \mu_j + \nu > 1, \end{cases}$$

$\mu_j := æ_j - \Delta_p^*(a_j) - \varepsilon$ and the constant c does not depend on z. Taking into account estimations (9), (10), and also inequalities

$$\alpha_j + \Delta_p(a_j) - \Delta_p^*(a_j) + \nu - 1 - 2\varepsilon > -1\,, \quad \beta_j + \Delta_p(a_j) - \Delta_p^*(a_j) - 2\varepsilon > -1\,,$$

which are fulfilled for sufficiently small ε, conclude, that the function Φ_0 satisfies the inequality of form (1).

Thus, Φ_0 is a particular solution of the nonhomogeneous Riemann boundary value problem. Notice, that for $æ < 0$ the function $\exp(\widetilde{p}(z)) \prod_{j=1}^{m} (z - a_j)^{-æ_j}$ has a pole of an order $-æ$ at infinity and Φ_0 is a solution of Riemann boundary value problem only when the conditions (5) are satisfied.

To complete the proof notice that in formula (6) the general solution of the nonhomogeneous Riemann boundary value problem is represented as a sum of the particular solution of this problem and the general solution of the homogeneous problem.

The following theorem, which is proved analogously to Theorem 3.1, does not contain conditions (2), (4) of Theorem 3.1, but it contains additional assumptions on the function G.

Theorem 3.2. *Let γ be a closed Jordan rectifiable curve satisfying the condition (1), where $0 < \nu \leq 1$. Suppose that the function G is represented as $G(t) = \exp(p(t))$, where $p \in H_T$, and satisfies the condition of the form (2) and estimations*

$$|G(t)| \geq c \prod_{j=1}^{m} |t - a_j|^{n_j} \quad \forall\, t \in \gamma \setminus T, \quad n_j \geq 0;$$

$$\int\limits_{[0,r_t]} \frac{\Omega_t(G, \gamma, \eta)}{\eta}\, d\theta_t(\eta) \leq c \prod_{j=1}^{m} |t - a_j|^{m_j} \quad \forall\, t \in \gamma \setminus T, \quad m_j \in \mathbb{R},$$

where the constant c does not depend on t, and, furthermore, the numbers $\Delta_p(a_j)$, $\Delta_p^*(a_j)$ are finite for all $j = \overline{1,m}$ (here $p(t) := \ln G(t)$ is understood as an arbitrary continuous on $\gamma \setminus T$ branch of this function). Suppose also that the function g is represented as $g = g^+ + g^-$, where $g^+ \in H_T^+$, and the function g^- is holomorphic in D^-, continuous on $\overline{D^-} \setminus T$ and satisfies the inequality

$$|g^-(z)| \leq c \prod_{j=1}^{m} |z - a_j|^{k_j} \quad \forall\, z \in D^-\,,$$

$$k_j > \Delta_p^*(a_j) - \Delta_p(a_j) + n_j + \max\{n_j - m_j - 1; -\nu\}\,,$$

where the constant c does not depend on z.

Then for $æ \geq -1$ the nonhomogeneous Riemann boundary value problem is solvable, but for $æ < -1$ it is solvable if and only if conditions (5) are satisfied. The general solution of the nonhomogeneous Riemann boundary value problem is given by formula (6).

As an application of Theorem 3.2 we have obtained a result on solvability of characteristic singular integral equation with Cauchy kernel.

References

1. F. D. Gakhov, *Boundary value problems*, Moscow: Nauka, 1977, 640 p. [in Russian]
2. N. I. Muskhelishvili, *Singular integral equations*, Moscow: Nauka, 1968, 511 p. [in Russian]
3. R. K. Seifullaev, Riemann boundary value problem on a nonsmooth open curve, *Mat. Sb.*, **112** (1980), no. 2, 147–161. [in Russian]
4. B. A. Kats, An exceptional case of the Riemann problem with an oscillating coefficient, *Izv. Vyssh. Uchebn. Zaved. Mat.* (1981), no. 12, 41–50. [in Russian]
5. E. A. Danilov, Dependence of the number of solutions of a homogeneous Riemann problem on the contour and the coefficient module, *Dokl. Acad. Nauk SSSR*, **264** (1982), no. 6, 1305–1308. [in Russian]
6. B. Gonzalez, J. Bory, The homogeneous Riemann boundary value problem on rectifiable open Jordan curves, *Science. Mat. Havana*, **9** (1988), no. 2, 3 – 9.
7. S. Plaksa, Riemann boundary value problem with an oscillating coefficient and singular integral equations on rectifiable curve, *Ukr. Math. J.*, **41** (1989), no. 1, 116–121.
8. K. Kutlu, On Riemann boundary value problem, *An. Univ. Timişoara: Ser. Matematiča-Informatiča*, **38** (2000), no. 1, 89 – 96.
9. D. Pena, J. Bory, Riemann boundary value problem on a regular open curve, *J. Nat. Geom.*, **22** (2002), no. 1, 1 – 17.
10. V. V. Salaev, Direct and inverse estimates for a singular Cauchy integral along a closed curve, *Mat. Zametki*, **19** (1976), no. 3, 365 – 380. [in Russian]

A NOTE ON THE MODIFIED CRANK-NICHOLSON DIFFERENCE SCHEMES FOR SCHRÖDINGER EQUATION

ALLABEREN ASHYRALYEV

Department of Mathematics, Fatih University, 34900 Buyukcekmece, Istanbul, Turkey

ALI SIRMA

Department of Mathematics, Gebze Institute of Technology, Gebze, Kocaeli, Turkey

In present paper the nonlocal boundary value problem

$$\begin{cases} i\frac{du}{dt} + Au = f(t),\ 0 < t < T, \\ u(0) = \alpha u(\lambda) + \mu, |a| < 1,\ 0 < \lambda \leq T \end{cases}$$

for Schrödinger equation in a Hilbert space H with the self-adjoint positive definite operator A is considered. The the second order of accuracy r-modified Crank- Nicholson difference schemes for the approximate solutions of this nonlocal boundary value problem are presented. The stability of these difference schemes is established. In practice, the stability inequalities for the solutions of difference schemes for Schrödinger equation are obtained. A numerical method is proposed for solving a one-dimensional Schrödinger equation with nonlocal boundary condition. A procedure of modified Gauss elimination method is used for solving these difference schemes. The method is illustrated by numerical examples.

Key Words: Schrödinger equation; Difference schemes; Stability

1. Introduction. Modified Crank-Nicholson Difference Schemes

It is known that various problems in physics lead to Schrödinger equation. Methods of solutions of the problems for Schrödinger have been studied extensively by many researchers (see, e.g., [1]-[5] and the references given therein). In the present paper the nonlocal boundary value problem

$$\begin{cases} i\frac{du}{dt} + Au = f(t),\ 0 < t < T, \\ u(0) = \alpha u(\lambda) + \mu, |a| < 1,\ 0 < \lambda \leq T \end{cases} \tag{1.1}$$

for Schrödinger equation in a Hilbert space H with the self-adjoint positive definite operator A is considered. In the papers [6]-[8] the second order of accuracy r-modified Crank- Nicholson difference schemes for the approximate solutions of parabolic equation in a Banach space E with nonsmooth data were presented. The main aim of this paper is to study r-modified Crank- Nicholson difference schemes for the approximate solutions of problem (1.1). It is assumed that $2\tau \leq \lambda$.The second order of accuracy r-modified Crank- Nicholson difference schemes

$$\begin{cases} i\frac{u_k-u_{k-1}}{\tau} + \frac{A}{2}(u_k+u_{k-1}) = \varphi_k,\ \varphi_k = f(t_k-\frac{\tau}{2}), t_k = k\tau,\ r+1 \leq k \leq N, \\ i\frac{u_k-u_{k-1}}{\tau} + Au_k = \varphi_k,\ \varphi_k = f(t_k-\frac{\tau}{2}), t_k = k\tau, 1 \leq k \leq r, \\ u_0 = \alpha(I+il_0A)u_l + \mu - il_0\varphi_l, r\tau \geq \lambda, \\ u_0 = \alpha u_l + \mu,\ r\tau < \lambda, \frac{\lambda}{\tau} \in Z^+, \\ u_0 = \alpha(I+il_1A)\frac{1}{2}(u_l+u_{l+1}) + \mu - il_1\varphi_l, r\tau < \lambda, \frac{\lambda}{\tau} \notin Z^+ \end{cases} \tag{1.2}$$

for the approximate solutions of this nonlocal boundary value problem are presented [6]. Z^+ denotes here the set $\{2, \cdots, n, \cdots\}$ and $l = [\frac{\lambda}{\tau}]$, $l_0 = \lambda - [\frac{\lambda}{\tau}]\tau$, $l_1 = \lambda - [\frac{\lambda}{\tau}]\tau - \frac{\tau}{2}$.

The stability of these difference schemes is established. In applications, the stability inequalities for the solutions of difference schemes for Schrödinger equation are obtained. A numerical method is proposed for solving a one-dimensional Schrödinger equation with nonlocal boundary condition. A procedure of modified Gauss elimination method is used for solving these difference schemes. The method is illustrated by numerical examples.

2. Theorem on Stability

By[6] ,

$$u_k = \begin{cases} R^k\xi - i\tau\sum_{j=1}^{k} R^{k-j+1}\varphi_j, k = 1, \cdots, r, \\ B^{k-r}R^r\xi - i\sum\limits_{j=1}^{r} B^{k-r}R^{r-j+1}\varphi_j\tau \\ -i\sum\limits_{j=r+1}^{k} B^{k-j}C\varphi_j\tau, k = r+1, \cdots, N \end{cases} \tag{2.1}$$

is the solution of the r-modified Crank-Nicholson difference schemes for the approximate solutions of Cauchy problem

$$\begin{cases} i\frac{u_k-u_{k-1}}{\tau} + \frac{A}{2}(u_k+u_{k-1}) = \varphi_k,\ \varphi_k = f(t_k-\frac{\tau}{2}), t_k = k\tau,\ r+1 \leq k \leq N, \\ i\frac{u_k-u_{k-1}}{\tau} + Au_k = \varphi_k,\ \varphi_k = f(t_k-\frac{\tau}{2}), t_k = k\tau, 1 \leq k \leq r, u_0 = \xi. \end{cases}$$

Here

$$R = (I - i\tau A)^{-1}, C = \left(I - i\frac{A}{2}\tau\right)^{-1}, B = \left(I + i\frac{A}{2}\tau\right)C.$$

For u_0, the three cases should be examined separately: (i) $r\tau \geq \lambda$, (ii) $r\tau < \lambda$ and $\frac{\lambda}{\tau} \in Z^+$, (iii) $r\tau < \lambda$ and $\frac{\lambda}{\tau} \notin Z^+$. Let $r\tau \geq \lambda$. Then, using formula (2.1) and the condition

$$u_0 = \alpha(I + il_0 A)u_l + \mu - il_0\varphi_l,$$

we get

$$\xi = \alpha(I + il_0A)R^l\xi - i\tau\alpha(I + il_0A)\sum_{j=1}^{l} R^{l-j+1}\varphi_j + \mu - il_0\varphi_l.$$

Since the operator $I - \alpha(I + il_0A)R^l$ has an inverse

$$\xi = T_\tau\left\{-i\tau\alpha(I + il_0A)\sum_{j=1}^{l} R^{l-j+1}\varphi_j + \mu - il_0\varphi_l\right\}, \tag{2.2}$$

where

$$T_\tau = \left(I - \alpha(I + il_0A)R^l\right)^{-1}.$$

Let $r\tau < \lambda$ and $\frac{\lambda}{\tau} \in Z^+$. Then, using formula (2.1) and the $u_0 = \alpha u_l + \mu$ condition, the following formula is obtained:

$$u_0 = \alpha\left(B^{l-r}R^r u_0 - i\sum_{j=1}^{r} B^{l-r}R^{r-j+1}\varphi_j\tau - i\sum_{j=r+1}^{l} B^{l-j}C\varphi_j\tau\right) + \mu.$$

Since the operator $I - \alpha B^{l-r}R^r$ has an inverse $(I - \alpha B^{l-r}R^r)^{-1}$, it follows that

$$u_0 = T_\tau\left\{i\alpha\tau B^{l-r}\sum_{j=1}^{r} R^{r-j+1}\varphi_j - i\alpha\tau\sum_{j=1}^{l-r} B^{l-r-j}C\varphi_{r+j} + \mu\right\}, \tag{2.3}$$

where

$$T_\tau = (I - \alpha B^{l-r}R^r)^{-1}.$$

Let $r\tau < \lambda$ and $\frac{\lambda}{\tau} \notin Z^+$. In this case if $r < \left[\frac{\lambda}{\tau}\right]$, then using formula (2.1) and the condition

$$u_0 = \alpha(I + il_1A)\frac{1}{2}(u_l + u_{l+1}) + \mu - il_1\varphi_l$$

the following formula is obtained

$$u_0 = \alpha(I + il_1A)\frac{1}{2}\left(B^{l-r}R^r u_0 - i\tau\sum_{j=1}^{r} B^{l-r}R^{l-j+1}\varphi_j - i\tau\sum_{j=r+1}^{l} B^{l-j}C\varphi_j\right)$$

$$+\alpha(I + il_1A)\frac{1}{2}\left\{B\left(B^{l-r}R^r u_0 - i\tau\sum_{j=1}^{r} B^{l-r}R^{l-j+1}\varphi_j\right.\right.$$

$$-i\tau\sum_{j=r+1}^{l}B^{l-j}C\varphi_j\Bigg)-i\tau C\varphi_{l+1}\Bigg\}+\mu-il_1\varphi_l.$$

Since the operator $\left(I-\alpha(I+il_1A)\frac{1}{2}(I+B)B^{l-r}R^r\right)$ has an inverse it follows that

$$u_0=T_\tau\Bigg\{-\alpha(I+il_1A)\frac{1}{2}i\tau\Bigg((I+B)\sum_{j=1}^{r}B^{l-r}R^{l-j+1}\varphi_j$$

$$+\sum_{j=r+1}^{l}B^{l-j}C\varphi_j+C\varphi_{l+1}\Bigg)+\mu-il_1\varphi_l\Bigg\},$$

where

$$T_\tau=\left(I-\alpha(I+il_1A)\frac{1}{2}(I+B)B^{l-r}R^r\right)^{-1}.$$

If $r=\left[\frac{\lambda}{\tau}\right]$, then in a similar manner one establishes

$$u_0=T_\tau\left\{-\alpha(I+il_1A)\frac{1}{2}i\tau\left((I+B)\sum_{j=1}^{r}R^{l-j+1}\varphi_j+C\varphi_{l+1}\right)+\mu-il_1\varphi_l\right\}.$$

So, for the solution of problem (1.2) we have the following formulas

$$u_k=\begin{cases}R^ku_0-i\tau\sum_{j=1}^{k}R^{k-j+1}\varphi_j,k=1,\cdots,r,\\[2ex] B^{k-r}R^ru_0-i\tau\sum\limits_{j=1}^{r}B^{k-r}R^{r-j+1}\varphi_j\\[2ex] -i\tau\sum\limits_{j=r+1}^{k}B^{k-j}C\varphi_j,k=r+1,\cdots,N\\[2ex] T_\tau\left\{-i\tau\alpha(I+il_0A)\sum_{j=1}^{l}R^{l-j+1}\varphi_j+\mu-il_0\varphi_l\right\},\\ k=0,r\tau\geq\lambda,\\[2ex] T_\tau\left\{i\alpha\tau B^{l-r}\sum_{j=1}^{r}R^{r-j+1}\varphi_j-i\alpha\tau\sum_{j=1}^{l-r}B^{l-r-j}C\varphi_{r+j}+\mu\right\},\\ k=0,r\tau<\lambda,\frac{\lambda}{\tau}\in Z^+,\\[2ex] T_\tau\left\{-\alpha(I+il_1A)\frac{1}{2}i\tau\left((I+B)\sum_{j=1}^{l}R^{l-j+1}\varphi_j+C\varphi_{l+1}\right)+\mu-il_1\varphi_l\right\},\\ k=0,r\tau<\lambda,\frac{\lambda}{\tau}\notin Z^+,r=\left[\frac{\lambda}{\tau}\right].\\[2ex] T_\tau\left\{-\alpha(I+il_1A)\frac{1}{2}i\tau\left((I+B)\sum_{j=1}^{r}B^{l-r}R^{l-j+1}\varphi_j\right.\right.\\ \left.\left.+\sum_{j=r+1}^{l}B^{l-j}C\varphi_j+C\varphi_{l+1}\right)+\mu-il_1\varphi_l\right\},r<\left[\frac{\lambda}{\tau}\right]\end{cases}\tag{2.4}$$

where

$$T_\tau = \begin{cases} \left(I - \alpha(I + il_0A)R^l\right)^{-1}, r\tau \geq \lambda, \\ (I - \alpha B^{l-r}R^r)^{-1}, r\tau < \lambda, \frac{\lambda}{\tau} \in Z^+, \\ \left(I - \alpha(I + il_1A)\frac{1}{2}(I + B)B^{l-r}R^r\right)^{-1}, r\tau < \lambda, \frac{\lambda}{\tau} \notin Z^+. \end{cases} \tag{2.5}$$

Theorem 3.1. The solution of the difference schemes (1.2) obeys the stability inequality

$$\max_{0 \leq k \leq N} \|u_k\|_H \leq C(\alpha) \left[\|\mu\|_H + \max_{1 \leq k \leq N} \|\varphi_k\|_H\right]. \tag{2.6}$$

Proof. Using the estimates

$$\|R\|_{H \to H} \leq 1, \|B\|_{H \to H} \leq 1, \|C\|_{H \to H} \leq 1 \tag{2.7}$$

and formula (2.4), we can obtain

$$\max_{1 \leq k \leq N} \|u_k\|_H \leq \left[\|u_0\|_H + T \max_{1 \leq k \leq N} \|\varphi_k\|_H\right]. \tag{2.8}$$

Using the spectral representation of the self-adjoint positive defined operators one can establishes

$$\|T_\tau\|_{H \to H} \leq C(\alpha). \tag{2.9}$$

The three cases should be examined separately: (i) $r\tau \geq \lambda$, (ii) $r\tau < \lambda$ and $\frac{\lambda}{\tau} \in Z^+$, (iii) $r\tau < \lambda$ and $\frac{\lambda}{\tau} \notin Z^+$. Let $r\tau \geq \lambda$. Then

$$\|T_\tau\|_{H \to H} \leq \sup_{\delta \leq \mu < \infty} \left| \left(1 - \alpha(1 + il_0\mu)\left(\frac{1}{1 - i\tau\mu}\right)^l\right)^{-1} \right|$$

$$\leq \sup_{\delta \leq \mu < \infty} \frac{1}{1 - \left|\alpha(1 + il_0\mu)\left(\frac{1}{1 - i\tau\mu}\right)^l\right|} \leq \sup_{\delta \leq \mu < \infty} \frac{1}{1 - |\alpha| \left|\frac{1 + il_0\mu}{1 - i\tau\mu}\right| \left|\frac{1}{1 - i\tau\mu}\right|^{l-1}} \leq \frac{1}{1 - |\alpha|}.$$

Let $r\tau < \lambda$ and $\frac{\lambda}{\tau} \in Z^+$. Then

$$\|T_\tau\|_{H \to H} \leq \sup_{\delta \leq \mu < \infty} \left| \left(1 - \alpha\left(\frac{1 + \frac{i\tau\mu}{2}}{1 - \frac{i\tau\mu}{2}}\right)^{l-r}\left(\frac{1}{1 - i\tau\mu}\right)^r\right)^{-1} \right|$$

$$\leq \sup_{\delta \leq \mu < \infty} \frac{1}{1 - \left|\alpha\left(\frac{1 + \frac{i\tau\mu}{2}}{1 - \frac{i\tau\mu}{2}}\right)^{l-r}\left(\frac{1}{1 - i\tau\mu}\right)^r\right|} \leq \sup_{\delta \leq \mu < \infty} \frac{1}{1 - |\alpha| \left|\frac{1 + \frac{i\tau\mu}{2}}{1 - \frac{i\tau\mu}{2}}\right|^{l-r} \left|\frac{1}{1 - i\tau\mu}\right|^r} \leq \frac{1}{1 - |\alpha|}.$$

Let $r\tau < \lambda$ and $\frac{\lambda}{\tau} \notin Z^+$. Then

$$\|T_\tau\|_{H\to H}$$

$$\leq \sup_{\delta\leq\mu<\infty}\left|\left(1-\alpha(1+il_1\mu)\frac{1}{2}\left(1+\frac{1+\frac{i\tau\mu}{2}}{1-\frac{i\tau\mu}{2}}\right)\left(\frac{1+\frac{i\tau\mu}{2}}{1-\frac{i\tau\mu}{2}}\right)^{l-r}\left(\frac{1}{1-i\tau\mu}\right)^{r}\right)^{-1}\right|$$

$$\leq \sup_{\delta\leq\mu<\infty}\frac{1}{1-\left|\alpha\left(\frac{1+il_1\mu}{1-\frac{i\tau\mu}{2}}\right)\left(\frac{1+\frac{i\tau\mu}{2}}{1-\frac{i\tau\mu}{2}}\right)^{l-r}\left(\frac{1}{1-i\tau\mu}\right)^{r}\right|}$$

$$\leq \sup_{\delta\leq\mu<\infty}\frac{1}{1-|\alpha|\left|\frac{1+il_1\mu}{1-\frac{i\tau\mu}{2}}\right|\left|\frac{1+\frac{i\tau\mu}{2}}{1-\frac{i\tau\mu}{2}}\right|^{l-r}\left|\frac{1}{1-i\tau\mu}\right|^{r}}\leq\frac{1}{1-|\alpha|}.$$

Now for estimate $\|u_0\|_H$ the three cases should be also examined separately: (i) $r\tau\geq\lambda, (ii)$ $r\tau<\lambda$ and $\frac{\lambda}{\tau}\in Z^+$, (iii) $r\tau<\lambda$ and $\frac{\lambda}{\tau}\notin Z^+$. Let $r\tau\geq\lambda$. Then by using formula (2.4), the triangle inequality and estimates (2.7), (2.9) the following estimate is obtained:

$$\|u_0\|_H\leq\|T_\tau\|_{H\to H}\left\{\tau|\alpha|\sum_{j=1}^{l}\left\|(I+il_0A)R^{l-j+1}\right\|_{H\to H}\|\varphi_j\|_H+\|\mu\|_H+l_0\|\varphi_l\|_H\right\}$$

$$\leq C(\alpha)\left\{\tau l|\alpha|\max_{1\leq j\leq l}\|\varphi_j\|_H+\|\mu\|_H+\tau\|\varphi_l\|_H\right\}\leq C(\alpha)\left\{(T+1)\max_{1\leq j\leq N}\|\varphi_j\|_H+\|\mu\|_H\right\}$$

$$=C_1(\alpha)\left\{\max_{1\leq j\leq N}\|\varphi_j\|_H+\|\mu\|_H\right\}.$$

The proof of estimate (2.6) for difference scheme (1.2) is based on the last estimate and estimate (2.8).

Let $r\tau<\lambda$ and $\frac{\lambda}{\tau}\in Z^+$,then by using formula (2.4), the triangle inequality and estimates (2.7), (2.9) the following estimate is obtained:

$$\|u_0\|_H\leq\|T_\tau\|_{H\to H}\left\{\tau|\alpha|\|B\|_{H\to H}^{l-r}\sum_{j=1}^{r}\|R\|_{H\to H}^{r-j+1}\|\varphi_j\|_H\right.$$

$$\left.+\tau|\alpha|\sum_{j=1}^{l-r}\|B\|_{H\to H}^{l-r-j}\|C\|_{H\to H}\|\varphi_{r+j}\|_H+\|\mu\|_H\right\}$$

$$\leq C(\alpha)\left\{\tau|\alpha|r\max_{1\leq j\leq r}\|\varphi_j\|_H+\tau|\alpha|(l-r)\max_{r+1\leq j\leq l}\|\varphi_j\|_H+\|\mu\|_H\right\}$$

$$\leq C(\alpha)\left\{T \max_{1\leq j\leq N} \|\varphi_j\|_H + \|\mu\|_H\right\} = C_2(\alpha)\left\{\max_{1\leq j\leq N} \|\varphi_j\|_H + \|\mu\|_H\right\}.$$

Let $r\tau < \lambda$ and $\frac{\lambda}{\tau} \notin Z^+$. If $r = \left[\frac{\lambda}{\tau}\right]$, then by using formula (2.4), the triangle inequality and estimates (2.7), (2.9) the following estimate is obtained:

$$\|u_0\|_H \leq \|T_\tau\|_{H\to H}\left\{|\alpha|\,\tau\left(\left\|(I+il_1A)\frac{1}{2}(I+B)\right\|_{H\to H}\sum_{j=1}^{l}\|R\|_{H\to H}^{l-j+1}\|\varphi_j\|_H\right.\right.$$

$$\left.\left.+\left\|(I+il_1A)\frac{1}{2}C\right\|_{H\to H}\|\varphi_{l+1}\|_H\right)+\|\mu\|_H+l_1\|\varphi_l\|_H\right\}$$

$$\leq C(\alpha)\left\{|\alpha|\,\tau l \max_{1\leq j\leq l+1}\|\varphi_j\|_H+\|\mu\|_H+l_1\|\varphi_l\|_H\right\}$$

$$\leq C(\alpha)\left\{(T+1)\max_{1\leq j\leq N}\|\varphi_j\|_H+\|\mu\|_H\right\} = C_3(\alpha)\left\{\max_{1\leq j\leq N}\|\varphi_j\|_H+\|\mu\|_H\right\}.$$

If $r < \left[\frac{\lambda}{\tau}\right]$, then by using formula (2.4), the triangle inequality and estimates (2.7), (2.9) the following estimate is obtained:

$$\|u_0\|_H \leq \|T_\tau\|_{H\to H}\left\{|\alpha|\,\tau\left(\left\|(I+il_1A)\frac{1}{2}(I+B)\right\|_{H\to H}\sum_{j=1}^{r}\|B\|_{H\to H}^{l-r}\|R\|_{H\to H}^{l-j+1}\|\varphi_j\|_H\right.\right.$$

$$\left.\left.+\frac{1}{2}\|(I+il_1A)C\|_{H\to H}\left(\sum_{j=r+1}^{l}\|B\|_{H\to H}^{l-r}\|\varphi_j\|_H+\|\varphi_{l+1}\|_H\right)\right)+\|\mu\|_H+l_1\|\varphi_l\|_H\right\}$$

$$\leq C(\alpha)\left\{|\alpha|\,\tau l \max_{1\leq j\leq l+1}\|\varphi_j\|_H+\|\mu\|_H+l_1\|\varphi_l\|_H\right\}$$

$$\leq C(\alpha)\left\{(T+1)\max_{1\leq j\leq N}\|\varphi_j\|_H+\|\mu\|_H\right\} = C_4(\alpha)\left\{\max_{1\leq j\leq N}\|\varphi_j\|_H+\|\mu\|_H\right\}.$$

Theorem 2.1 is proved.

Note that if the stability estimate of Theorem 2.1 and the passing to the limit for $\tau \to 0$ is considered, one can recover Theorem of papers [2], [3] on the stability of the nonlocal boundary value problem (1.1) .

In the next section abstract Theorem 2.1 is applied in the investigations of two nonlocal boundary value problems.

3. Applications

First, the nonlocal boundary value problem for one dimensional Schrödinger equation

$$\begin{cases} iu_t - (a(x)u_x)_x + \delta u = f(t,x), & 0 < t < T, \ 0 < x < 1, \\ u(0,x) = \alpha u(\lambda, x) + \mu(x), & 0 \leq x \leq 1, \ 0 < \lambda \leq T, \\ u(t,0) = u(t,1), \ u_x(t,0) = u_x(t,1), & 0 \leq t \leq T \end{cases} \tag{3.1}$$

is considered. Problem (3.1) has a unique smooth solution $u(t,x)$ for the smooth $a(x) \geq a > 0$ $(x \in (0,1))$, $\mu(x)(x \in [0,1])$ and $f(t,x)$ $(t \in [0,T], x \in (0,1))$ functions and $\delta > 0$ constant.This allows us to reduce the mixed problem (3.1) to the nonlocal boundary problems (1.1) in a Hilbert space $H = L_2[0,1]$ with a self-adjoint positive definite operator A defined by (3.1).

The discretization of problem (3.1) was carried out in two steps. In the first step the grid set

$$[0,1]_h = \{x = x_m = hm, 0 \leq m \leq M, Mh = 1$$

is defined. To the differential operator A generated by the problem (3.1) we assign the difference operator A_h^x by the formula

$$A_h^x u_x^h = \left\{ -\left(a_r(x) u_{\bar{x}_r}^h\right)_{x_r,m} + \delta u_m \right\}_1^{M-1} \tag{3.2}$$

acting in the space of grid functions $u^h(x) = \{u_m\}_0^M$, satisfying the conditions $u_0 = u_M, u_1 - u_0 = u_N - u_{N-1}$ .With the help of A_h^x we arrive at the nonlocal boundary-value problem

$$\begin{cases} i\frac{du^h(t,x)}{dt} + A_h^x u^h(t,x) = f^h(t,x), 0 < t < T, \ x \in [0,1]_h, \\ u^h(0,x) = \alpha u^h(\lambda, x) + \mu^h(x), \ 0 < \lambda \leq T, x \in [0,1]_h \end{cases} \tag{3.3}$$

for an infinite system of ordinary differential equations.

In the second step problem (3.3) is replaced, by the difference scheme (1.2)

$$\begin{cases} i\frac{u_k^h(x)-u_{k-1}^h(x)}{\tau}+\frac{A_h^x}{2}(u_k^h(x)+u_{k-1}^h(x))=\varphi_k^h(x), \\ \varphi_k^h(x)=f^h(t_k-\frac{\tau}{2},x), t_k=k\tau,\ r+1\le k\le N, x\in[0,1]_h \\ \\ i\frac{u_k^h(x)-u_{k-1}^h(x)}{\tau}+A_h^x u_k^h(x)=\varphi_k^h(x), \\ \varphi_k^h(x)=f^h(t_k-\frac{\tau}{2},x), t_k=k\tau, 1\le k\le r, x\in[0,1]_h \\ \\ u_0^h(x)=\alpha(I_h+il_0A_h^x)u_l^h(x)+\mu^h(x)-il_0\varphi_l^h(x), r\tau\ge\lambda, x\in[0,1]_h \\ \\ u_0^h(x)=\alpha u_l^h(x)+\mu^h(x),\ r\tau<\lambda, \frac{\lambda}{\tau}\in Z^+, x\in[0,1]_h \\ \\ u_0^h(x)=\alpha(I_h+il_1A_h^x)\frac{1}{2}(u_l^h(x)+u_{l+1}^h(x))+\mu^h(x)-il_1\varphi_l^h(x), \\ r\tau<\lambda, \frac{\lambda}{\tau}\notin Z^+, x\in[0,1]_h. \end{cases} \tag{3.4}$$

To formulate the result, one needs to introduce the space $L_{2h}=L_2([0,1]_h)$ of the all grid functions $\varphi^h(x)$ defined on $x\in[0,1]_h$, equipped with the norm

$$\left\|\varphi^h\right\|_{L_{2h}}=\left(\sum_{x\in[0,1]_h}\left|\varphi^h(x)\right|^2 h\right)^{1/2}.$$

Theorem 3.1. Let τ and $|h|$ be a sufficiently small numbers.Then the solutions of difference scheme (3.4) satisfy the following stability estimate:

$$\max_{0\le k\le N}\left\|u_k^h\right\|_{L_{2h}}\le C(\alpha)\left[\left\|\mu^h\right\|_{L_{2h}}+\max_{1\le k\le N}\left\|\varphi_k^h\right\|_{L_{2h}}\right]. \tag{3.5}$$

Here $C(\alpha)$ does not depend on τ, h, $\mu^h(x)$ and $\varphi_k^h(x), 1\le k\le N$.

The proof of Theorem 3.1 is based on the abstract Theorem 2.1 as well as the symmetry properties of the difference operator A_h^x defined by the formula (3.2) in L_{2h}.

Second, let Ω be the unit cube in the $n-$dimensional Euclidean space R^n $(0<x_k<1,\ 1\le k\le n)$ with boundary S and $\widetilde{\Omega}=\Omega\cup S$. In $[0,T]\times\widetilde{\Omega}$ the nonlocal boundary-value problem for multidimensional Schrödinger equation

$$\begin{cases} iu_t-\sum_{r=1}^n(a_r(x)u_{x_r})_{x_r}=f(t,x),\quad 0<t<T,\quad x\in\Omega, \\ \\ u(0,x)=\alpha u(\lambda,x)+\mu(x),\ x\in\widetilde{\Omega}, \\ \\ u(t,x)=0,\ x\in S,\ 0\le t\le T \end{cases} \tag{3.6}$$

is considered. The problem (3.6) has a unique smooth solution $u(t,x)$ for the smooth

$a_r(x)\ge a>0$ $(x\in\Omega)$, $\mu(x)(x\in\widetilde{\Omega})$ and $f(t,x)$ $(t\in[0,T],\ x\in\Omega)$ functions. This allows us to reduced mixed problem (3.6) to the nonlocal boundary value problem

(1.1) in a Hilbert space $H = L_2(\Omega)$ of the all integrable functions defined on Ω, equipped with the norm

$$\|f\|_{L_2(\Omega)} = \left\{ \int \cdots \int_{x\in\Omega} |f(x)|^2 \, dx_1 ... dx_n \right\}^{\frac{1}{2}}$$

with a self-adjoint positive definite operator A defined by (3.6).

The discretization of problem (3.6) was carried out in two steps. In the first step the grid sets

$$\widetilde{\Omega}_h = \{x = x_m = (h_1 m_1, \cdots, h_n m_n), m = (m_1, \cdots, m_n) \, ,$$

$$0 \le m_r \le N_r, h_r N_r = L, r = 1, \cdots, n\} \, ,$$

$$\Omega_h = \widetilde{\Omega}_h \cap \Omega, S_h = \widetilde{\Omega}_h \cap S$$

are defined. To the differential operator A generated by the problem (3.6) we assign the difference operator A_h^x by the formula

$$A_h^x u_x^h = -\sum_{r=1}^{n} \left(a_r(x) u_{\bar{x}_r}^h \right)_{x_r, j_r} \tag{3.7}$$

acting in the space of grid functions $u^h(x)$, satisfying the conditions $u^h(x) = 0$ for all $x \in S_h$.With the help of A_h^x we arrive at the nonlocal boundary-value problem

$$\begin{cases} i\frac{du^h(t,x)}{dt} + A_h^x u^h(t,x) = f^h(t,x), 0 < t < T, \ x \in \widetilde{\Omega}_h, \\ u^h(0,x) = \alpha u^h(\lambda, x) + \mu^h(x), \ 0 < \lambda \le T, x \in \widetilde{\Omega}_h \end{cases} \tag{3.8}$$

for an infinite system of ordinary differential equations.

In the second step problem (3.3) is replaced, by the difference schemes (1.2)

$$\begin{cases} i\frac{u_k^h(x) - u_{k-1}^h(x)}{\tau} + \frac{A_h^x}{2}(u_k^h(x) + u_{k-1}^h(x)) = \varphi_k^h(x), \\ \varphi_k^h(x) = f^h(t_k - \frac{\tau}{2}, x), t_k = k\tau, \ r+1 \le k \le N, x \in \widetilde{\Omega}_h \\ \\ i\frac{u_k^h(x) - u_{k-1}^h(x)}{\tau} + A_h^x u_k^h(x) = \varphi_k^h(x), \\ \varphi_k^h(x) = f^h(t_k - \frac{\tau}{2}, x), t_k = k\tau, 1 \le k \le r, x \in \widetilde{\Omega}_h \\ \\ u_0^h(x) = \alpha(I_h + i l_0 A_h^x) u_l^h(x) + \mu^h(x) - i l_0 \varphi_l^h(x), r\tau \ge \lambda, x \in \widetilde{\Omega}_h \\ \\ u_0^h(x) = \alpha u_l^h(x) + \mu^h(x), \ r\tau < \lambda, \frac{\lambda}{\tau} \in Z^+, x \in \widetilde{\Omega}_h \\ \\ u_0^h(x) = \alpha(I_h + i l_1 A_h^x)\frac{1}{2}(u_l^h(x) + u_{l+1}^h(x)) + \mu^h(x) - i l_1 \varphi_l^h(x), \\ r\tau < \lambda, \frac{\lambda}{\tau} \notin Z^+, x \in \widetilde{\Omega}_h. \end{cases} \tag{3.9}$$

To formulate the result, one needs to introduce the space $L_{2h} = L_2(\Omega_h)$ of the all grid functions $\varphi^h(x) = \{\varphi(h_1 m_1, \cdots, h_n m_n)\}$ defined on Ω_h, equipped with the norm

$$\left\|\varphi^h\right\|_{L_2(\Omega_h)} = \left(\sum_{x\in\Omega_h} \left|\varphi^h(x)\right|^2 h_1 \cdots h_n\right)^{1/2}.$$

Theorem 3.2. Let τ and $|h| = \sqrt{h_1^2 + \cdots + h_n^2}$ be a sufficiently small numbers. Then the solutions of difference scheme (3.4) satisfy the following stability estimate:

$$\max_{0\leq k\leq N} \left\|u_k^h\right\|_{L_2(\Omega_h)} (\alpha) \left[\left\|\mu^h\right\|_{L_2(\tilde{\Omega}_h)} + \max_{1\leq k\leq N} \left\|\varphi_k^h\right\|_{L_2(\tilde{\Omega}_h)}\right]. \tag{3.10}$$

Here $C(\alpha)$ does not depend on τ, h, $\mu^h(x)$ and $\varphi_k^h(x), 1 \leq k \leq N$.

The proof of Theorem 3.2 is based on the abstract Theorem 2.1 as well as the symmetry properties of the difference operator A_h^x defined by the formula (3.7) in L_{2h}.

4. Numerical Analysis

In this section, the numerical solutions of the nonlocal-boundary value problem

$$\begin{cases} i\frac{\partial u(t,x)}{\partial t} - \frac{\partial^2 u(t,x)}{\partial x^2} + u(t,x) = \exp(it\pi^2)\sin \pi x, \ 0 < t, x < 1, \\ u(0,x) = \frac{1}{2}u(1,x) + \varphi, \ \varphi = \left(1 - \frac{1}{2}\exp(i\pi^2)\right)\sin \pi x, \quad 0 \leq x \leq 1, \\ u(t,0) = u(t,1) = 0, \ 0 \leq t \leq 1 \end{cases} \tag{4.1}$$

for Schrödinger equation by using r-modified Crank-Nicholson difference schemes (1.2) are investigated. The exact solution of this problem is

$$u(t,x) = \exp(it\pi^2)\sin \pi x.$$

For the approximate solution of problem (4.1), the set $[0,1]_\tau \times [0,1]_h$ of a family of grid points depending on the small parameters τ and h

$$[0,1]_\tau \times [0,1]_h = \{(t_k, x_n) : t_k = k\tau, \ \ 1 \leq k \leq N-1, \ \ N\tau = 1, \\ x_n = nh, \ \ 1 \leq n \leq M-1, Mh = 1\}$$

is defined. Applying (1.2) for the approximate solutions of the problem (4.1), we get the following r-modified Crank-Nicholson difference schemes

$$\begin{cases} i\frac{u_n^k-u_n^{k-1}}{\tau}-\frac{1}{2}\left(\frac{u_{n+1}^k-2u_n^k+u_{n-1}^k}{h^2}+\frac{u_{n+1}^{k-1}-2u_n^{k-1}+u_{n-1}^{k-1}}{h^2}\right)+\frac{1}{2}u_n^k+\frac{1}{2}u_n^{k-1}=f(t_k-\frac{\tau}{2},x_n), \\ r+1\le k\le N-1, 1\le n\le M-1, \\ \\ i\frac{u_n^k-u_n^{k-1}}{\tau}-\frac{u_{n+1}^k-2u_n^k+u_{n-1}^k}{h^2}+u_n^k=f(t_k-\frac{\tau}{2},x_n), 1\le k\le r,\ 1\le n\le M-1, \\ \\ u_n^0=\frac{1}{2}u_n^N+\varphi(x_n),\ 1\le n\le M-1;\ u_0^k=u_M^k=0,\ 0\le k\le N, \\ \\ f(t,x)=\exp(it\pi^2)\sin\pi x, \varphi(x)=(1-\frac{1}{2}\exp(i\pi^2))\sin\pi x. \end{cases} \tag{4.2}$$

So, we have $(N+1)\times(M+1)$ system of linear equations. This system one can write in the following equivalent form

$$\begin{cases} au_{n+1}^{k-1}+au_{n+1}^k+bu_n^{k-1}+cu_n^k+au_{n-1}^{k-1}+au_{n-1}^k=\varphi_n^k, \\ r+1\le k\le N,\ 1\le n\le M-1, \\ \\ du_{n+1}^k+eu_n^{k-1}+fu_n^k+du_{n-1}^k=\varphi_n^k, \\ 1\le k\le r,\ 1\le n\le M-1 \\ \\ u_n^0-\frac{1}{2}u_n^N=\varphi(x_n),\ 1\le n\le M-1, \\ \\ u_0^k=u_M^k=0,\ 0\le k\le N, \end{cases} \tag{4.3}$$

where

$$a=-\frac{1}{2h^2},\quad b=\left(-\frac{i}{\tau}+\frac{1}{h^2}+\frac{1}{2}\right),\quad c=\frac{i}{\tau}+\frac{1}{h^2}+\frac{1}{2},$$

$$d=-\frac{1}{h^2},\quad e=-\frac{i}{\tau},\quad f=\frac{i}{\tau}+\frac{2}{h^2}+1.$$

Then (4.3) can be written as the matrix form

$$\begin{cases} A\,U_{n+1}+B\,U_n+C\,U_{n-1}=D\varphi_n, \quad 0\le n\le M-1, \\ \\ U_0=\vec{0},\ U_M=\vec{0}, \end{cases} \tag{4.4}$$

where

$$\varphi_n^k=\begin{cases} (1-\frac{1}{2}\exp(i\pi^2)\sin\pi x_n,\ k=0, \\ f(t_k-\frac{\tau}{2},x_n),\ 1\le k\le N \end{cases}$$

$$\varphi_n = \begin{bmatrix} \varphi_n^0 \\ \varphi_n^1 \\ \dots \\ \varphi_n^N \end{bmatrix},$$

$$A = \begin{bmatrix} 0 & d & 0 & 0 & \dots & 0 & 0 & 0 & \dots & 0 & 0 & 0 \\ 0 & 0 & d & 0 & \dots & 0 & 0 & 0 & \dots & 0 & 0 & 0 \\ \dots & \dots & \dots & \dots & \dots & \dots & \dots & \dots & \dots & \dots & \dots & \dots \\ 0 & 0 & 0 & 0 & \dots & d & 0 & 0 & \dots & 0 & 0 & 0 \\ 0 & 0 & 0 & 0 & \dots & a & a & 0 & \dots & 0 & 0 & 0 \\ 0 & 0 & 0 & 0 & \dots & 0 & a & a & \dots & 0 & 0 & 0 \\ 0 & 0 & 0 & 0 & \dots & 0 & 0 & a & \dots & 0 & 0 & 0 \\ \dots & \dots & \dots & \dots & \dots & \dots & \dots & \dots & \dots & \dots & \dots & \dots \\ \dots & \dots & \dots & \dots & \dots & \dots & \dots & \dots & \dots & \dots & \dots & \dots \\ 0 & 0 & 0 & 0 & \dots & 0 & 0 & 0 & \dots & a & a & 0 \\ 0 & 0 & 0 & 0 & \dots & 0 & 0 & 0 & \dots & 0 & a & a \\ 0 & 0 & 0 & 0 & \dots & 0 & 0 & 0 & \dots & 0 & 0 & 0 \end{bmatrix},$$

$$B = \begin{bmatrix} e & f & 0 & 0 & \dots & 0 & 0 & 0 & \dots & 0 & 0 & 0 \\ 0 & e & f & 0 & \dots & 0 & 0 & 0 & \dots & 0 & 0 & 0 \\ \dots & \dots & \dots & \dots & \dots & \dots & \dots & \dots & \dots & \dots & \dots & \dots \\ 0 & 0 & 0 & 0 & \dots & f & 0 & 0 & \dots & 0 & 0 & 0 \\ 0 & 0 & 0 & 0 & \dots & b & c & 0 & \dots & 0 & 0 & 0 \\ 0 & 0 & 0 & 0 & \dots & 0 & b & c & \dots & 0 & 0 & 0 \\ 0 & 0 & 0 & 0 & \dots & 0 & 0 & b & \dots & 0 & 0 & 0 \\ \dots & \dots & \dots & \dots & \dots & \dots & \dots & \dots & \dots & \dots & \dots & \dots \\ \dots & \dots & \dots & \dots & \dots & \dots & \dots & \dots & \dots & \dots & \dots & \dots \\ 0 & 0 & 0 & 0 & \dots & 0 & 0 & 0 & \dots & b & c & 0 \\ 0 & 0 & 0 & 0 & \dots & 0 & 0 & 0 & \dots & 0 & b & c \\ 1 & 0 & 0 & 0 & \dots & 0 & 0 & 0 & \dots & 0 & 0 & \frac{-1}{2} \end{bmatrix},$$

and

$$C = A,$$

$$D = \begin{bmatrix} 1 & 0 & 0 & \dots & 0 & 0 \\ 0 & 1 & 0 & \dots & 0 & 0 \\ 0 & 0 & 1 & \dots & 0 & 0 \\ \dots & \dots & \dots & \dots & \dots & \dots \\ 0 & 0 & 0 & \dots & 1 & 0 \\ 0 & 0 & 0 & \dots & 0 & 1 \end{bmatrix},$$

$$U_s = \begin{bmatrix} U_s^0 \\ U_s^1 \\ \ldots \\ U_s^{N-1} \\ U_s^N \end{bmatrix}, \quad s = n-1,\ n, n+1.$$

To solve this difference equation we have applied a procedure of modified Gauss elimination method for difference equation with respect to n with matrix coefficients. Hence, we seek a solution of the matrix equation in the following form

$$U_n = \alpha_{n+1}U_{n+1} + \beta_{n+1}, \quad n = M-1, \cdots, 2, 1, 0, \tag{4.5}$$

where α_j $(j = 1, \cdots, M-1)$ are $(N+1) \times (N+1)$ square matrices and β_j $(j = 1, \cdots, M-1)$ are $(N+1) \times 1$ column matrices. Since

$$U_0 = 0 = \alpha_1 U_1 + \beta_1, \tag{4.6}$$

we have

$$\alpha_1 = \begin{bmatrix} 0 & 0 & 0 & \ldots & 0 & 0 \\ 0 & 0 & 0 & \ldots & 0 & 0 \\ 0 & 0 & 0 & \ldots & 0 & 0 \\ \ldots & \ldots & \ldots & \ldots & \ldots & \ldots \\ 0 & 0 & 0 & \ldots & 0 & 0 \\ 0 & 0 & 0 & \ldots & 0 & 0 \end{bmatrix},$$

and

$$\beta_1 = \begin{bmatrix} 0 \\ 0 \\ \ldots. \\ 0 \\ 0 \end{bmatrix}$$

Using the equality

$$U_s = \alpha_{s+1}U_{s+1} + \beta_{s+1}, (\text{for } s = n,\ n-1)$$

and the matrix equation

$$AU_{n+1} + B\,U_n + CU_{n-1} = D\varphi_n,$$

we can write

$$[A + B\alpha_{n+1} + C\alpha_n\alpha_{n+1}]U_{n+1} + [B\beta_{n+1} + C\alpha_n\beta_{n+1} + C\beta_n] = D\varphi_n.$$

The last equation is satisfied if we select

$$A + B\alpha_{n+1} + C\alpha_n\alpha_{n+1} = 0,$$

$$[B\beta_{n+1} + C\alpha_n\beta_{n+1} + C\beta_n] = D\varphi_n, \quad 1 \le n \le M-1.$$

From that it follows

$$\alpha_{n+1} = -\left(B + C\alpha_n\right)^{-1} A, \tag{4.7}$$

$$\beta_{n+1} = \left(B + C\alpha_n\right)^{-1} \left(D\varphi_n - C\beta_n\right), n = 1, 2, 3, \cdots, M - 1.$$

Using formulas (4.7) and we can compute α_n and β_n. After that we obtain U_n, $1 \leq n \leq M - 1$, starting from

$$U_M = \vec{0},$$

and using recursive equation

$$U_n = \alpha_{n+1} U_{n+1} + \beta_{n+1}, \ n = M - 1, ..., 2, 1.$$

Now, we will give the results of the numerical analysis. In order to get the solution of (4.2) we use MATLAB programs. The numerical solutions are recorded for different values of $N = M$ and u_n^k represents the numerical solutions of these difference schemes at (t_k, x_n). For their comparison, the errors computed by

$$E = \max 1 \leq k \leq N 1 \leq n \leq M \left| u(t_k, x_n) - u_n^k \right|.$$

Tables 1 give the error analysis between the exact solution and the solutions derived by difference schemes. Table 1 is constructed for $N = M = 60, 80,$ and100 respectively.

Table 1

Method	N=M=60	N=M=80	N=M=100
M.C.N. r=1	0.0310	0.0175	0.0112
M.C.N. r=2	0.0413	0.0234	0.0150
M.C.N. r=3	0.0522	0.0297	0.0191
1^{st}order accuracy	0.4704	0.3866	0.3267

Note that the second order of accuracy r- modified Crank-Nicholson difference schemes are more accurate comparing with the first order of accuracy difference scheme.

References

1. M.E. Mayfield, *Non-reflective boundary conditions for Schroedinger's equation,* PhD Thesis,University of Rhode Island (1989)
2. D.G. Gordeziani and G.A. Avalishvili, *Time- nonlocal problems for Schrödinger type equations: I. Problems in abstract spaces,* Differential and Equations **41**(2005), no.5, 703–711
3. D.G. Gordeziani and G.A. Avalishvili, *Time- nonlocal problems for Schrödinger type equations: II. Results for specific problems,* Differential and Equations **41**(2005) no.6, 852-859
4. H. Han, J. Jin, X. Wu, *A finite difference method for the one-dimensional Schrödinger equation on unbounded domain,* Computers and Mathematics with Applications **50** (2005), 1345-1362
5. X. Antoine, C. Besse and V. Mouysset, *Numerical schemes for the simulation of the two-dimensional Schrödinger equation using non-reflecting boundary conditions,* Mathematics of Computation **73** (2004), no.248, 1779-1799
6. A. Ashyralyev, *An estimation of the convergence for the solution of the modified Crank-Nicholson difference schemes for parabolic equations with nonsmooth initial data,* Izv. Akad. Nauk Turkmen. SSR Ser. Fiz.-Tekhn. Khim. Geol. Nauk (1989), no.1, 3–8. (Russian).
7. A. Ashyralyev, S. Piskarev and S. Wei, *On well-posedness of the difference schemes for abstract parabolic equations in $L_p([0,1],E)$ spaces,* Numerical Functional Analysis and Optimization **23**(2002), no.7-8, 669–693
8. R. Rannacher, *Discretization of the heat equation with singular initial data,* Z. Angew. Math. Mech. 62(1982), no.5, 346–348.

PART B

OPEN PROBLEMS

SOME OLD (UNSOLVED) AND NEW PROBLEMS AND CONJECTURES ON FUNCTIONAL EQUATIONS OF ENTIRE AND MEROMORPHIC FUNCTIONS

CHUNG-CHUN YANG

The Hong Kong University of ScienceTechnology, Hong Kong
mayang@ust.hk

Introduction

The purpose for posing these old and new problems and conjectures arised in studying the Nevanlinna's value distribution theory and its applications is to stimulate peers' interest to study further of the theory and properties of entire and meromorphic functions of one complex variable. Also be reminded that one can carry the studies to similar problems and conjectures entire and meromorphic functions of several complex variables as well as that of p-adic complex variables. For simplicity , here we shall address the problems and conjectures for (non-constant) entire or meromorphic functions of one complex variable only, and most of them were proposed by the present author and his co-workers earlier. We refer the reader to [3] for the Nevanlinna's theory and its associated notations. Problems and conjectures will be stated in three different topics : 1.Factorization and fixed points of entire functions, 2. Dynamics of two permutable entire functions, and 3. Functional equations of Diophantine type over functions field. Sources, definitions, and results relevant to these problems and conjectures will be stated or cited, with references.

1. Factorizations and fixed points of entire functions

Definition. Let F, f and g be non-constant entire functions. The expression or composition: $F = f(g)$ (or fog) is called a factorization of F, with f and g the left and right factor, respectively. F is called a prime (pseudo-prime) function iff whenever F has a factorizaion : $F = f(g)$ implies that either f or g is a linear function (polynomial), see, e.g.,[1]. There have been many sufficient conditions or criteria to judge whether or not a certain function is prime or pseudo-prime, but no necessary conidtion or criterion for such a verification has been found yet.

Question 1.1. What is a necessary condition for an entire function to be prime or pseudo-prime ?

Conjecture 1.2 ([1], p.181). Let F be a pseudo-prime entire function and P a non-constant polynomial. Then $F(P)$ remains to be pseudo-prime.

Note that one can exhibit some examples to show that, in general, $P(F)$ may not be psudo-prime.

Definition Let f, F, g, G, h, H, k and K denote entire functions. h is called a (right)factor of H if and only if $H = k(h)$ and will be noted as $h|H$. And K is called a greatest common (right factor) of F and G iff $K|F$, $K|G$ and if h is any other common factor of F and G, then $K|h$. K is called a least common multiplier of F and G iff $F|K$ and $G|K$, and if H is any other function satisfying $H|F$ and $H|G$, then $K|H$. It has been shown [2] that any collection of entire functions $E = \{F_\alpha\}$, under a slightly general definition of a (right) factor, there always exists a common factor for E, which may be a linear function. However, by example of two functions: $\exp(z)$ and $\exp(iz)$ in [9], it follows that a common multiplier of two functions may not exist.

Question 1.3. What is a necessary and condition for two entire functions F and G to be relatively prime ? That is the only common factor of F and G is a linear function.

It has been known that $F(z) = z + p(z)\exp\alpha(z)$ is a prime function, where p is a polynomial ($\not\equiv 0$) and α a non-constant entire function. Which immediately shows that if neither f nor g is linear, then a transcendental function of the form $f(g)$ must have infinitely many fixed points. In fact, it had been a conjecture for long while that such a function $f(g)$ must have infinitely many fixed points, see, e.g., [8]. As a further study of the quantitative estimate of the numer of fixed points of composite functions, the following conjecture was raised:

Conjecture 1.4 ([7], [8]). Let f be a transcendental meromorphic function, g a nonlinear function , $a(z)$ a small function of f. Then

$$N\left((r, 1/(f(g) - a)\right) \neq o(1)\ T(r, f(g)) \quad \text{as} \quad r \to \infty.$$

2. Dynamics of two permutable entire functions

It has been a well known result that if two polynomials p, q are permutable, then p, q have the same Julia and Fatou sets. Thus far,the same conclusion holds for two permutable transcendental entire functions f, g that are of bounded type, that is the sets of finite singularities (critical values and asymptotic values) of both f and g are bounded , see, e.g.,([6]).

Question 2.1 . What are the relations between two permutable transcendental entire functions f and g other than the following two situations : 1. $f = ag + b$, with a, b constants and 2. $f = h_n$ and $g = h_m$ for some transcendental entire function h, where h_k denotes kth-iteration of h.

Remark. One may find the example in [6] which shows f and g are permutable but may satisfy neither situations 1 and 2 above.

3. Functional equations of Diophantine type

Conjecture 3.1 ([10]). If a Diophantine equation : $F(x, y) = 0$, where F is an irreducible polynomial of degree higher than 3 , with rational numbers as the coefficients, has none or finitely many rational solutions, then the correspondening equation $F(f, g) = 0$ has none or finitely many non-constant transcendental meromorphic solutions f and g. Here the two pairs of solutions (f, g) and $(f(h), g(h))$ for any non-constant entire h are said to be equivalent.

Conjecture 3.2. Let P denote a non-constant polynomial. The only transcendental entire solutions for the equation of the form : $f^2(z) + p(z)g^2(z) = 1$ are pairs :$(\pm f, \pm g)$, with $f = \cos\sqrt{p(z)}$ and $g = \sin\sqrt{p(z)}/\sqrt{p(z)}$.

Definition. Let $P(z, f, g)$ be a polynomial in f and g, with entire or meromorphic functions of z as the coefficients. A pair of meromorphic solutions (f, g) of the equation $P(z, f, g) = 0$ is called admissible if all the coefficients of the equation are small functions of f and g. See [4] for some results on the existence of admissible solutions of functional equations of the form :

$$f^n + a_1 f^{n-m} + b_1 = c(g^n + a_2 g^{n-m} + b_2),$$

where $a_i (i = 1, 2), b_i (i = 1, 2)$ and c are meromorphic functions and none of them is identically zero.

Note, in the above result, n is assumed to be greater than 8, with $m \geq 2$ and $n > 2m+3$. It seems to be quite difficult to deal with the above functional equations when the degree of f or g is less than 8.

Conjecture 3.3([4]). Let p, q be two polynomials both having at least three distinct zeros, and $a(z)$ be any non-constant meromorphic function. Then the functional equation: $p(f) = a(z)q(g)$ has no admissible solutions f and g.

Conjecture 3.4 ([5]). The functional equation: $f^2 = g^3 + b(z)g + c$, where b is a non-constant meromorphic function and c a constant, has no pair of admissible solutions.

References

1. Chi-Tai Chuang and Chung-Chun Yang, Fix-points and factorization of meromorphic functions, World Scientific , 1990 .
2. A. Eremenko and L.A. Rubel, The arithmetic of entire functions under composition, Advances in Mathematics, 124 (1996) ,334-354 .
3. W.K. Hayman, Meromorphic Functions, Oxford Press, 1964 .
4. Ping Li and Chung-Chun Yang, Admissible solutions of functional equations of Diophantine type, Preprint.
5. Ping Li and Chung-Chun Yang, Decompositions of meromorphic functions over small functions fields , This Proceedings .
6. Liang-Wen Liao and Chung-Chun Yang, On the Julia sets of two permutable entire functions, Rocky Mountain J. of Math. , vol. 35 no. 5 2006, p.1659-1674 .
7. Chung-Chun Yang, On the zeros of $f(g(z)) - \alpha(z)$, Proc. Nankai Conference, International Press, 1994, 261-269 .
8. Chung-Chun Yang and Jian-Hua Zheng, On the fix-points of composite meromorphic functions, J. D'Analyse Mathematique, vol. 68 (1998), 55-93 .
9. Chung-ChunYang, On common right factor and common right multiplier of two entire functions, Chinese J. of Contemporary Math., vol. 18, no. 3, 1997, 347-375 .
10. Chung-Chun Yang and Ping Li, Some further results on the functional equation $P(F) = Q(G)$, Value distribution theory and related topics, ACAA series, Kluwer 2004, p. 219-231 .

AN OPEN PROBLEM ON THE BOHR RADIUS

LEV AIZENBERG

Department of Mathematics,
Bar-Ilan University, 52900
Ramat-Gan, Israel
aizenbrg@math.biu.ac.il

Keywords: Hausdorff operator, Fourier transform, real Hardy space

Let $G \subset \mathbb{C}$ be any domain. A point $p \in \partial G$ is called a point of convexity if $p \in \partial \tilde{G}$. Here $\tilde{G}$ is the convex hull of G. A point of convexity p is called regular if there exists a disk $U \subset G$ so that $p \in \partial U$.

In[1] the following result is proved.

Theorem 1. If the function

$$f(z) = \sum_{k=0}^{\infty} c_k z^k$$

is such that $f(U_1) \subset G$, with $\tilde{G} \neq \mathbb{C}$ and U_1 the unit disk, then for $|z| < 1/3$ the inequality

$$\sum_{k=0}^{\infty} |c_k z^k| < \operatorname{dist}(c_0, \partial \tilde{G})$$

is valid. The constant $1/3$ cannot be improved if ∂G contains at least one regular point of convexity.

Open problem: if $\tilde{G} \neq \mathbb{C}$ and no regular point of convexity exists in ∂G, figure out whether it is possible or not to improve the constant $1/3$ in the above theorem. A similar problem concerns multidimensional analogs of that theorem (see[1]).

References

1. L. Aizenberg, 'Generalization of results about the Bohr radius for power series', submitted to Studia Mathem.

OPEN PROBLEMS ON HAUSDORFF OPERATORS

ELIJAH LIFLYAND

Department of Mathematics, Bar-Ilan University,
Ramat-Gan, 52900, Israel
E-mail: liflyand@math.biu.ac.il

After 2000 an interest in Hausdorff operators grew. Here we give a list of open problems in the subject along with history and prerequisites.

Keywords: Rower series, Hausdorff operator, Fourier transform, Hardy space.

The aim of these notes is to present open problems concerning Hausdorff operators. The idea appeared during the International Conference on Complex Analysis and Potential Theory in Istanbul, Turkey, September 8-14, 2006. The author's talk on recent results in the subject and further discussions with participants of the conference confirmed that there is an interest in these problems.

We will give minimal information on the topic, enough to formulate the problems and make the paper more or less self-contained. Further details can be found in the relevant literature from the list of references.

Hausdorff means, the Cesàro means among them, are known long ago in connection with summability of number series. We are not going to touch this subject, though we do not exclude that there are interesting open problems there.

1. Power series. Cesàro means for power series from the Hardy space H^1 in the unit disk were considered by A. Siskakis. The idea is to substitute the coefficients a_k in the expansion $f(z) = \sum_{k=0}^{\infty} a_k z^k$, with $f \in H^1$, for their Cesàro means $k^{-1}\sum_{p=0}^{k-1} a_p$:

$$\sum_{k=0}^{\infty}\left(k^{-1}\sum_{p=0}^{k-1} a_p\right) z^k; \qquad (1)$$

for an elegant proof of the boundedness of the corresponding operator in H^1, see Ref.[14]

Later, already after appearance of the paper Ref.[9] in 2000, general Hausdorff matrices were considered in Ref.[5] (we are also aware of continuation of the latter work by P. Galanopoulos and M. Papadimitrakis) as follows.

Let Δ be the forward difference operator defined on scalar sequences $a = (a_n)_n$

by $\Delta a_n = a_n - a_{n+1}$ and set $C_{n,k} = \binom{n}{k}\Delta^{n-k}, a_k \quad k \le n$. The Hausdorff matrix $H(a)$ is defined by $(H(a))_{ij} = \begin{cases} 0, & i < j \\ C_{ij}, & i \ge j \end{cases}$. It induces two operators on spaces of power series which are formally given by

$$H(a)\left(\sum_0^\infty a_n z^n\right) = \sum_{n=0}^\infty \left(\sum_{k=0}^n C_{n,k} a_k\right) z^n \tag{2}$$

and

$$A(a)\left(\sum_0^\infty a_n z^n\right) = \sum_{k=0}^\infty \left(\sum_{n=k}^\infty C_{n,k} a_n\right) z^k. \tag{3}$$

Necessary and sufficient conditions on $a = a(\mu)$ are found in order that $H(a)$ and $A(a)$ induce bounded linear operators on the Hardy spaces H^p, $1 \le p < \infty$.

Problem 1.
a) *Find multivariate versions of all aforementioned results.*
b) *Study partial sums of (1) as well as those of (2) and (3).*

2. Cesàro means for Fourier transforms. The Fourier transform setting for the Cesàro means was considered in Ref.;[7] this is a partial case of general Hausdorff operators (4) below by letting $\varphi(x)$ to be the indicator of $[0,1]$. . For a special interesting case $p \le 1$ a complete solution for the Cesàro means was given in Ref.[12]

Problem 2. *Study the boundedness of the Hausdorff operators (4) in H^p for* $p \le 1$.

3. One-dimensional Hausdorff operators. In fact, general Hausdorff means of a Fourier-Stieltjes transform were introduced in Ref.,[6] but only in L^1. In the real Hardy space on $\mathbb{R}$, for the Hausdorff operator defined, by means of $\varphi \in L^1(\mathbb{R})$, as

$$(\mathcal{H}f)(x) = (\mathcal{H}_\varphi f)(x) = \int_{\mathbb{R}} |u|^{-1}\varphi(u) f(x/u)\, du \tag{4}$$

the boundedness of this operator taking $H^1(\mathbb{R})$ into $H^1(\mathbb{R})$ was proved in Ref.[9]

Observe that, as in Ref.,[6] in Refs.[1]-[3] and 13 similar problems are considered for the Hausdorff operators defined by suitable measures, while (4) is the partial case of absolutely continuous measures. Here we restrict ourselves to the latter case for the sake of brevity and convenience; we are aiming at a different generality, more transparent in a simpler setting. Extensions to arbitrary measures go through as in the cited papers.

Similarly to the case of power series, the point is as follows. Since, generally speaking, the inverse formula

$$f(x) = (2\pi)^{-1} \int_{\mathbb{R}} \widehat{f}(t) e^{ixt} dt$$

does not take place for $f \in L^1(\mathbb{R})$ as well as for $f \in H^1(\mathbb{R})$, where $\hat{f}$ is the Fourier transform of f; expected is that

$$\int_{\mathbb{R}} (\mathcal{H}f)\widehat{\ }(y) e^{ixy} dy$$

behaves better and characterizes f properly, in a sense. Though these operators were considered in various spaces, the most important is the case of Hardy spaces; see, for example, Refs.[9] or 12.

Some open problems that will be given below in the multivariate setting are also open for dimension one. Here we present a very general problem which first of all deserves investigation on the real axis.

Problem 3. *Prove or disprove that the real Hardy space $H^1(\mathbb{R})$ can be characterized via Hausdorff operators in the sense that if (4) for an integrable function f is in $H^1(\mathbb{R})$ then f is necessarily in $H^1(\mathbb{R})$. If not (this is very probable), a counterexample should be given.*

4. Multidimensional Hausdorff operators. In the multidimensional case the situation is, as usual, more complicated. The Cesàro means in Ref.[8] and special Hausdorff means in Ref.[10] were considered in dimension 2 only for the so-called product (mixed) Hardy space $H^{11}(\mathbb{R} \times \mathbb{R})$. In the recent paper Ref.[16] a slight extension was made in the same direction. In Ref.[1] the boundedness in $H^1(\mathbb{R}^n)$ of the Hausdorff type operator

$$(\mathcal{H}_\mu f)(x) = \int_{\mathbb{R}} |u|^{-n} f(x/u)\, d\mu(u),$$

where $x \in \mathbb{R}^n$, was proved. But defining the operator by one-dimensional averaging does not seem to be natural for the multivariate case.

Similar to Refs.[3] and 13, we generalize (4) by defining a Hausdorff type operator as

$$(\mathcal{H}f)(x) = (\mathcal{H}_\Phi f)(x) = (\mathcal{H}_{\Phi,A} f)(x) = \int_{\mathbb{R}^n} \Phi(u) f\big(xA(u)\big)\, du, \tag{5}$$

where $A = A(u) = (a_{ij})_{i,j=1}^n = \big(a_{ij}(u)\big)_{i,j=1}^n$ is the $n \times n$ matrix with the entries $a_{ij}(u)$ being measurable functions of u. This matrix may be degenerate at most on a set of measure zero; $xA(u)$ is the row n-vector obtained by multiplying the row n-vector x by the matrix A.

By this, each Hausdorff operator is defined by a couple (Φ, A) satisfying at least the condition

$$\|\Phi\|_{L_A} = \int_{\mathbb{R}^n} |\Phi(u)|\,|\det A^{-1}(u)|\,du < \infty,$$

which provides the boundedness of the operator in $L^1(\mathbb{R}^n)$.

In Ref.[13] the boundedness of such operators in $H^1(\mathbb{R}^n)$ is proved for a very special case of diagonal matrices A with all entries on the diagonal equal to one another.

But, defining $\Phi \in L_A^*$ by

$$\|\Phi\|_{L_A^*} = \int_{\mathbb{R}^n} |\Phi(u)|\,\|A(u)\|^n du < \infty,$$

with $\|A\| = \|A(u)\| = \max_j(|a_{1j}(u)| + ... + |a_{nj}(u)|)$, we proved in Ref.[11] that the boundedness on $H^1(\mathbb{R}^n)$ for general matrices A is provided by

$$\Phi \in L_{A^{-1}}^*. \tag{6}$$

Since the dual approach is used in Ref.,[11] the key ingredient in the proof is a lemma on the behavior in u of the BMO-norm of $f(xA(u))$. This also allows us to get conditions for the boundedness of the operator in $BMO(\mathbb{R}^n)$. Moreover, similar results are obtained for the adjoint operator as well.

These lead to

Problem 4.

a) *Prove (or disprove) the sharpness of the condition (6) for the boundedness of the Hausdorff operator in $H^1(\mathbb{R}^n)$.*

b) *Find a different proof of the boundedness in $H^1(\mathbb{R}^n)$ of the Hausdorff operator satisfying (6).*

5. Mixed Hardy spaces. Besides "usual" H^1 space of importance are the so-called product (mixed) Hardy spaces considered in detail by R. Fefferman (see, e.g., Ref.[4]). Results on Hausdorff operators in very special mixed Hardy spaces are mentioned above.

Problem 5. *Find connections between (Φ, A) Hausdorff operators and mixed Hardy spaces in terms of belonging of Φ to a space defined by A.*

6. Partial integrals. It is known (see Ref.[3]) that

$$(\mathcal{H}f)\widehat{\ }(y) = \int_{\mathbb{R}^n} \Phi(u)\,|\det A^{-1}(u)|\,\widehat{f}\big(y(A^{-1})^T(u)\big)\,du, \tag{7}$$

where $(A^{-1})^T$ is the transpose of A^{-1}. Taking

$$\mathcal{H}_N f(x) = \int_{|x|\le N} (\mathcal{H}f)\hat{}(y) e^{i\langle x,y\rangle} dy,$$

using (7), and applying Fubini's theorem we arrive at

$$\mathcal{H}_N f(x) = \int_{\mathbb{R}^n} \Phi(u) |\det A^{-1}|\, du \int_{\mathbb{R}^n} f(v)\, dv \int_{|y|\le N} e^{-i\langle x - vA^{-1}(u), y\rangle} dy.$$

The last integral is well-known (see, e.g., a formula from the proof of Theorem 3.3 from Chapter IV in Ref.[15]), and we get

$$(2\pi N)^{-n/2}\mathcal{H}_N f(x)$$
$$= \int_{\mathbb{R}^{2n}} \Phi(u) |\det A^{-1}|\, f(v)\, |x - vA^{-1}(u)|^{-n/2} J_{n/2}(N|x - vA^{-1}(u)|)\, du\, dv,$$

where $J_{n/2}$ is the Bessel function of order $n/2$, with f either from L^1 or from H^1. In dimension one it is of the following form: with $\phi \in L^1$

$$\mathcal{H}_N f(x) = \int_{R^2} \phi(u) f(t) \frac{\sin N(x - ut)}{y - ut}\, du\, dt.$$

Problem 6. *Study $\mathcal{H}_N f$ or maybe $\mathcal{H}_N^* f = \sup_N |\mathcal{H}_N f|$.*

7. More problems. Certain problems worth being studied become apparent in discussions during the conference, first of all with O. Martio. One of such problems is to consider the spaces defined by general singular operators rather than those defined by the Hilbert or Riesz transforms.

We mention also the question the author was asked after his talk about compactness of Hausdorff operators.

References

1. K.F. Andersen, *Boundedness of Hausdorff operators on $L^p(\mathbb{R}^n), H^1(\mathbb{R}^n)$, and $BMO(\mathbb{R}^n)$*, Acta Sci. Math. (Szeged) **69** (2003), 409–418.
2. G. Brown and F. Móricz, *The Hausdorff and the quasi Hausdorff operators on the spaces L^p, $1 \le p < \infty$*, Math. Inequal. Appl. **3** (2000), 105–115.
3. G. Brown and F. Móricz, *Multivariate Hausdorff operators on the spaces $L^p(\mathcal{R}^n)$*, J. Math. Anal. Appl. **271** (2002), 443–454.
4. R. Fefferman, *Some recent developments in Fourier analysis and H^P theory and product domains. II*, Function spaces and applications, Proc. US-Swed. Semin., Lund/Swed., Lect. Notes Math. **1302** (1988), 44–51.
5. P. Galanopoulos and A.G. Siskakis, *Hausdorff matrices and composition operators*, Ill. J. Math. **45** (2001), 757–773.
6. C. Georgakis, *The Hausdorff mean of a Fourier-Stieltjes transform*, Proc. Am. Math. Soc. **116** (1992), 465–471.

7. D.V. Giang and F. Móricz, *The Cesàro operator is bounded on the Hardy space H^1*, Acta Sci. Math. **61** (1995), 535–544.
8. D.V. Giang and F. Móricz, *The two dimensional Cesàro operator is bounded on the multi-parameter Hardy space $\mathcal{H}^1(\mathbb{R} \times \mathbb{R})$*, Acta Sci. Math. **63** (1997), 279–288.
9. E. Liflyand and F. Móricz, *The Hausdorff operator is bounded on the real Hardy space $H^1(\mathbb{R})$*, Proc. Am. Math. Soc. **128** (2000), 1391–1396.
10. E. Liflyand and F. Móricz, *The multi-parameter Hausdorff operator is bounded on the product Hardy space $H^{11}(\mathbb{R} \times \mathbb{R})$*, Analysis **21** (2001), 107–118.
11. A. Lerner and E. Liflyand, *Multidimensional Hausdorff operator on the real Hardy space*, to appear in J. Austr. Math. Soc.
12. A. Miyachi, *Boundedness of the Cesàro Operator in Hardy spaces*, J. Fourier Anal. Appl. **10** (2004), 83–92.
13. F. Móricz, *Multivariate Hausdorff operators on the spaces $H^1(\mathbb{R}^n)$ and $BMO(\mathbb{R}^n)$*, Analysis Math. **31** (2005), 31–41.
14. A.G. Siskakis, *The Cesàro operator is bounded on H^1*, Proc. Am. Math. Soc. **110** (1990), 461–462.
15. E. M. Stein and G. Weiss, *Introduction to Fourier Analysis on Euclidean Spaces*, (Princeton Univ. Press, Princeton, N.J. 1971).
16. F. Weisz, *The boundedness of the Hausdorff operator on multi-dimensional Hardy spaces*, Analysis **24** (2004), 183–195.

AUTHOR INDEX